THEORY, DETERMINATION AND CONTROL
OF PHYSICAL PROPERTIES OF FOOD MATERIALS

SERIES IN FOOD MATERIAL SCIENCE: VOLUME 1

THEORY, DETERMINATION AND CONTROL OF PHYSICAL PROPERTIES OF FOOD MATERIALS

Edited by

CHOKYUN RHA

Department of Nutrition and Food Science,
Massachusetts Institute of Technology, Cambridge, Mass., U.S.A.

D. REIDEL PUBLISHING COMPANY

DORDRECHT-HOLLAND / BOSTON-U.S.A.

Library of Congress Catalog Card Number 74–76484

ISBN 90 277 0468 6

Published by D. Reidel Publishing Company,
P.O. Box 17, Dordrecht, Holland

Sold and distributed in the U.S.A., Canada, and Mexico
by D. Reidel Publishing Company, Inc.
306 Dartmouth Street, Boston,
Mass. 02116, U.S.A.

Printed in The Netherlands by D. Reidel, Dordrecht

*Dedicated to Dr George W. Scott Blair
to commemorate his pioneering work in
food rheology*

TABLE OF CONTENTS

PREFACE

In recent years, the importance of material science, or the understanding of the physical properties of food materials in the progress of food engineering, has become more recognized. Increasing numbers of basic and applied studies in this area appear in numerous journals and literature scattered around various disciplines. This 'Series in Food Material Science' is planned to survey, collect, organize, review and evaluate these studies. By doing so, it is hoped that this series will be instrumental in bringing about a better understanding of the physical properties of food materials, better communication among scientists, and rapid progress in food engineering, science and technology.

This volume, *Theory, Determination and Control of Physical Properties of Food Materials*, Volume 1 of the 'Series in Food Material Science', contains basic principles, methods and instrumental methods for determination and application of the modification of physical properties.

In this book, noted investigators in the subjects have pooled their knowledge and made it available in a condensed form. Every chapter is selfcontained with most of them starting with a review or introduction, including the viewpoint of the author. These should offer a beginner a very general introduction to the subjects covered, make the scientists and technologists in the field aware of current progress and allow the specialists a chance to compare different viewpoints.

The editor's experience in the organizing and teaching a graduate level course in physical properties of food and biological materials and related research suggests a need for a book of this kind which could serve as a text or reference.

With some minor editorial changes, this volume was derived mainly from the lectures delivered at a special summer program on 'Theory, Determination and Control of Physical Properties of Food Materials' offered jointly by the Food and Agricultural Engineering Department, Cooperative Extension Service and Division of Continuing Education at the University of Massachusetts.

Thanks are due to the contributors for their willingness to participate, for careful preparation of the manuscripts and for their patience in coping with the long delays.

Thanks are also due to the publisher, D. Reidel Publishing Company for their accommodativeness.

The editor appreciates the most effective secretarial assistance of Mrs Roberta Zidik.

Amherst, Massachusetts
Winter, 1973

LIST OF CONTRIBUTORS

G. W. Scott Blair, Oxford, England.

Malcolm C. Bourne, Professor, Department of Food Science and Technology, New York State Agricultural Experiment Station, Geneva, New York.

Fergus M. Clydesdale, Associate Professor of Food Science, Dept. of Food Science and Technology, University of Massachusetts, Amherst, Mass.

Stevenson W. Fletcher, Head Dept. of Hotel and Restaurant Management, Associate Professor of Food Engineering, Food and Agricultural Engineering Department, University of Massachusetts, Amherst, Mass.

Frederick J. Francis, Professor and Head, Dept. of Food Science and Technology, University of Massachusetts, Amherst, Mass.

Julian F. Johnson, Dept. of Chemistry and Institute of Materials Science, University of Connecticut, Storrs, Conn.

Ronald Jowitt, Reader in Food Engineering, National College of Food Technology, University of Reading, Weybridge, Surrey, England.

Marcus Karel, Professor of Food Engineering, and Deputy Head, Dept. of Nutrition and Food Science, Massachusetts Institute of Technology, Cambridge, Mass.

Amihud Kramer, Professor of Horticulture and Food Science, Department of Horticulture, University of Maryland, College Park, Maryland.

Theodore P. Labuza, Professor, Department of Food Science and Industries, University of Minnesota, St. Paul, Minnesota.

John R. Martin, Research Department, The Foxboro Company, Foxboro, Mass.

Stanley Middleman, Professor of Chemical Engineering and Polymer Science, University of Massachusetts, Amherst, Mass.

Roger Porter, Professor and Head of Polymer Science Program, University of Massachusetts, Amherst, Mass.

ChoKyun Rha, Associate Professor of Food Process Engineering, Dept. of Nutrition and Food Science, Massachusetts Institute of Technology, Cambridge, Mass.

Philip Sherman, Professor, Queen Elizabeth College, University of London, London, England

Peter W. Voisey, Head, Research Service Section, Engineering Research Service, Canada Department of Agriculture, Ottawa, Canada.

Lester Whitney, Associate Professor of Food Machinery, Food and Agricultural Engineering Dept., University of Massachusetts, Amherst, Mass.

INTRODUCTION

CHOKYUN RHA

Dept. of Nutrition and Food Sciense, Massachusetts Institute of Technology, Cambridge, Mass., U.S.A.

Food is undoubtedly the most important essential for human life, and the food industry one of the largest. However, the disciplines related to food products and food processing are generally not recognized as technologically oriented or advanced subjects equivalent to other sciences and engineering. This may be because the development in food engineering, science and technology has not been as spectacular as in some of the other fields. Yet, remarkably the modest progress made in food processing, for instance the processing method for frozen orange juice concentrate, development of instant coffee or whipped cream substitute in an aerosol, can directly improve the quality of everyday life of the average person. This is a good incentive to strive toward a more rapid progress in the field.

In order to achieve faster progress, food engineering and technology should rely less on superficial knowledge resulting from the traditional familiarity with food, inherent conceptual understanding of food, accumulation of uncoordinated empirical data or uncontrolled observations. What is required is a well thought out systematic study of the material properties of food systems. Understanding of the primary physical properties, such as viscosity and thermal conductivity, and complex properties, such as texture and structure properties, can be achieved through careful defining of the system, collecting of consistent and accurate data and rational analysis of experimental results. Only thorough understanding of the physical properties of food materials will lead to improvement of the existing products and processes, and development of the new products and processes through effective application of the engineering and technological principles. For this reason this volume is written on the theory, determination and control of physical properties of food materials.

The first four chapters of this volume are devoted to rheology, survey, principles, determination, and applications. A relatively large portion is directed to rheology since this is by far the most important and interesting fundamental property because of many deviations from ideality and complexity caused by biological anomalities of food materials. Texture and mechanical properties take an even larger portion of this book. This is because the texture properties are unique properties of food and play an important role in acceptability of food; also these properties are a complex attribution of several primary physical properties. Understanding of these properties is even more at the infantile stage than other physical properties of food but interestingly, more work has been attempted because of the need for evaluation which will relate to organoleptic quality. Chapter V, VI and VII deal with definition, measurement and instrumentation and the following two chapters discuss the mechanical properties in relation to processing and handling.

Chapter X covers sorption phenomena with respect to principles, measurement and related to food deterioration. Chapter XI discusses the properties controlling mass transfer which plays an important role in food processing and acceptability. The physical properties, mainly viscosity, influencing the evaluation of emulsion based food is found in Chapter XII. Colorimetry is presented from Chapters XIII to XVI. Chapter XVII is a review on thermal properties including some discussion on heat transfer. A short discussion on aerodynamic and hydrodynamic properties, electrical properties, and physical characteristics can be found in Chapter XVIII. The last chapter gives a good example of how vast information, if available, can be compiled for later recall and utilization.

CHAPTER I

SURVEY OF THE RHEOLOGICAL STUDIES
OF FOOD MATERIALS

GEORGE W. SCOTT BLAIR
Iffley, Oxford, England

1. Rheology of Foodstuffs – An Historical Survey

Rheology as a separately organized branch of physics was inaugurated at a meeting held at Columbus, Ohio on April 29, 1929. A committee of twelve scientists had been appointed by the Third Plasticity Symposium to set up a permanent organization. A Society of Rheology was founded, a constitution was drawn up and plans were made for a preliminary conference to be held in December of the same year in Washington, D.C. A journal to be called the *Journal of Rheology*, later simply *Rheology*, was to be started but unfortunately, it lived only a few years. The Society was originally planned to be International but this did not come about and it is now the American Society of Rheology. The writer had the good fortune to be present and indeed to preside at one of the sessions of this Conference in Washington. A striking feature was the diversity of interest represented.

Neither at the Conference nor in the abstracts in the earlier numbers of the journal were foodstuffs mentioned as far as can be recorded. This does not mean that no work had been published on foodstuffs before 1929. In the first textbook on rheology by Bingham [1], published in 1922, the author also mentioned work on milk and cream as early as 1908. Syrups, flours, gelatine and olive oil were also referred to. Although little known at the time, systematic work on rheology of flour dough was being done in Budapest.*

Kosutány [2], described an apparatus designed by Rejtő. Strips of dough, rectangular in cross-section, were stretched on a series of low friction metal rollers. Many later workers, for example Schofield and Scott Blair [3], in England; Bailey [4] in Minnesota; and Issoglio [5] in Italy 'improved' on this method using cylindrical test pieces and mercury baths in place of the rollers. The former was indeed an improvement but it is interesting that in fairly recent work, Reiner & Lerchenthal in Israel [6], have returned to the rollers. This is mainly because of the danger of the mercury vapor poisoning and partly perhaps because the mercury must be neither too clean, in which case the test piece floats to the side of the trough, nor, more obviously too dirty. Another Hungarian, Hankoczy [7] was a pioneer in developing the method for measuring the work done during the kneading of the dough in a mechanical mixer, a method later commercial-

---

* Before the First World War, 'Budapest' was part of the Austro-Hungarian Empire and was spelt 'Buda-Pesth'.

ChoKyun Rha (ed.), Theory, Determination and Control of Physical Properties of Food Materials, 3–6.

ized by Brabender in Germany and still used today. In earlier times, flour doughs and suspensions in water were probably the most widely tested materials. It was first thought that the viscosity of the suspension of flour gave a measure of the baking quality. Then better correlation was obtained from the viscosity-concentration relation. The writer's first paper on rheology was concerned with this [8]. But it was soon realized that since it was the doughs that are actually handled by the baker the early Hungarian workers were wise to study these rather than the dilute suspensions. Unfortunately, up to recent times [6], there has been far too little study of the correlations between rheological properties of doughs and the quality as assessed by bakers and others. Quite often the process of preparing the test piece destroys the very structure that one sets out to measure. For far too brief a period the distinguished German psychologist, David Katz, was in England studying the problem and he published a classical paper [9] in which he discussed the relationship between the baker as to what he appears to have assessed when he handled the dough and what the rheologist measures. It is essential to make reliable and reproducible rheological measurements but it is rash to judge the value of foodstuffs, or their intermediates, solely in terms of such measurements unless it can be shown that the parameters relate to quality.

In a study of an admittedly limited population of French flours, Scott Blair and Potel [10] found no correlation between bakers' and millers' scores and the data from the Chopin extensimeter or the alviograph which measures the size and internal pressure of a bubble of air blown in a disk of dough. Milk and other dairy products also appear early in the rheological literature. Bingham quotes papers as early as 1895, 1902 and 1905. The viscosity of milk and even its temperature coefficient have not found much practical application. See an excellent review article by Cox [11] and another by Cox *et al.* [12].

A long program of work was carried out at the British National Institute for Research in Dairying on the rheology of cheese curd and butter though the industry has made little use of the results, and the latest revolutions in cheese making procedures have put many of them out of date. A good summary of the work up to 1952 is given by Baron [13]. Much of this work was concerned with psycho-physical aspects [14]. At about the same time, a series of books was published in Holland on the application of rheology to many materials. But because of a foolish prejudice against the name rheology, this did not appear in any of the titles. The writer edited and wrote a part of the book dealing with foodstuffs [15]. The materials dealt with were as follows: starch, cereals, dairy products, honey, and miscellaneous foods. A chapter by R. Harper on the psychological aspects was also included. The rheology of muscle was studied from early times. *In vivo*, this is a highly complex field; *in vitro*, this is rather less difficult, and clearly it has a bearing on the organoleptic properties of meat. A pioneer in this field is Dr Bate-Smith [16] though of course much work has been done since 1948. Even dead meat is not at all an easy material for the rheologist. He has to cut test pieces more or less at random or separate out the individual fibers. Penetrometers have often been used and the various imitative chewing machines are perhaps rather too recent for discussion in a historical survey. But mention should be made at this point of a novel experiment on many foodstuffs devised by Dr Drake in Sweden [17] in which he records and amplifies the sounds made by the jaws of subjects chewing.

The properties of fruits and vegetables have been widely studied since the 30's but the writer has found little in the literature before that period. The invention of the tenderometer perhaps marks the start of the modern era in this field. (See Martin *et al.* [18].)

Honey has received much attention especially in Great Britian where the thixotropic honey from ling-heather, *Caluna-vulgaris* is much prized and highly priced. The presence of large air bubbles in a clear gel medium give it an attractive appearance. Only one other floral source, *Leptospermum scoparium*, in New Zealand appears to show these characteristics. It is difficult to get this latter honey from a pure floral source, whereas in some high land in England, in Devon and Yorkshire, practically no other plant is available for the bees at the appropriate time of year. Pryce-Jones made many studies along these lines and he succeeded in producing thixotropy in Newtonian honeys by the addition of certain proteins. Perhaps the most suitable of his many papers for reference would be those published in 1936, [19] and 1944, [20]. An early paper on the consistency of eggs should be mentioned because of the ingenuity of the method. The egg was attached to a swinging torsional pendulum and the damping which depends on the internal consistency, was measured. (See Wilcke [21].)

Not much work on the rheology of chocolate would seem to have been published before the paper of Clayton *et al.* in 1937 [22]. This is somewhat surprising in view of the amount of work that these authors could quote on other physical and chemical measurements. Katz [23] would seem to be a pioneer in measuring the compressibility and tensile strength of bread. Work on biscuits and cake followed later.

A good summary of early work on the rheology of fish is given by Charnley and Bolton [24]. The penetrometer is the favorite instrument.

It is inevitable that a review of this kind must be incomplete. Many food products have not been mentioned and it may well be that some pioneer papers have been missed. But, it is not intended to list recent work; for this, recent books and journals must be consulted. Especially to be recommended is the comparatively new *Journal of Texture Studies*. But, it is hoped that the present short article may be of some historical interest showing how the science of rheology, which is applied in so many fields, has made its way into the range of foodstuff research.

References

1. Bingham, E. C.: *Fluidity and Plasticity*. McGraw-Hill Book Co., New York, 1922.
2. Kosutány, T.: *Der ungarische Weizen und das ungarisches Mehl*, Verlag Malnarok Lapja, Budapest, 1907.
3. Schofield, R. K. and Scott Blair, G. W.: *Proc. Roy. Soc.* (A) **138**, 707 (1932); **141**, 72 (1933); **160**, 87 (1937).
4. Bailey, C. H.: *Wheat Studies of the Food Res. Inst.* **16**, 243, (1940).

5. Issoglio, G.: *R. Acad. Agric. Torino* **76**, 4, (1933); *Annali di Chimica Applicata* **25**, 274 (1935).
6. Reiner, M. and Lerchenthal, C. H.: Haifa Technion Report, 1967.
7. Hankoczy, E. V.: *Z. ges. Getreidewesen* **12**, 57 (1920).
8. Scott Blair, G. W., Watts, G., and Denham, H. J.: *Cereal Chem.* **4**, 63 (1927).
9. Katz, D.: *Cereal Chem.* **14**, 382 (1937).
10. Scott Blair, G. W. and Potel, P.: *Cereal Chem.* **14**, 257 (1937).
11. Cox, C. P.: *J. Dairy Res.* **19**, 72 (1952).
12. Cox, C. P., Hosking, Z., and Posener, L. N.: *J. Dairy Res.* **26**, 182 (1959).
13. Baron, M.: *The Mechanical Properties of Cheese and Butter*, Dairy Industries, London, 1952.
14. Harper, R.: *Brit. J. Psychol. Monogr. Suppl.* **18** (1952).
15. Scott Blair, G. W. (ed.): *Foodstuffs: Their Plasticity, Fluidity and Consistency*, North-Holland Publ. Co., Amsterdam, 1953.
16. Bate-Smith, E. C.: Chap. in E. M. Mrak and G. F. Stewart's (eds.), *Advances in Food Research*, Vol. I, Academic Publ. Inc., New York, 1948.
17. Drake, B.: *J. Food Sci.* **28**, 233 (1963).
18. Martin, W. McK., Lueck, R. H., and Sallee, E. D.: *Canning Age* **19**, 146 (1938).
19. Pryce-Jones, J.: *Bee World* (Aug. 1936).
20. Pryce-Jones, J.: *Scottish Beekeeper* (Sept. 1944).
21. Wilcke, H. L.: Iowa State Coll. Res. Bull. No. 194 (1936).
22. Clayton, W., Back, S., Johnson, R. I., and Morse, J. F.: *J. Soc. Chem. Ind.* **61**, 196 (1937).
23. Katz, J. R.: *Z. ges. Getreide Mühl.u.Bäck.* **20**, 181 and 206 (1933).
24. Charnley, F. and Bolton, R. S.: *J. Fish. Res. Bd., Canada* **4**, 162 (1938).

CHAPTER II

THEORIES AND PRINCIPLES OF VISCOSITY

CHOKYUN RHA

Dept. of Nutrition and Food Science, Massachusetts Institute of Technology, Cambridge, Mass., U.S.A.

Viscosity is a transport phenomenon. Viscosity is the transport of momentum due to a velocity gradient. The beginning of the phenomenological studies of viscosity goes back to the ancient Greeks and later the Romans characteristically applied what they had learned in practical ingenious ways [1]. Modern theories of viscosity of liquids are based on continuum mechanics and molecular theory.

Since Maxwell's prediction of viscosity based on the kinetic theory, many theories of viscosity were developed. Although it has been several decades since the first development of the theory of viscosity, obtaining the numerical values for the theoretical equations, or confirmation of the theories by experimental results has been confined to only a few monoatomic or diatomic fluids. The application of the theories of viscosity to the complex poly-atomic associating food components is still far away. Nevertheless, these theories help us to visualize the viscosity phenomena at the molecular level and with molecular forces.

1. Theories of Viscosity

The theories of viscosity will be discussed here rather superficially. The three main theories of viscosity discussed here are molecular theory, the rate process theory and diffusional theory.

1.1. MOLECULAR THEORY

The molecular theory attempts to relate the viscosity to two factors. One is the distribution of the nearest molecules surrounding a given molecule and the other, the intermolecular potential between this pair of molecules.

The molecules in a liquid are coupled to each other by intermolecular attraction. Therefore, if two adjacent layers of fluid move with different velocities, each will tend to drag the other due to this intermolecular attraction. This drag will dissipate the velocity gradient within the liquid. As a result of the drag exerted on one another by molecules in adjacent layers of the fluid moving with different velocities, the molecular structure of the fluid becomes deformed. The configuration of the molecules surrounding a given molecule effected by such drag is more ellipsoidal than the usual radial distribution. Principle axes of this ellipsoidal distribution are determined by the local velocity gradient, and the degree of deformation of the molecular structure determines the magnitude of the coefficient of viscosity [2, 3, 4].

ChoKyun Rha (ed.), Theory, Determination and Control of Physical Properties of Food Materials, 7–24.
Copyright © 1975 by D. Reidel Publishing Company, Dordrecht-Holland. All Rights Reserved.

Starting from this concept, Born and Green [4] gave an approximate solution for the viscosity by the following equation:

$$\eta = 0.48 \frac{r}{v} [m\phi_a(r)]^{1/2} \exp\left[-\frac{1}{kT}\phi(r) \right]$$

η = viscosity
r = distance between the centers of two adjacent molecules;
v = molecular volume;
m = mass per molecules;
$\phi_a(r)$ = attractive component of the pair potential cr^{-6};
$\phi(r)$ = mutual potential energy of a pair of molecules at distance r;
k = Boltzman constant;
T = absolute temperature.

1.2. THE RATE-PROCESS THEORY

The rate-process theory is similar to very familiar rate theory governing the kinetics of chemical reactions. The key aspect of this theory is that a molecule has to pass over a potential barrier in order to flow. Figure 1 shows the cross-section of an idealized liquid illustrating the fundamental rate process involved in viscous flow [5]. In order to move into an unoccupied site, a molecule passes through a potential energy bottle-

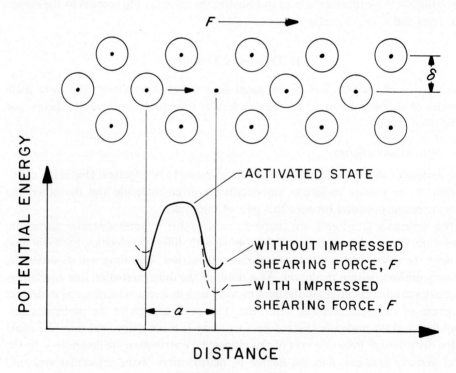

Fig. 1.   Activation for flow.

neck which is lowered when shear force is imposed [6, 7]. The assumption that a molecule passing through its neighbors can be considered as passage over a potential as such permits it to be treated by the theory of absolute reaction rate.

Figure 2 shows that two neighboring molecules form a pair instantaneously upon collision. This double molecule passes through its neighbors by rotation when there is free space in the adjacent layer. The free space in the adjacent layer is created by local

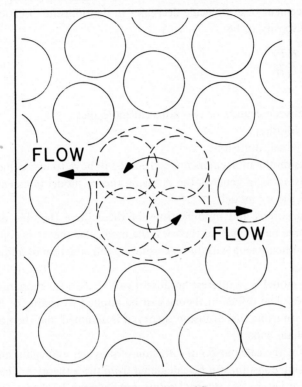

Fig. 2. Rate-process of flow.

density difference due to temperature fluctuation. Following the rotation, the pair of molecules dissociates. When no external stress is applied, 'forward' and 'backward' motion would be equally frequent. However, when shearing force is applied the potential barrier will be deformed so as to favor the 'forward' move as illustrated in Figure 1.

From these considerations, the expression for viscosity is found to be [8]:

$$\eta = \left(\frac{v_f^{1/3}}{v_L}\right)(2\pi m k T)^{1/2} \exp\left(\frac{E}{kT}\right).$$

v_f = fluctuation volume per molecule, i.e. fluctuation volume within which the center of gravity of the molecule carries out its thermal motion;
v_L = molecular volume in liquid;

k = Boltzman constant;
T = absolute temperature;
E = activation energy.

1.3. DIFFUSION THEORY

For diffusion theory, Simha [9] suggested that under application of external shear force, the self diffusion of molecules increases in the direction of shear field, and naturally the displacement of a molecule decreases in the opposite direction to the shear. This gave the viscosity to be:

$$\eta = \frac{KTr^2}{vD_s}$$

where:

r = distance between centers of two adjacent molecules;
v = molecular volume;
D_s = coefficient of self diffusion.

If in self diffusion the moleculars are considered to have to overcome the activation energy for diffusion as in general, the diffusion theory model in essence is similar to the rate-process theory model.

Numerous variations and modifications of these three theories were developed. Among these three basic types of theories, the molecular theory is considered to best explain the flow phenomena without necessitating introduction of additional assumptions or concepts.

The diffusion model is of interest because it can be used for moderately large polyatomic molecules. The diffusion theory can be applied to chains of 6 to 50 carbon atoms in length. In using this diffusion theory, it is assumed that the flow is caused by the motion of chain ends.

The theories for the calculation of diffusion of polymer end chain have been developed. The diffusion constant can be calculated from these theories. The motion of the chain end is retarded by (1) internal rotation which can be determined by thermodynamic measurements, and (2) by external barrier which is calculated from the interaction of the chain ends as independent molecules. Viscosity of hexadecane was calculated from this method. The calculated values are in agreement with the measured values [8].

2. Effect of Molecular Structure on Viscosity

These theories, although rigorous, at present can be applied to only extremely simple monoatomic fluids and the methods for applying these theories to materials of more general interest should be developed. The understanding of the relations between the viscosity and molecular structure is very important and so pressing that rightly, there is a general agreement that physically sound semi-empirical property correlations would be more immediately useful. Such an approach could lead to semi-quantitative estimates of viscosity related to structure and permit generalization of available information. The effect of molecular structure on the viscosity is discussed below and based mainly on Bondi's review [8].

2.1. NON-ASSOCIATING POLYATOMIC LIQUIDS

The use of the concept of central forces gives reasonable values of the pair potential for many of the small molecules such as diatomic and triatomic molecules. In non-associating polyatomic molecules, in general, the molecules interact primarily through their outer atoms. Hydrocarbons are a good example. Because of their relatively simple structure, a quantitative estimation of the interaction can be made by separately considering each set of interactions; the C—H and C—H, C—H and C—C, and C—C and C—C [10]. This approach, again, can be only useful in simple, non-associating molecules.

2.2. POLAR LIQUIDS

When the permanent dipoles exist in the molecules of a liquid, the liquid has a higher intermolecular force. This higher intermolecular force will hinder or prevent the free external rotation of the molecule. This is because the higher intermolecular interaction might lead to more or less permanent association of neighboring molecules. However, the effect of typical dipoles on viscosity is shown to be very small [8]. But, one exception, a very important one, is that of the hydrogen-bonding types of dipoles which are very common in food materials. Effect of H-bonding on viscosity becomes more significant when the association extends over many molecules. It is even more significant when the association is strong enough for the formation of dimers, trimers or polymers.

In the case of fatty acids, the hydrogen bonding is completely neutralized between two molecules and there is no appreciable effect on viscosity [11]. On the other hand, if the molecular association is more continuous, the continuous association has a more important bearing on the viscosity. It is especially so when this hydrogen bond definitely prefers a given configuration as in the case of water. This type of liquid cannot be treated rigorously in order to calculate viscosity because of the increase in intermolecular potential, as well as complexity of determining the radial distribution of neighboring molecules.

However, fortunately quantitatively, the viscosity reflects the strength of the hydrogen bonds in liquids. This quantitative relation is shown by comparing the energies of vaporization with apparent energy of activation for flow [8].

Hydrogen bonding is very weak compared with covalent bond. Taking water as an example, because all food material contains some water and most 70% or more, the O—H bond within a single molecule is stronger than hydrogen bonding. However, hydrogen bonding is strong enough to form the continuous tetrahedral structure of water as illustrated in Figure 3.

From the mean configuration of the surrounding molecules of a given molecule, it is likely that the distribution deviates from the more theoretically comfortable radial to ellipsoidal. As the molecular configuration of water is distorted, the hydrogen bonding between molecular groups becomes stronger. The shear force acting on the molecular groups tends to bond the groups more strongly, thus causing the force resulting motion to vary with the force causing the motion. This is considered a probable reason for fluid friction being different from sliding friction between solids [11].

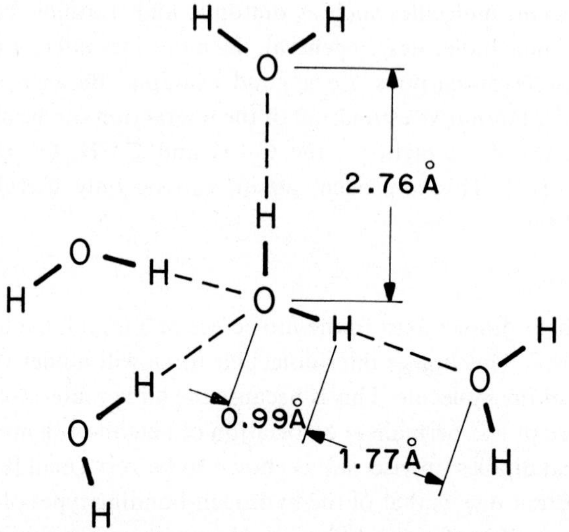

Fig. 3.   Tetrahedral structure of four water molecules formed by hydrogen bonding.

When a material with many OH and H groups, such as many types of sugars, are dissolved in water, the hydrogen bonding increases; thus the increase in viscosity. In looking into the mechanism of flow Eyring's group reasoned that in short chain alcohols, passing of a molecule past another involves the breaking of hydroxyl bonds [13].

2.3. HOMOLOGOUS POLYMER

In polymer homologues with repeating unit increasing in one direction, generally viscosity increases with degree of polymerization as shown in Figure 4. Figure 6 in Chapter III also shows the increase in viscosity, i.e. slope of shear stress vs. shear rate. Increase in viscosity with increasing degree of polymerization is not linear at the low degree of polymerization (less than 50 D.P. in general). This is because the significant contribution of the different interactions of the end groups is the successive dilution of the terminal group effect taking place up to linearity.

2.4. ORDINARY HOMOLOGOUS

In ordinary homologues, the increase in the repeating units will dilute the effect of functional groups; and when repeating units become large, then all homologous series will reach the same viscosity. Figure 5 is more typical of the homologous series effects on viscosity.

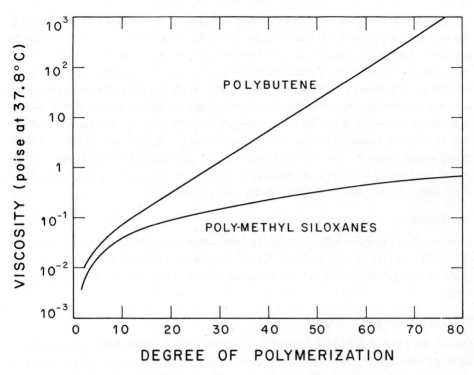

Fig. 4. Increase in viscosity with increase in degree of polymerization.

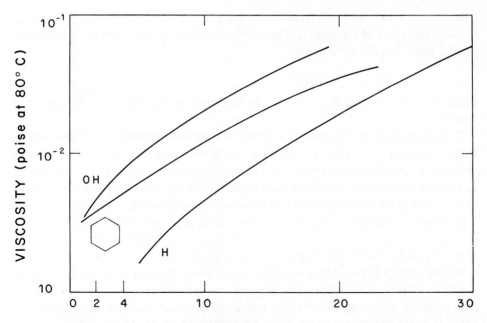

Fig. 5. Increase in viscosity with increase in non-functional repeating unit.

One series of compounds with OH which indicate strong dipole or hydrogen bonding increase in viscosity with an increase in M.W. This increase is much similar to paraffins. The other compound with benzene ring shows a viscosity vs. M.W. approaching that of paraffin of equal chain length at high M.W. The reason for this is that for dipole or H-bond interaction, overwhelming strength of dipole-dipole compared with the dipole-CH_2 interaction is manifested in intermolecular potential. For weakly interacting groups such as the one containing benzene ring, the interaction between two functional groups and the interaction between a functional group and CH_2 group are of the same order. Therefore, as M.W. increases, the probability of interaction between CH_2 groups increases and therefore it is reasonable that viscosity of these series reaches that of hydrocarbons with similar M.W.

2.5. ISOMERS

Isomerization causes no effect on the viscosity unless it causes the shielding of powerful dipoles or OH groups which bring the change in intermolecular potential. This again shows the importance of the influence of intramolecular flexibility on the viscosity.

2.6. HOMOMORPHS

One of the most striking features of the viscosity of liquids is that the molecular of similar shape has a similar viscosity regardless of their composition. This is of course provided that there is no difference in strong dipoles and molecular flexibility.

2.7. EFFECT OF TEMPERATURE AND PRESSURE ON VISCOSITY

The discussion here will only supplement the effect of temperature and pressure on viscosity presented in Chapter III. The relationship between the viscosity of liquid and boiling point was found to be [5]:

$$\eta = hne^{3.8\,T_b/T}$$

where h = Plank's constant, n = number of particles/volume, and T_b = boiling point at 1 atm.

From experimental determination of the temperature dependence of viscosity at constant pressure, enthalpy of activation can be obtained. For most substances the enthalpy of activation ranges from $\frac{1}{3}$ to $\frac{1}{4}$ of the energy of vaporization. For associating liquids it is much larger than for normal liquids. For long polymer molecules, the heat of activation is found to be around 10 kcal mole^{-1} independent of chain length. This corresponds to a flow unit between 20 to 40 carbon atoms in a molecule. From this it is postulated that long chain molecules move one coil at a time similar to the movement of a snake, rather than one whole molecule at once.

Moore *et al.* [14] found that at low temperatures, the motion of the molecule is mostly translational. At intermediate temperatures rocking or vibration becomes more important. At high temperatures, the molecules have free rotation about their long axes. At low temperatures, flow takes place through cooperation of two or more molecules, but at high temperatures, the flow is a unimolecular process, and perhaps even fractional as discussed above.

The temperature dependency of viscosity at constant volume can show us something of the role of holes in the flow process such as the concept in rate process theory. Considering the fact that the number of holes in a liquid would be almost independent of temperature at constant volume, change in viscosity with temperature at constant pressure would be very small [5]. This is true for normal liquids but not so for associated liquids. The viscosity of associated liquids is affected considerably by temperature. The reason for this temperature effect was considered to be caused by the breaking of the bonds between associated molecules in order to make smaller units which can flow easier.

From pressure dependency of viscosity at constant temperature, the volume of activation or the additional volume required for activation can be determined. For normal liquids this is approximately $\frac{1}{6}$ of the volume/molecule.

2.8. Viscosity of solution

Eyring and coworker's empirical relationship for the variation in viscosity with composition is shown below.

$$\log \eta = x_1 \log \eta_1 + x_2 \log \eta_2$$

where x_1, x_2 = mole fraction.

In true solution where particle size is smaller than 10^{-6} mm, the viscosity exhibit ideal, Newtonian characteristics. Terminology and definition of the flow characteristics can be found in the latter part of this chapter as well as in Chapter IV. Viscosity of solutions depends on the viscosity of solvent and solute and concentration of solvent.

Viscosity of sucrose solutions at various concentrations is given by Bingham and Jackson (Table I). Viscosity of sucrose solutions increases with concentration and

TABLE I

Viscosity of sucrose solutions in centipoises[a]

Temp., °C	Percentage sucrose by weight			Temp., °C	Percentage sucrose by weight		
	20	40	60		20	40	60
0	3.818	14.82	–	50	0.974	2.506	14.06
5	3.166	11.60	–	55	0.887	2.227	11.71
10	2.662	9.830	113.9	60	0.811	1.989	9.87
15	2.275	7.496	74.9	65	0.745	1.785	8.37
20	1.967	6.223	56.7	70	0.688	1.614	7.18
25	1.710	5.206	44.02	80	0.592	1.339	5.42
35	1.336	3.776	26.62	85	0.552	1.226	4.75
40	1.197	3.261	21.30	90	–	1.127	4.17
45	1.074	2.858	17.24	95	–	1.041	3.73

[a] From *International Critical Tables*, **5**, 23 (1917); Bingham, E. C. and Jackson, R. F.: *Bur. Standard Bul.* **14**, 59, 1919 [15].

effect of concentration is greater at higher concentration. Many types of food solids, in general, increase the viscosity upon addition. However, not all the solutions increase in viscosity with solute concentration. In case of alcohol and water solutions, the viscosity increases with increase in alcohol content, then further addition decreases the viscosity (Table II). Interestingly the maximum viscosity of alcohol-water mixture is at between 40 to 50% alcohol concentration which is close to some of the popular alcoholic beverages.

TABLE II

Viscosity of alcohol-water mixtures in centipoises[a]

Temp. °F	Concentration of ethanol, percent by weight									
	10	20	30	40	50	60	70	80	90	100
32	3.311	5.32	6.94	7.14	6.58	5.75	4.76	3.69	2.73	1.77
41	2.58	4.06	5.29	5.59	5.26	4.63	3.91	3.12	2.31	1.62
50	2.18	3.16	4.05	4.39	4.18	3.77	3.27	2.71	2.10	1.47
59	1.79	2.62	3.26	3.53	3.44	3.14	2.77	2.31	1.80	1.33
68	1.54	2.18	2.71	2.91	2.87	2.67	2.37	2.01	1.61	1.20
77	1.32	1.82	2.18	2.35	2.40	2.24	2.04	1.75	1.42	1.10
86	1.16	1.55	1.87	2.02	2.02	1.93	1.77	1.53	1.28	1.00
95	1.01	1.33	1.58	1.72	1.72	1.66	1.53	1.36	1.15	0.91
104	0.91	1.16	1.37	1.48	1.50	1.45	1.34	1.20	1.04	0.83

[a] From *Food Texture*, Samuel A. Matz, The Avi Publishing Company, Inc. 1972

2.9. VISCOSITY OF DISPERSED SYSTEM

Viscosity of emulsion is discussed in Chapter XII with appropriate mathematical equations. Often the general approach to emulsion can be applied to suspensions as well.

According to Scott Blair [16], for concentrated suspensions Hatscheck proposed the viscosity to be

$$\eta_{sus} = \frac{1}{1 - \sqrt[3]{\phi K}},$$

where ϕ is the ratio of volume suspended to total volume. This equation further includes voluminosity factor, K, which takes into account the swelling due to the solvent attaching to the suspension and increasing the volume of particle.

Roscoe studied suspensions and proposed, for suspension of uniform size particles in high concentration,

$$\eta_{sus} = \eta_0 (1 - 1.35 \, \phi)^{-2.5}$$

and for diversified size particles in high concentration

$$\eta_{sus} = \eta_0 (1 - \phi)^{-2.5},$$

where η_0 is the viscosity of solvent. As indicated by these equations, the high concentration suspension deviates from Newtonian. One of the reasons for the deviation is self 'crowding factor'. When the concentration of suspension increases, there is an increase in the volume of the dispersed phase. This increase in the volume of the dispersed phase is accompanied by the decrease in volume of the continuous phase.

This self crowding factor is incorporated in an equation to calculate viscosity of suspension as follows [16]:

$$\eta_{sus} = \eta_0 K\phi \,(1 - Q\phi),$$

where Q = self crowding factor.

Another factor causing the anomaly is the electrical charge. When there are electrical charges present, they can increase viscosity of a dispersion by mutual interaction, and it is called electro-viscous effects.

In colloidal suspensions, suspended particles are strongly solvated. Solvation involves the strong binding of solvent molecules, usually in a layer not more than one molecule thick. The combined unit, the particle plus the solvent layer, as a whole, acts as a single kinetic unit in solution. This increase in volume, if not accounted for, will cause deviation from the viscosities calculated.

The degree of solvation for spherical colloidal particles can be indicated by the ratio of hydrodynamic volume to dry specific volume. Hydrodynamic volume can be obtained by finding the value required to confirm the appropriate equation [18]. Some numerical values for the ratios of hydrodynamic to dry specific volumes listed in the literature [12] are:

	1.6	for sucrose in water
	0.9	for egg albumin in water
	6	for isoelectric gelatin at 40°C
and	30	for isoelectric gelatin at 30°C

For sucrose and albumin the ratio is close to one. However, extremely large values introduced by high solution viscosities are due to cause other reasons besides solvation. It is caused by:

(1) Intermolecular network of solute molecules, possibly holding solvent molecules in between. The intermolecular network may be covalent bonding but more often it is H-bonding and also often van der Waals forces.

(2) Non-spherical shape of colloidal particles.

The asymmetric particle requires increased extra energy for flow because of the rotation of the particle induced by the velocity gradient, and interaction between the molecules, or intermolecular network such as covalent bonding and hydrogen bonding.

The expression for calculating viscosity of dilute suspension using the ratio of long axis to short axis has also been developed [12]. So far, only the dispersion without

electrical charge was discussed. However, in actuality, most of the dispersions, especially food dispersions, have electrical charges. These electrical charges can increase
the intrinsic viscosity of a dispersion by mutual interaction and lead to electro-viscous
effects. The electrical charge effects have been related to the viscosity of suspension
through use of such factors as electrokinetic potential, dielectric constant of the dispersion medium and specific electric conductivity of the suspension.

2.10. VISCOSITY PHENOMENA

In practice, what is actually observed is the velocity gradient during the flow due to
viscous effect. The concept of viscosity is demonstrated by the frictional resistance of
a fluid in motion to the applied shear stress. The nature of this resistance is shown in
Figure 6. Chapter IV also discusses the flow under applied shear, and the similar

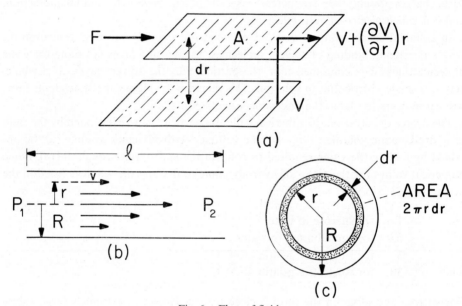

Fig. 6.   Flow of fluids.

illustration is given in Figure 1 of that chapter. When a fluid is placed between two
parallel plates and force is applied to one plate, or when fluid is placed in a tube and a
pressure is applied, the fluid adjacent to the plate assumes the velocity of the wall.
For instance, the fluid between these two planes will have velocity ranging from the
velocity of the moving plate, $V + (\partial V/\partial r)\, dr$ at the wall to zero velocity adjacent to
the stationary plate as in Figure 6a. In a tube, fluid at the center has maximum velocity
and the fluid next to the wall has zero velocity as in Figure 6b. The velocity of fluid
depends on the shear stress (τ) applied, and the distance from the moving plates (dr)
or distance from the center of the tube (r) and the characteristics of the fluid which
is viscosity (η). The shear stress (τ) is the shear force applied per unit area which is

F/A for two parallel plates and $(P_2 - P_1)(R/2l)$ for the tube where F is force applied and P_2 and P_1 are the pressures at two different positions l distance apart in the tube, and R is the radius of the tube.

2.11. Effect of Rate of Shear on Viscosity

For ideal fluids, the relationship between the change in velocity with respect to distance, which is shear rate, and the force applied per unit area, shear stress is given by a simple equation,

$$\tau = \eta \left(-\frac{dV}{dr} \right).$$

This linear relationship holds for Newtonian or ideal fluids but more often fluids are not Newtonian.

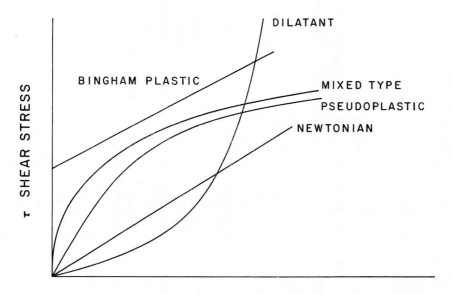

Fig. 7.   Flow behavior of fluids.

Figure 7 shows shear stress-shear rate relationship of various types of fluids. Newtonian fluids show straight line passing through origin.

Bingham plastic has yield stress but is linear. Viscosity of pseudo-plastic liquid decreases with increased rate of shear and viscosity of dilatant fluid increases at higher shear rate. Not too many food products are found to be dilatant. Many of the food materials fall into mixed type having some pseudoplastic characteristics with a yield value. When viscosity is not constant and varies with shear, the ratio of shear stress to rate of shear at a given point is usually called apparent viscosity.

For a more general relationship between shear stress and shear rate, or to include

TABLE III

Summary of flow characteristics of various types of fluids

Flow characteristics of fluids	Typical fluids having the flow characteristics	Value of b	Value of S	Value of C	Examples of Foods
Newtonian	Thin solution	Viscosity $b > 0$	$S = 1$	$C = 0$	Clarified juices, such as apple juice, cranberry juice, oils, confectionary syrup
Pseudoplastic	Emulsions, suspensions	Apparent viscosity $b > 0$	$0 < S < 1$	$C = 0$	Vegetable soup, chowder
Bingham plastic	Thick solution, colloids	Plasticity constant $b > 0$	$S = 1$	Yield stress $C > 0$	French dressing, tomato catsup, fudge sauce
Mixed type	Thick solution with suspended particles of irregular shape	Proportionality constant $b > 0$	$0 < S < 1$	Yield stress $C > 0$	Sandwich spread, jelly, marmalade
Dilatant	Nearly saturated or supersaturated suspensions	Proportionality constant $b > 0$	$1 < S < \infty$	$C = 0$	Sausage slurry, homogenized peanut butter

all types of flow behavior, it is common practice to use an empirical general equation, power law equation,

$$\tau = b(\gamma)^s + C,$$

where γ = rate of shear, dV/dr; b = proportionality constant; s = dimensionless power law constant; C = yield stress.

Table III gives the summary of flow characteristics of various types of fluids expressed in terms of power law equation with examples in food.

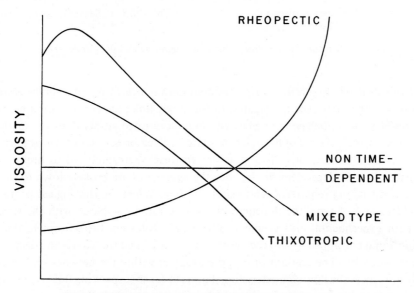

Fig. 8. Time dependency of viscosity.

2.12. EFFECT OF TIME OF SHEAR ON VISCOSITY

For an ideal fluid, the viscosity will remain constant when the rate of shear remains constant, regardless of the period the fluid has been exposed to the shear. But in many heterogeneous or structured systems, the flow properties change with time of shear. Figure 8 shows the types of time dependency exhibited in viscosity. Thixotropic fluids decrease in viscosity as the time goes on, and rheopectic fluids increase with time. Then there are fluids which increase in viscosity, initially, then decrease later and vice versa. Of course, there are some fluids which show the combination of these behaviors. An example of this will be discussed in the following section of this chapter.

2.13. GENERAL FACTORS AFFECTING THE VISCOSITY

In complex material, there are numerous factors which could lead to anomalies of the flow properties. For instance, in suspension, the individual particles may attract each other and form flocs. The flocs can, in turn, group together to form larger aggre-

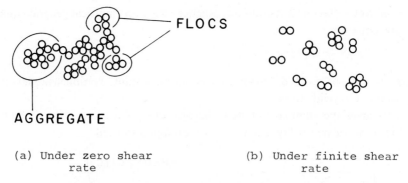

(a) Under zero shear   (b) Under finite shear
      rate   rate

Fig. 9.   Structure of suspensions breaking aggregate and flocs under shear.

gates as illustrated in Figure 9a. Forces maintaining aggregates are not usually as strong as those forming flocs. The aggregates, in turn, may link together to form a network. The non-Newtonian behavior or plasticity in concentrated suspensions is usually accompanied by some degree of flocculation of aggregated or net-work formed particles.

At any rate of shear, there is a tendency to form aggregates and to break down aggregates. In time dependent fluids, once the aggregates are broken down to smaller flocs, reconstructing requires a relatively long time. When the rheological properties are independent of time, the flocculation must be rapid compared with the time required for experimental measurement. When aggregates are broken down, the suspending liquid entrapped within the network is released and this causes the decrease in apparent viscosity. The amount of liquid entrapped within the network or aggregate greatly influences the flow characteristics. This is because suspending fluid is entrapped in the larger aggregates while freed in small aggregates, and the larger aggregates are more likely to be able to absorb energy due to its flexibility and elasticity. Also in smaller groupings, there is less support from adjacent aggregates and therefore, greater ease of movement and flow.

When suspension becomes more concentrated, the surface area of the material increases and the film of fluid coating the particles becomes thinner. In this case, the film acts as a lubricant between particles and the probability of particles coming in contact with each other increases. As the particles come together they adhere to each other forming aggregates. The number and size of the aggregates formed depends on the contact area and frequency of collisions during the flow. When there is sufficient free fluid left to prevent contact between particles, the force required for flow is mainly to shear fluids. But when concentration is high enough to form aggregates, the force is also expended to shear and possibly to break the aggregates [12]. When the concentration of suspended particles increases further beyond this point, the fluid phase becomes discontinuous due to direct contact of many particles. There is a rapid increase in consistency and therefore, dilatancy.

A fluid can exhibit perfectly different time dependencies at different shear rates or with different experimental methods or instruments. For instance, an elastic gel under very low rate of shear may show rheopectic effect. The same gel when sheared at a much higher rate can be thixotropic. This is because gel structure is continuously formed while the shearing breaks the structure and what is observed is the status of the equilibrium structure network in a much similar way as in suspensions discussed above. It is very interesting to note that the properties which are responsible for non-ideal shear dependency are, through same mechanics, also responsible for time dependency such as thixotropic or rheopectic nature.

The factors which were discussed above, such as molecular structures, molecular weight, electrical charges, interaction between molecules or particles, etc. must affect the viscosity of food materials. However, because of the complex nature and many compounds present, isolation of a certain responsible factor which attribute to a given viscosity phenomenon is difficult. Still some crude generalization can be made from the experimental studies of flow behavior.

One of the food materials which has been exposed to a good measure of rheological study is honey. Most of honey is Newtonian but two particular types of honey show anomalities. These honeys show thixotropic behavior with gelatinous consistency [21]. It was elucidated that the gelatinization which takes place when heather honey is allowed to rest undisturbed is due to the presence of about 1 to 2% of a characteristic protein. Heating of these honeys to about 149°F is said to accentuate the gelling behavior of the honey by denaturation of protein [22]. On the other hand, large numbers of honeys are dilatant. These honeys contain dextran, $(C_6H_{10}O_5)_n$, in which n is near 8000. Once the dextran was removed from honeys, they became true Newtonian [21, 22]. What is interesting is that even in this relatively thick and complex fluid, the rheological peculiarity is not governed by the highest concentration solute, sugar in this case, but rather by high molecular weight substances such as protein and dextran although they are at the concentration level of only a few percent. This perhaps suggests the predominant role high molecular weight material plays on viscosity.

Many fruit and vegetable juices and purees containing dissolved long chain molecules and suspending particles, are non-Newtonian, and often pseudo-plastic. Viscosity of tomato juice depends primarily on the structure and composition of cell walls. Whittenberger and Nutting [23] realized that the insoluble solids have a stronger effect on viscosity than soluble pectin. The tomato cell walls contain amorphous cellulose which is different from the crystalline cellulose in wood or cotton. This amorphous tomato cellulose alone can make suspensions which are much thicker than most of the common thickening agents [22]. This amorphous cellulose is affected by electrolytes, and the viscosity of tomato juice can be kept at a relatively low level by either natural or added electrolytes. When electrolytes, such as soluble pectins, organic acids and mineral salts are removed from the tomato juice, it will thicken to almost a gel. However, removal of non-electrolytes such as sucrose, glycol, and ethanol would have no effect on the viscosity. As the electrolytes decrease, the cell walls swell suggesting the neutralization effect of electrolytes on the cell wall.

These outstanding viscosity characteristics can be a good example for speculating the relative importance of the role which many factors affecting the flow behavior play. For instance, does tomato juice really show the stronger effect on viscosity of insoluble solids than soluble solids including the high molecular weight solubles, dramatic ability of electrolytes to change the viscosity, and increasing viscosity with increase in hydrodynamic volume?

References

1. Scott Blair, George W.: *A Survey of General and Applied Rheology*, Sir Isaac Pitman & Sons, London, 2nd ed. 1949.
2. Green, H. S.: *The Molecular Theory of Fluids*, Interscience, New York, 1952.
3. Kirkwood, J. G., Buff, F. P., and Green, M. S.: *J. Chem. Phys.* **17**, 988 (1949).
4. Born, M. and Green, H. S.: *Proc. Roy. Soc., London* **A190**, 455 (1947).
5. Hirschfelder, Joseph O., Curtis, C. F., and Bird, R. Byron: *Molecular Theory of Gas and Liquids*, John Wiley & Sons, New York, 1954.
6. Ewell, R. E. and Eyring, H.: *J. Chem. Phys.* **5**, 726 (1937).
7. Glasstone, S., Laidler, K. J., and Eyring, H.: *The Theory of Rate Process*, McGraw-Hill, New York, 1947.
8. Bondi, A.: 'Theories of Viscosity' in *Rheology* vol. 1 (ed. by F. R. Eirich), Academic Press, Inc., New York, 1956.
9. Simha, R.: *J. Chem. Phys.* **7**, 202 (1939).
10. Muller, A.: *Proc. Roy. Soc., London* **A154**, 624 (1936).
11. Bondi, A.: *J. Chem. Phys.* **14**, 591 (1946).
12. Charm, S. E.: *Adv. Food Res.* **11**, Academic Press, Inc., New York, 1962.
13. Ewell, R. E. and Eyring, H.: *J. Chem. Phys.* **5**, 726 (1937).
14. Moore, R. J., Gibby, P., and Eyring, H.: *J. Phys. Chem.* **57**, 172 (1953).
15. Bingham, E. C. and Jackson, R. F.: *International Critical Tables*, **5**, 23 (1917); *Bur. Standard Bul.* **14**, 59 (1919).
16. Scott Blair, G. W.: *Elementary Rheology*, Academic Press, London and New York, 1969.
17. Roscoe, R.: *Br. J. Appl. Phys.* **3**, 267 (1952),
18. Alexander, A. E. and Johnson, P.: *Colloid Science*, vol. 1, p. 361, The Clarendon Press, Oxford, 1949.
19. Charm, Stanley E.: *Fundamentals of Food Engineering*, Avi Publishing Co., Inc., Westport Conn., 1962.
20. Perce-Jones, J.: 'The Rheology of Honey', in G. W. Scott Blair (ed.), *Foodstuffs: Their Plasticity, Fluidity and Consistency*, Interscience Publishers, New York, 1953.
21. Matz, Samuel A.: *Food Texture*. Avi Publishing Co., Inc., Westport, Conn., 1962.
22. Whittenberger, R. T. and Nutting, G. C.: *Food Technology* **2**, 19 (1957).

CHAPTER III

DETERMINATION OF VISCOSITY OF FOOD SYSTEMS

JULIAN F. JOHNSON

Dept. of Chemistry and Institute of Materials, Science,
University of Connecticut, Storrs, Conn. 06268, U.S.A.

JOHN R. MARTIN

Research Dept., The Foxboro Co., Foxboro, Mass. 02035, U.S.A.

and

ROGER S. PORTER

Polymer Science and Engineering Program, University of Massachusetts,
Amherst, Mass. 01002, U.S.A.

1. Introduction

Viscosity is an important physical property of food systems. It is related to flow properties of fluids which in turn control manufacturing operations in many cases. This chapter describes a variety of methods for determining viscosities of liquids and gases as a function of temperature, pressure and shear. As only fluids are considered, viscoelastic measurements, normal stress effects and elongational flow are excluded. The various theories of viscosity and the interpretation of flow characteristics of suspensions have been reviewed in Chapter II and elsewhere [1–3] and are therefore also omitted.

If an external stress is exerted on a fluid, flow will occur. The rate of flow will be decreased by internal frictional forces. The viscosity of the fluid is a measure of the resistance to flow. Dynamic viscosity, η, is defined as the tangential force on unit area of one of two parallel planes at unit distance apart with the space between the planes filled with the fluid and one of the planes moving relative to the other with unit velocity in its plane. The unit of η is the poise with dimensions $(ML^{-1}T^{-1})$. A more readily measured viscosity is the kinematic viscosity, γ. This is defined by the equation:

$$\gamma = \frac{\eta}{\varrho},\tag{1}$$

where ϱ is the density of the fluid.
The dimensions of γ are (L^2T^{-1}) and the unit is the stoke.

For simple fluids usually referred to as Newtonian fluids the rate of flow or shear rate is directly proportional to the applied shear stress. Viscosity can be expressed as:

$$\eta = \frac{\tau}{\dot{\gamma}} \left\{ \begin{array}{l} \text{shear stress, dyne cm}^{-2} \\ \text{shear rate, reciprocal seconds} \end{array} \right\}.\tag{2}$$

Thus for Newtonian fluids the viscosity is independent of the shear stress and, by definition, it is not for non-Newtonian fluids. In general low molecular weight liquids tend to be Newtonian over a wide range of shear stresses whereas high molecular weight polymers are usually non-Newtonian.

At low rates of shear, fluids are in laminar flow. The internal motion of the fluid is smooth. As the rate of shear increases there comes a critical point above which the flow changes from laminar to turbulent. In turbulent flow the internal motion contains vortices and other irregularities. The flow regime and the point at which turbulence sets in is usually given in terms of the Reynolds number, a dimensionless quantity. Methods of calculating the Reynolds number, Re, are given for specific viscometers.

The viscosity of fluids changes markedly with temperature so good temperature control is required for precise viscosity measurements. Similarly the viscosity of liquids changes with pressure. The effects of temperature, pressure and shear are described in more detail in later sections.

2. Viscosity of Liquids

2.1. MEASUREMENT

2.1.1. *Capillary Viscometers*

A commonly used and convenient method for measuring the viscosity of liquids is by use of a glass capillary viscometer. An example of such a viscometer is shown in Figure 1. In principle, the time, t, for a given volume of liquid, V, as defined by the etched lines 1 and 2, to flow through the capillary of length, L, and radius, r, is measured. If the density of the liquid, ϱ, is known the viscosity may be calculated from the Poiseuille equation:

$$\eta = \frac{\pi r^4 t P}{8LV},\tag{3}$$

where η is the absolute viscosity and P, the applied pressure differential is defined by:

$$P = hg\varrho,\tag{4}$$

where h is the average height of the liquid in the tube and g is the acceleration due to gravity. Viscometers have been used in this manner to determine absolute viscosities. Usually, however, the absolute dimensions of a capillary are difficult to determine precisely. Therefore, the flow time for a liquid of known viscosity is determined and a viscometer constant C is measured such that the viscosity and C are defined by the equations:

$$\gamma = Ct \qquad C = \frac{h\pi r^4 g}{8LV}.\tag{5}$$

This equation is not exact. A more exact form often used is:

$$\gamma = Ct - \frac{B}{t}.\tag{6}$$

Here, the constant B represents a term based on the geometry of the viscometer to correct for the kinetic energy effects. This 'constant' is a complex one and, in reality, is not a constant. It may be measured by calibrating with two standards of known different viscosities or computed by means of the formula cited in footnote b in Table I.

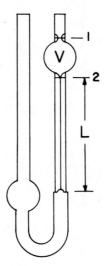

Fig. 1. Ostwald viscometer

Typically capillary viscometers are designed so that at flow times of 200 s or more the kinetic energy correction ranges from 1% downward so that precise determination of B is not required. Similarly the design features make it unnecessary to correct results for a number of other effects such as entrance effects, surface tension, drainage, curvature of capillary, etc.

Glass capillary viscometers have been designed in a variety of configurations. Figure 2 shows comparative sizes of a number of commonly used viscometers. In general the various types have originated to make measurements on a wide variety of liquids ranging from transparent to opaque, with viscosities from 0.5 to 100000 centistokes. Good temperature control, $\pm 0.01\,°C$, and timing to at least 0.1 s are required. Thermostating is readily achieved by a variety of commercially available liquid constant temperature baths. Normally the viscometer is immersed in the constant temperature bath. Temperatures range from -180 to $+500\,°C$. Table I gives viscometer characteristics for the capillary viscometers illustrated in Figure 2.

The shear rate in a capillary is not constant. It varies from zero in the center of the tube to a maximum value at the wall given by:

$$S_M = \frac{4V}{\pi r^3 t}.$$

(7)

The maximum value, S_M, expressed in units of reciprocal seconds, occurs at the wall of the capillary. For a Newtonian liquid, that is, one whose viscosity is independent of

TABLE I
Viscometer characteristics [6]

Designation	FitzSimons	Ubbelohde	Fenske-Cannon	Zeitfuchs	Dean-Ruh	Cannon Master	Fenske-Cannon	Zeitfuchs-Lantz	Zeitfuchs cross arm
Type		Suspended Level		Normal Flow	Normal Flow			Reverse Flow	
Recommended use	Transparent liquids						Opaque liquids		Both transparent and opaque
Useful range, centistokes	0.5-1200	1-100000	0.8-5600	0.6-5000	0.6-10000	0.5-6000	0.8-10000	1-15000	0.5-100000
Sample size, ml	12	15	8	10	15	8	13[a]	13	2
Required bath depth, cm	32	30	20	36	30	60	36	27	28
Capillary length, cm	12	9	7	15	14.5	45	15	20	21
Viscometer length, cm	28	9	25	39	29.5	63	29	29	28
Value of C temperature dependent	No	No	Yes	No	No	Yes	Yes	No	No
Value of B[b]	1.25	2.8	1.7	1.4	1.4	0.2	1.8	1.4	0.07
Adjustment of liquid head	Automatic	Automatic	Manual	Automatic	Automatic	Manual	Manual	Automatic	Automatic
Rerun possible	Yes	Yes	Yes	Yes	Yes	Yes	No	No	No
Calibration starting with water	Yes	Yes	No	Yes	Yes	Yes	No	Yes	Yes
Fill while in bath	Yes	Yes	No	Yes	Yes	No	No	Yes	Yes
Time to reach 210 °F, minutes	1-2	10	10	10	10	10	10	10	5
Clean in bath	Yes	Yes	Yes	Yes	No	Yes	Yes	Yes	Yes

[a] Can be constructed to use 5-ml sample.

[b] Calculated from $B = mV/8\pi L \times 100$, where V = volume in ml, L = length in cm, $m = 1$.

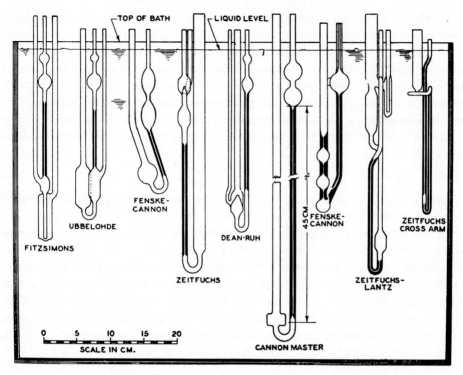

Fig. 2. Comparative sizes of capillary viscometers.

shear rate, the shear rate profile is that of a parabola; and the average value, S_A, throughout the capillary is:

$$S_A = \frac{8V}{3\pi r^3 t}. \tag{8}$$

In reality, there is no interest in shear rate unless the viscosity is varying with shear rate. So, Equation (8) is an approximation to the actual average shear rate. Another viscometer constant that is of interest is the Reynolds number. This is defined by:

$$\text{Reynolds Number} = \frac{2Rv\varrho}{\eta}. \tag{9}$$

The term, v is the velocity of the liquid in the capillary in centimeters per second. The significance of the Reynolds number is that it is an approximate measure of the point where flow changes from laminar to turbulent conditions. At Reynolds numbers below 1500–2000 in smooth-walled tubes or capillaries, the desired laminar flow conditions prevail. Reynolds numbers can also be defined for other geometries and for non-Newtonian fluids. The transition point between flow regimes is, of course, different for these other geometries. Reference [3] discusses methods of defining Re for non-Newtonian fluids.

Capillary viscometers operating under gravity flow are limited to liquids with viscosities of less than about 100000 centistokes or roughly 1000 poise. This is an extreme limit reached by using very large capillaries and long flow times. A more practical estimate would be 100 poise. To measure higher viscosities variable pressure capillaries may be used, thus giving a wide range of P in Equation (3). There are a variety of methods of varying pressure, use of compressed gases, pumps, mechanical drives, etc. It is necessary to measure the pressure precisely for determining viscosity either by use of Equation (3) or (5). If pressure is a variable Equation (5) must be modified to include this term:

$$\gamma = CtP. \tag{10}$$

One such variable pressure capillary viscometer will be described. Although details of operation differ it will serve as an example for the various types. Figure 3 is a schematic diagram of the capillary and pressure measuring device in an extrusion rheometer (Instron Engineering Corp., Canton, Mass.). The viscometer operates as follows. The capillary is electrically thermostated. Above this is a barrel containing the fluid to be measured. A machined screw crosshead is advanced at a constant rate. This drives a plunger resting on the fluid thus forcing the fluid through a capillary. The pressure required for flow at the constant rate is measured by a compression load cell and recorded. By use of a variety of plunger speeds and capillary sizes a wide range of viscosities and shear rates may be covered. Viscosities from about 5 to5×10^6 poise may be measured.

As the pressures encountered in the barrel may reach 25000 psi compressibility corrections for the fluid can be significant. Other corrections related to entrance effects, capillary lengths to diameter ratios, etc., may be necessary but will not be discussed in detail here. The detailed practice of viscometry is given in several standard references (See, for example, references [1, 4, 5]).

The precision and accuracy of the capillary viscometers may be increased by the use of automatic timing and recording devices. A summary of the areas of utility of the various viscometers is given in Table II.

TABLE II
Areas of utilization of viscometers

Capillary viscometers – gravity operation
Viscosities of Newtonian fluids – 0.5–10000 centistokes, − 180–500 °C. Inexpensive, simple to operate, good precision. Can be used to study viscosity as a function of shear over limited, low shear rate region. Suitable for dilute solution viscosities of macromolecules.

Capillary viscometers – gravity operation – automatic operation
Same viscosity and shear range as above. Give higher precision. Expensive. Operate unattended. Find most applications in area of dilute solution viscosities of macromolecules.

Capillary viscometers – variable pressure
Viscosities of Newtonian and non-Newtonian fluids 0.01 − 5 × 10⁶ poise. Wide temperature range. Variety of designs range from moderately to very expensive. Can be used to study viscosity as a function of shear from very low to very high shear stresses. Precision good to moderate.

2.1.2. *Cone and Plate Viscometers*

For measurements of viscosities in the moderate to high range and for measurements of viscosity as a function of shear, the cone and plate design has several attractive features. Figure 4 illustrates the viscometer configuration. The cone is driven at a

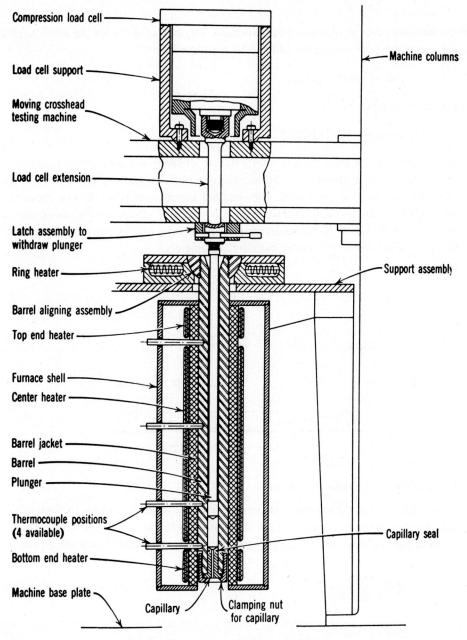

Fig. 3.   Cross section of an Instron rheometer.

known speed giving an angular velocity Ω. The resulting torque, M, is usually measured by some type of strain gauge on the fixed plate. For small, $<3°$, angles of the cone the viscosity is given by:

$$\eta = \frac{3\alpha M}{2\pi R^3 \Omega} \tag{11}$$

the cone angle, α, is in radians in this equation. The rate of shear for the small angles used is approximately constant and can be calculated from:

$$\text{rate of shear} = \Omega/\alpha .$$

Shear stress is given by:

$$\text{shear stress} = 3M/2\pi R^3 .$$

The constant rate of shear makes interpretation of shear viscosity plots straightforward. By varying the speed, cone angle and cone radius a wide variety of viscosities and shear rates may be measured. The required sample size is small. The geometry makes precise thermostating cumbersome but it can be done, usually using air or inert gas thermostats. At high stresses the temperature rise can introduce serious errors. This is discussed under concentric cylinder viscometers.

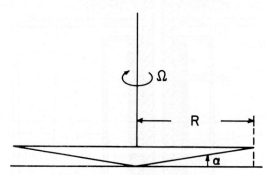

Fig. 4.   Cone and plate viscometer.

2.1.3. Concentric Cylinder Viscometers

Another major class of rotational viscometers is the concentric cylinder configuration. An example is shown in Figure 5. If the length of the cylinder is l, the radii of the inner and outer cylinders R_i and R_o respectively, the angular rotation Ω, and the torque M then:

$$\eta = \frac{M}{4\pi\Omega l}\left(\frac{1}{R_i^2} - \frac{1}{R_o^2}\right) \tag{12}$$

$$\text{rate of shear} = \frac{2\pi R_i^2 R_o^2}{R_x^2(R_o^2 - R_i^2)} \tag{13}$$

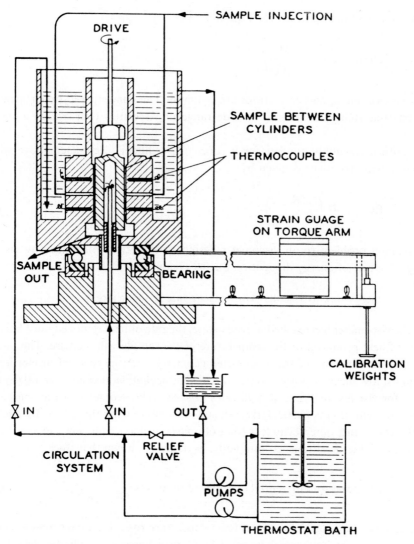

Fig. 5. Section through recording high shear concentric cylinder viscometer.

for any radius R_x. For clearances between cylinders that are small compared to the cylinder radius i.e. $R_i \approx R_o$ and $R_o - R_i \ll R_i$, the shear rate is approximately constant and given by:

$$\text{rate of shear} = \frac{2\pi R_{AV}^2}{(R_o^2 - R_i^2)}, \tag{14}$$

where $R_{AV} = (R_o + R_i)/2$

$$\text{shear stress} = \frac{M}{2\pi R_x l} \tag{15}$$

or, for the conditions specified for Equation (14):

$$\text{shear stress} = \frac{M}{2\pi R_{AV} l}. \tag{16}$$

For large clearances, end corrections are required. The concentric cylinder arrangement has the ability to measure a wide range of viscosities over a wide range of shear.

The critical Reynolds number for a concentric cylinder viscometer above which turbulence will take place is given by:

$$\text{Re}_{\text{crit}} \simeq 41 \left(\frac{R_o}{R_o - R_i} \right)^2. \tag{17}$$

A common problem to all viscometer designs is the error introduced by viscous heating. The rate of energy dissipation due to viscous flow is given by:

$$\text{energy due to viscous flow} = \text{shear rate} \times \text{shear stress}.$$

When the viscometer has reached a steady state, the rate of heat removed must balance the rate of heat produced or the temperature of the sample will increase. The rate of heat removed depends on the viscometer geometry and the thermal properties of the fluid. At low shear rates and stresses the effect is negligible. It can be several degrees or more for viscous materials at high rates of shear. The geometry of the concentric cylinder configuration is particularly well suited to minimizing this effect. By cooling both the inner and outer cylinders for example, the maximum temperature rise is decreased by a factor of four. Additionally, as it can be shown that since:

$$\text{Temperature differential in center of fluid is} \sim \frac{1}{(R_o - R_i)^2}, \tag{18}$$

the use of very small gap widths gives low temperature rises. Figure 6 shows viscosity shear measurements on normal hydrocarbons on a concentric cylinder viscometer with cylinder F and $4A$ having radial clearances of 2.77 and 1.32×10^{-4} cm, respectively. The measured viscosities are constant over the entire shear range measured indicating an absence of significant temperature effects.

2.1.4. Industrial and Miscellaneous Viscometers

The three types of viscometers described in the previous sections are the most widely used. There are a number of other viscometers which find specialized uses.

The falling ball or rolling ball viscometers measure the time for a ball to fall through a liquid or to roll down an inclined plane. These are particularly useful in measurements of viscosity as a function of pressure. Similarly the time for a bubble to rise through a liquid is a measure of viscosity.

Sliding plate viscometers have been applied to very viscous materials, for example asphalt.

Measurement of the rate of damping of a rapidly vibrating reed can be used to measure the product of viscosity times density. Process continuous viscometers using this principle are commercially available.

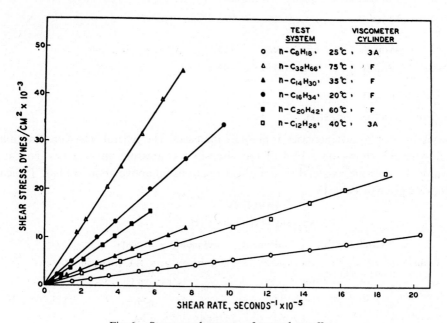

Fig. 6.   Stress vs. shear rate of normal paraffins.

There are a large number of viscometers basically of the capillary type that are used for specific tests tied to particular materials. Most of these viscometers are carefully defined dimensionally. Usually they utilize a relatively short capillary and therefore have large kinetic energy corrections that are ignored. Some of these are the Saybolt, Engler, Zahn cups, etc. Fortunately their use is rapidly decreasing and measurements are made in terms of actual viscosities rather than flow times through an arbitrary design.

2.2. THE EFFECT OF TEMPERATURE

The viscosities of liquids are strongly temperature dependent with the viscosity decreasing as temperature increases. There are a few exceptions to this directional change but they usually involve changes in molecular association. For many liquids the change in viscosity as a function of temperature may be represented over quite extensive temperature ranges by:

$$\log \eta = A + \frac{B}{T},$$ (19)

where A and B are constants and T is the absolute temperature. This relationship is valid for simple liquids and solutions including high molecular weight polymers and their solutions. Some representative values are given in Table III.

TABLE III

Viscosity of liquids in centipoises

$T\,^\circ C$	0°	25°	50
Water	1.79	0.90	0.55
Benzene	0.90	0.61	0.44

2.3. The effect of pressure

The viscosity of liquids increases as pressure increases. The general behavior of normal liquids usually shows on a plot of log viscosity vs. pressure an area first concave towards the pressure axis and then a linear relationship above about 3 kbar. Typical values are given in Table IV.

TABLE IV

Viscosity in centipoises [7]

Pressure in kbar	n-Pentane	n-Butane	Propane
0 (extrapolated)	0.21	0.16	0.11
2	0.73	0.56	0.38
4	1.54	1.06	0.70
6	2.82	1.85	1.08
8	4.77	2.93	1.57

2.4. The effect of shear: non-newtonian liquids

Newtonian liquids as previously defined and illustrated in Figure 6 are characterized by viscosities that are independent of the shear rate or shear stress. Many high molecular weight materials and solutions of high molecular weight polymers exhibit a decrease in viscosity as shear is increased. This is usually attributed to shear induced alignment of the molecules in the direction of flow to produce a decreased resistance to flow. When alignment of the molecules is complete a second Newtonian region is reached where the viscosity is again independent of shear. This general behaviour is illustrated in Figure 7. In theory such behavior should be observed for all molecules except perhaps spherical nondeformable molecules. In practice it is difficult to reach sufficiently high shear rates, without introducing temperature gradients in the liquids, to observe non-Newtonian behavior in materials other than high molecular weight polymers and solutions of such polymers. For polymers and polymer solutions non-Newtonian behavior has been studied extensively. Figure 8 shows typical results on a 11.5% solution of polyisobutene, molecular weight 2 200 000 in cetane [8].

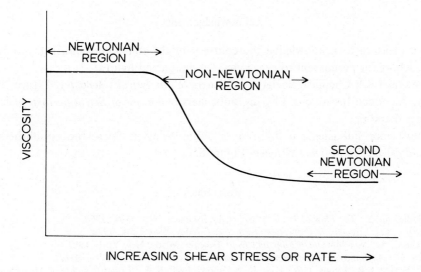

Fig. 7.   Generalized plot of viscosity vs shear rate or shear stress.

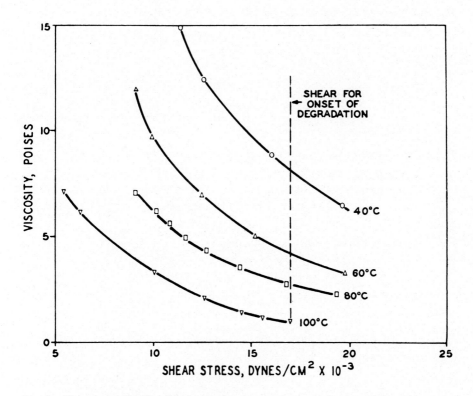

Fig. 8.   Viscosity of 11.5% polyisobutene in cetane. Polymer molecular weight $= 2.2 \times 10^6$.

Acknowledgements

It is a pleasure to acknowledge the courtesy of the following publishers, journals and authors for permission to reproduce the designated figures:

The American Chemical Society, publishers of *Analytical Chemistry*, Figure 2.

The American Institute of Physics, publishers of *Review of Scientific Instruments*, Figures 5 and 6.

Interscience Publishers, a division of John Wiley & Sons, Inc., publishers of *Journal of Applied Polymer Science*, Figure 8.

References

1. Middleman, S.: *The Flow of High Polymers*, Interscience, New York, 1968.
2. Ferry, J. D.: *Viscoelastic Properties of Polymers*, Wiley, New York, 1970.
3. Skellard, A.: *Non-Newtonian Flow and Heat Transfer*, Wiley, New York, 1967.
4. Barr, G.: *A Monograph of Viscometry*, Oxford University Press, London, 1931.
5. Van Wazer, J. R., Lyons, J. W., Kim, K. Y., and Colwell, R. E.: *Viscosity and Flow Measurement*, Interscience, New York, 1963.
6. Johnson, J. F., LeTourneau, R. L., and Matteson, R.: *Anal. Chem.* **24**, 1505 (1952).
7. Babb, Jr., S. E. and Scott, G. J.: *J. Chem. Phys.* **40**, 3666 (1964).
8. Porter, R. S. and Johnson, J. F.: *Appl. Polym. Sci.* **3**, 107 (1960).

CHAPTER IV

ADVANCES IN POLYMER SCIENCE AND ENGINEERING: APPLICATIONS TO FOOD RHEOLOGY

STANLEY MIDDLEMAN

University of Massachusetts, Amherst, Mass., U.S.A.

1. Introduction

Flow is an essential step in the production and fabrication of many foods, and so the flow properties of materials constitute a body of information essential to the economic design of food process equipment and operation. Rheology is the science of flow, and while it thus includes all classes of fluids (gases, liquids, slurries, etc.) we usually use the term 'rheology' to refer to non-Newtonian materials, of which polymeric materials are the most important class, in an economic sense.

This chapter discusses some aspects of polymer rheology that are, in part, review of basic points [1] and, in part, descriptions of new research that will be important as rheologists try to make progress. The progress in polymer rheology can be utilized for food materials. It is worthwhile for food scientists and engineers to give serious attention to progress that has been made in the field of polymer science and engineering.

The central problem in rheology is easily stated: given the deformation undergone by a material, what are the internal stresses experienced by 'particles' of the fluid? An equivalent statement is: Given the forces acting on a material, how does that material deform? We can organize such a discussion along the lines suggested by several questions: What motivates the study of rheology? What specific information is sought, and how is it obtained? How is the information used?

With regard to motivation, there is, of course, the challenge of fitting complex, and in some ways fascinating, rheological phenomena into a general physico-mathematical framework. To most of us, however, motivation lies in the economic importance of engineering design based on accurate and complete understanding of rheological phenomena.

With regard to information sought about a material, the more basic question is: "How do we describe flow?"

A completely general description is so complex, if achievable at all, that we make progress in terms of *models* involving simple flows.

A "simple flow' is easy to define: it is one which can be exactly described. The classical example is the 'simple shear flow', shown schematically in Figure 1. Here it is suggested that a fluid is confined between two rigid surfaces infinite in extent. (The *infinite* restriction is met so long as the separation H is quite small in comparison to other dimensions of the system.) What is suggested in the figure is that the lower surface is stationary and the upper surface moves at uniform speed **V** parallel to the lower surface. A force **F** is required to maintain such a motion, and one observes,

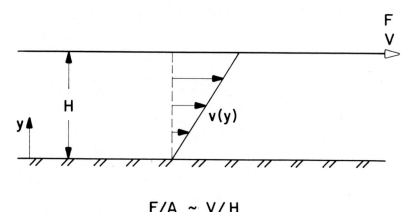

F/A ~ V/H

Fig. 1.   Shear flow.

for a Newtonian fluid, that the force per unit area in the plane of motion, F/A, is proportional to the speed V and inversely proportional to the spacing H:

$$F/A \sim V/H. \tag{1}$$

For this simple flow F/A is the shear stress τ, and V/H is the velocity gradient, or shear rate:

$$V/H = dv/dy = \dot{\gamma}. \tag{2}$$

The relationship between shear stress and shear rate provides the simplest definition of viscosity η:

$$\tau = \eta \frac{dv}{dy} = \eta\dot{\gamma}. \tag{3}$$

While this definition is undoubtedly well known to most readers, one still can raise the practical question: "Can such a flow be achieved in the laboratory with some desired degree of accuracy?" The answer, of course, is: "Yes – with care in design and execution."

Figure 2 shows several configurations of simple shear flows. Most commercial instruments are based on one of the flow geometries illustrated here. In all of these cases there are *end* effects (due to finite geometrical restrictions) and *flow* effects (due to departure from the assumed *laminar steady* flow. The quantitative use of these flows requires a knowledge of their limitations, most of which have been carefully worked out in the past ten years. A means of checking the accuracy of viscosity measurements generally lies in the use of a versatile instrument and a knowledge of proper methods

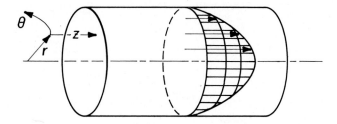

Capillary Flow

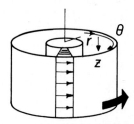

Cylindrical Couette Flow

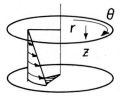

Torsional Flow

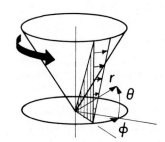

Cone and Plate Torsion

Fig. 2.   Shear flow.

of data treatment. Figure 3 is basically a pressure drop-flow rate curve, obtained using several different sizes of capillaries. The ordinate and abscissa are chosen so as to produce a *single* curve if the data are free of experimental artifact. Another internal check on viscosity data follows if we use two distinct instruments (Figure 4) such as a capillary and rotating coaxial cylinders. Here the data are reduced to shear stress *vs*

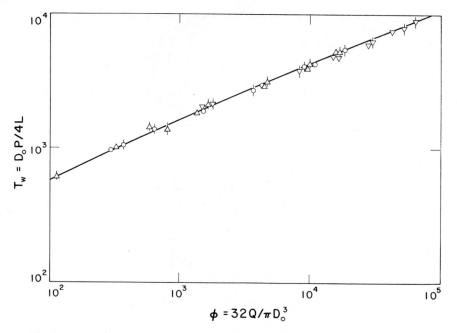

Fig. 3.   Pressure drop vs. flow rate for aqueous solution of carboxymethylcellulose.

shear rate, a curve which *should* be characteristic of the material, and independent of the instrument.

Viscosity behavior is very well understood and documented, yet one finds some laboratories doing a poor job through ignorance of modern instrumentation and theory.

Typical viscosity data have the appearance shown in Figure 5 for a *molten-polymer* at various temperatures, and in Figure 6 for a *polymer solution* at various concentrations. In each case we see the typical behavior: at low shear rates the viscosity becomes independent of shear rate, i.e., the fluid is Newtonian. At high shear rates (and the criterion of 'high' clearly depends on temperature and concentration) non-Newtonian behavior is observed.

One major distinction between synthetic polymer rheology and food rheology is in the more common appearance of a 'yield stress' in food materials. Figure 7 shows a comparison of a Newtonian fluid, typical non-Newtonian fluid, and a 'Bingham plastic', a material that exhibits a yield stress. The distinctions are best displayed in an arithmetic plot of τ' vs $\dot{\gamma}$, rather than in logarithmic plotting (as in Figures 5 and 6) of η vs $\dot{\gamma}$.

The physical significance of the yield stress τ_Y is quite simple: the material behaves as a solid (i.e., it does not flow) until the applied stress exceeds τ_Y. As an example of data on yield stresses for foods, Figure 8 shows yield stress values correlated with viscosity values in a variety of food types. [2] Note that a range of four orders of magnitude is displayed here.

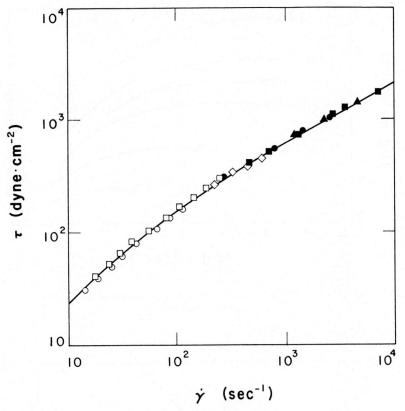

Fig. 4.   Shear stress vs. shear rate from coaxial cylinder (open symbols) and
capillary (filled symbols) viscometers.

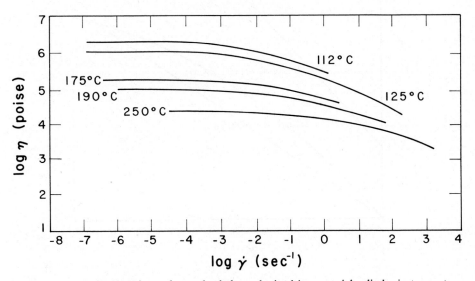

Fig. 5.   Viscosity data for molten polyethylene obtained in a coaxial cylinder instrument.

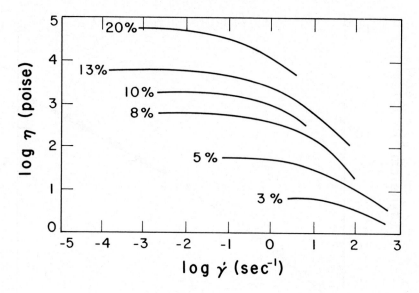

Fig. 6. Viscosity data for solutions of polyisobutylene in decalin at 25°C obtained in a cone and plate viscometer.

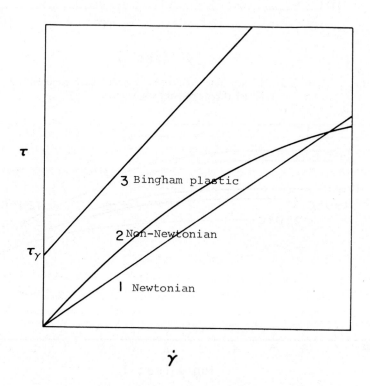

Fig. 7. Flow characteristics of fluid.

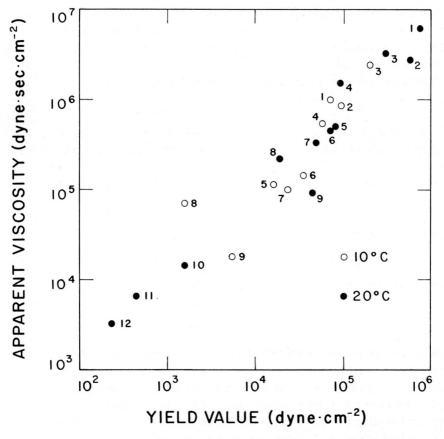

Fig. 8.   Apparent viscosity vs. yield stress for several food products (Reference [2]).

While a yield stress is measurable, its significance in processing depends on conditions of operation. For example, if shear rates in typical processing equipment exceed $10\,\text{s}^{-1}$, which is probably a reasonable lower bound, then shear stresses within the material, given approximately by the product of apparent viscosity times shear rate, are about one hundred times as large as the yield stress, for the materials of Figure 8. In such a case, yield stress may not be a significant parameter with regard to process design.

It is sometimes convenient to have a simple mathematical model of the shear stress-shear rate relationship. Equation (3), for example, defines the Newtonian fluid. A model which, with three parameters, gives a good fit of non-Newtonian data over a wide range of shear rates is the Ellis model (1):

$$\tau = \frac{\eta_0 \dot{\gamma}}{1 + \left(\dfrac{\tau}{\tau_{1/2}}\right)^{(1-n)/n}} \tag{4}$$

(Note that the shear stress appears on both sides of the equation.) η_0 is called the 'zero-shear viscosity', and $\tau_{1/2}$ is a material constant which represents the shear stress at which the viscosity has fallen to half its zero-shear value. The third parameter, n, is the 'power law index'.

Over a limited range of shear rate the 'power law' is often used:

$$\tau = K\dot{\gamma}^n. \tag{5}$$

The parameter K is called the 'consistency index'. Application of the power law to a variety of foodstuffs, and tables of n and K values, are given in a paper by Holdsworth [3].

Finally we note two popular 'yield stress' models, that of Casson [4]:

$$\tau^{1/2} = \tau_Y^{1/2} + K_1\dot{\gamma}^{1/2} \tag{6}$$

and that of Charm [5]:

$$\tau = \tau_Y + K\dot{\gamma}^n \tag{7}$$

In the Charm model, if $n=1$ and $K=\eta_0$, we have the 'Bingham plastic'.

2. Elasticity

While viscosity is often the major fluid property of interest in process design, we know that most polymeric materials, and many food materials, are *viscoelastic*. What does the term viscoelasticity mean, in terms that have some relevance to processing behavior? There are two ways to discuss this point: one in terms of elastic, or *normal* stresses; the other in terms of time dependent behavior, stress relaxation.

Let us look at the simple shear flow again, as in Figure 9. In addition to the shearing stresses that act *in the plane* of the deformation there can be a set of stresses that act in planes perpendicular (normal) to the deformation. These are called, then, *normal stresses*; they would not be observed in a Newtonian fluid subject to a simple shear flow.

Assuming, for the moment, the capability to measure normal stress components, how do we differentiate fluids with respect to normal stress behavior? With respect to viscous behavior, recall that we consider a ratio of shear stress to shear rate:

$$\tau/\dot{\gamma} = \eta. \tag{8}$$

Why a simple ratio? Because that ratio is observed to be a constant (at a given temperature, pressure, composition) for Newtonian fluids, and approaches constancy at low shear rate for most polymeric fluids.

To quantify normal stresses, and so to differentiate fluids with regard to elasticity, we again lean on experience. What we observe is that normal stresses are proportional to the *square* of shear rate, at low shear rates in polymeric fluids. We commonly work in terms of normal stress *differences*, such as $\tau_{11} - \tau_{22}$, and $\tau_{22} - \tau_{33}$, since, for incompressible fluids, we could always add an arbitrary hydrostatic pressure to the entire system without changing the flow. The stress *difference* is free of hydrostatic pressure. Thus we define a normal stress function

$$\psi_{12} = \frac{\tau_{11} - \tau_{22}}{\dot{\gamma}^2} \tag{9}$$

the behavior of which is shown in Figure 10. The limiting behavior of this function at low shear rates, that is, $\Psi_{12}^0 = \dot{\gamma} \xrightarrow{\lim 0} \Psi_{12}$, is a parameter which, like the zero-shear viscosity, differentiates among various fluids, in this case with regard to normal stress behavior.

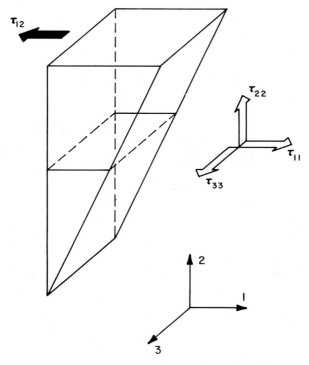

Fig. 9.   Coordinate for normal stresses.

Normal stresses can be measured in a variety of ways, and a large literature in this field exists [1]. In Figure 11 is shown one simple configuration, the cone-and-plate. The rise of fluid in the manometer tubes allows a measure of the stress distribution. Note that if the fluid were newtonian, centrifugal effects would cause an opposite stress distribution. In fact, centrifugal effects are a major limitation in the use of this configuration.

Now let us examine elasticity from the viewpoint of stress relaxation. Stress relaxation is a classical experiment in rubber elasticity theory, and in the study of elastic solids. The basic experiment is one in which a change of some kind (an input) is imposed on a material, and the response (an output) is observed. Typical examples

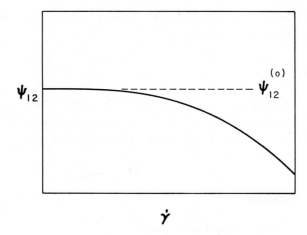

Fig. 10.   Normal stress function, ψ_{12}.

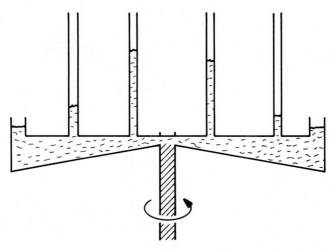

Fig. 11.   The normal stress distribution in cone and plate shear indicated by fluid level in the manometer tubes.

are creep following a sudden loading, or stress in response to constant speed stretching. Time scales of the response are of the order of minutes to days when one works with rubbery materials.

In the case of *fluids*, time scales (or relaxation times) are of the order of milliseconds, and one can not perform classical tests. In recent years we have put some effort into developing experiments suitable to the measurement of very short time viscoelastic phenomena.

Figure 12 shows one such experiment, the measurement of the swelling which accompanies the efflux of a viscoelastic liquid from a capillary or die. What do we determine in such an experiment?

Figure 13 shows the manner in which stress changes as the fluid enters and leaves the capillary. We note that while this is a steady state flow, to the laboratory observer, it is a transient flow *to the fluid*. Furthermore, the time scales of these changes are very small, of the order of milliseconds.

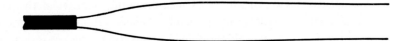

Fig. 12.   Swelling in the efflux of a viscoelastic fluid from a capillary.

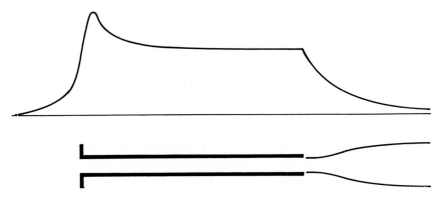

Fig. 13.   Normal stress distribution in capillary extrusion.

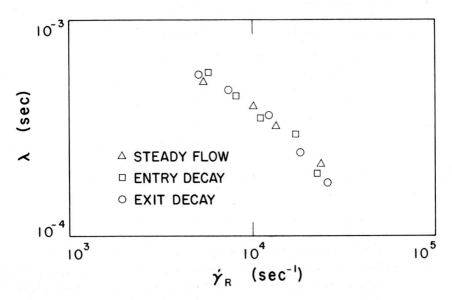

Fig. 14.   Relaxation times measured by a jet-swelling experiment.

There are two relaxation regions accessible in such an experiment. In the entrance region the fluid is subject to a suddenly imposed shear rate, and we can measure the response of the normal stresses by using capillaries of varying length.

In the exit region the shear stress acting on the fluid suddenly disappears as the fluid leaves the end of the capillary, and the relaxation of normal stress is followed. In both cases, normal stresses are measured by measuring the diameter of the jet, and using a theoretical analysis based on the principle of conservation of momentum [1].

In Figure 14 some of our experimental results for relaxation times are shown. Three independent experiments are shown: the 'entry' and 'exit' experiments are based on the previous picture of stress relaxation after entering and leaving the capillary. The data marked 'steady flow' refer to a third independent experiment, which is a steady state experiment, and which is based, in part, on some results of theoretical continuum mechanics. The agreement among the three sets of data is quite encouraging. Note that we are measuring relaxation times that are fractions of a millisecond.

3. Elongational Flows

To this point we have talked exclusively of *shearing* flows, and of the viscosity and normal stress coefficients that describe how a fluid responds to shear. However, there is a second class of flow – the elongational flow. Figure 15 shows one mode of elongational flow: extension of one end of a specimen. This is a type of flow that, except at the finite boundaries, is completely free of shear.

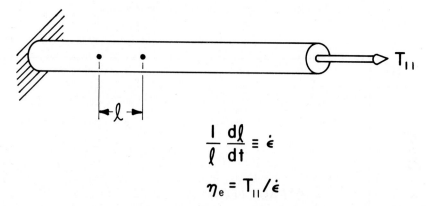

$$\frac{1}{\ell}\frac{d\ell}{dt} \equiv \dot{\epsilon}$$

$$\eta_e = T_{11}/\dot{\epsilon}$$

Fig. 15. Elongational viscosity.

Figure 15 indicates how we define a material property, the *elongational viscosity*, relevant to such a flow. We define a strain rate, as shown, and the axial stress T_{11} associated with this elongational deformation.

The elongational viscosity is defined as the ratio of stress to strain rate, at constant strain rate. It is known from theory, as well as from observation, that Newtonian fluids exhibit an elongational viscosity that is three times the shear viscosity. The important point is that the elongational viscosity is not an elastic parameter, *per se*, but is exhibited by Newtonian (inelastic) materials as well. But the manner in which elongational viscosity depends on strain rate may be strongly affected by the elastic nature of the material.

There is quite a bit of controversy today about the nature of elongational viscosity in polymeric materials; specifically, how does it depend on strain rate? Figure 16 shows some recent experimental data [6], obtained with a rubbery elastomeric material. The lower curve, the shear viscosity, shows the usual non-Newtonian viscosity. The upper curve, the elongational viscosity, gives the appearance of Newtonian behavior.

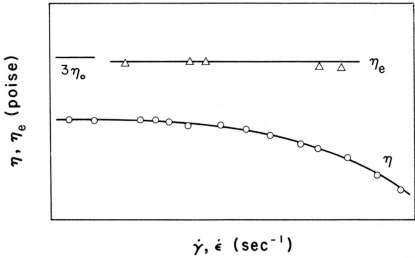

Fig. 16.   Shear viscosity and elongational viscosity of a rubbery material. (Data from reference [6]).

The elongational viscosity is extremely difficult to measure in polymer *solutions*. We are currently examining a method based on the growth of a bubble within a larger body of fluid. Bubble growth can be shown to be an elongational flow, in the sense that it is without shear. High speed movies of bubble growth, accompanied by measurement of pressure within the bubble, should allow us to determine models of viscoelastic fluids suitable to description of elongational flow. The importance of elongational flow in processing will be dealt with further below.

4. Continuum Mechanics

That branch of mathematical physics called 'Continuum mechanics' has played a central role in the progress made in the field of rheology, especially in the past decade. The role of continuum mechanics is four-fold:

(1) *Organization.* – By first establishing the fundamental mechanical principles, and giving them a concise mathematical formulation, research in rheology is organized.

(2) *Delineation of appropriate variables.* – Should viscosity depend on the kinetic energy of a fluid? Can questions like this be answered prior to experiment? To a great extent, yes, with the result that experimental work is better directed.

(3) *Definition of, and interrelationships among, material functions.* – There *are* improper ways to define material functions. Many of the material functions are interrelated and it is possible to develop some internal checks on data much like those used in checking the thermodynamic consistency of data.

(4) *Development of constitutive equations.* – Constitutive equations give mathematical expression to the relationship between deformation and stress. Continuum mechanics has provided a rational basis for development of such equations.

Because continuum mechanics is mathematical, often it is regarded as irrelevant to real behavior. It is also true that many continuum mechanics are abstract as to have no clear relationship to reality.

However, it is impossible to understand the value of simple approximate approaches to physics (regardless of which branch we discuss) without having some feelings for what is exact. Furthermore, the nature of continuum mechanics must be appreciated before it can be more effectively incorporated into the progress in rheology.

5. Some Applications of Rheology in Food Processing

Mixing is usually a significant operation at some stage of processing. Often at least one, and sometimes several, of the materials to be mixed are of relatively high viscosity. Yet mixing of viscous materials is a poorly understood operation.

In conducting model experiments, attempting to develop computer simulations, and carrying out pilot scale operations, all in the area of mixing of viscous fluids, the basic question is: In what way do rheological properties determine the ease of mixing of two viscous fluids? The viscosity ratio is expected to be the principle variable that determines the nature of flow in a mixing device.

A second area of study is that of fiber spinning, which for years has been a means of producing a variety of synthetic polymeric materials of commercial importance. It is known that the shape of the extrudate depends strongly on the fluid properties, and that mechanisms exist for distorting the surface of the fiber. These mechanisms, which are mainly in the nature of instabilities, are believed to be associated with the viscoelastic properties of the melt or solution.

In recent years fiber spinning from protein solutions is becoming an increasingly important means of producing textured foods. In the spinning region many phenomena occur: heat and mass transfer, coagulation, and stretching being the most important. Stretching, during spinning, is an elongational flow of the type discussed above. Our studies of protein fiber spinning show some indication that spinning stability depends on the elongational behavior of the spinning dope.

References

[1] Middleman, S.: *The Flow of High Polymers*, Interscience, New York, 1968.
[2] Tanaka, M., DeMan, J. M., and Voisey, P. W.: *J. Texture Studies* **2**, 306 (1971).
[3] Holdsworth, S. D.: *J. Texture Studies* **2**, 393 (1971).
[4] Casson, N.: in C. C. Mill (ed.), *Rheology of Disperse Systems*, Pergamon Press, New York, 1959
[5] Charm, S. E.: *I.E.C. Proc. Des. Dev.* **2**, 62 (1963).
[6] Stevenson, J. F.: *AIChE J.* **18**, 540 (1972).

CHAPTER V

FOOD TEXTURE – DEFINITION, MEASUREMENT AND RELATION TO OTHER FOOD QUALITY ATTRIBUTES

AMIHUD KRAMER

Dept. of Horticulture, University of Maryland, College Park, Md., U.S.A.

1. Introduction

Certainly many difficulties, misunderstandings, and actual conflicting conclusions can be avoided when terms are defined precisely so that everyone employing the same term in a specific field is referring to the same things. It would appear therefore that the logical beginning would be with Webster's Dictionary. Unfortunately, however, this very useful compendium lists seven distinct definitions, none of them applying specifically to food. Certainly the English language is flexible enough so that the same word or term may have more than one meaning, but for specific scientific use, the term should be defined with precision. The one dictionary definition that appears to have some application: "The disposition or manner of the union of the particles of a body or substance." Even this definition, however, would seem to apply more accurately as a definition, or a description of the Universe, than as a property of food.

It would appear therefore that the food field requires its own definition of the word texture which is not as yet included in Webster's Dictionary.

The problem of defining texture as a major component of sensory food quality arose when during the 1920's there developed a gradual awareness that sensory quality of foods does not consist of a single well-defined attribute, but is a composite of any number of attributes which are perceived by the human senses individually and are then integrated by the brain into a total, or overall, impression of quality. Among the first to recognize the advantage of such an analytical approach to quality evaluation were those in government agencies, particularly the Department of Agriculture, who were responsible for the development of grades and standards of quality for various raw and processed food products (USDA-CMS).

As late as 1940, however, Lee, writing on the "quality determination of vegetables", was reporting exclusively on what we would now consider textural quality. To this day, we encounter not only lay consumers, but food scientists describing a specific attribute and assigning to it the totality of quality. In a recent conversation with a flavor physiologist, for example, I was impressed by his conviction that sensory quality of foods consists of nothing more than 'flavor', with texture included as contributing to flavor quality in some undefined manner that might be described by some term such as mouth-feel. Similarly, a geneticist indicated his total lack of understanding why a new strain of peach of 'excellent quality' was not being produced and marketed commercially. Upon further interrogation, it was found that he totally ignored the light color and mushy texture of the peach flesh and was referring to quality strictly on the basis of sweetness and fruity aroma.

2. Classification

Smith (1947) was among the first to list more specific properties of quality, as distinct parameters contributing to overall quality. Of the nine parameters listed (size, viscosity, thickness, texture, consistency, turbidity, color, succulence, and flavor), it is interesting to note that no less than five would probably be included under the general term of texture by many food technologists today.

Kramer (1955) proposed that sensory quality of foods, being a psycho-physical phenomenon, should be systemized or classified in accordance with the senses by which the various attributes of quality are perceived by the consumer. He, therefore, classified sensory quality under the three major senses: appearance as sensed by the eye, flavor as sensed by the papillae on the tongue and the olfactory epithelium of the nose, and kinesthetics (borrowed from Crocker, 1945) or texture as sensed by the nerve endings that subserve muscle. Of 61 commodities for which the U.S.D.A. Standards of Quality were listed in the *Canning Trade Almanac* in 1965, only *absence of defects* and *color* were generally included at weights ranging from 0.15 to 0.6 of the total score – both factors listed by Kramer under the *appearance* category. *Flavor* was listed as a component of quality for only 25 commodities, and *texture* for just 4. The other 57 commodities however, were graded on the basis of such terms as *character, consistency, tenderness* or *maturity*, practially all of which may be classified under the general (or primary) term of texture.

At that time, therefore, there was some reluctance to assign the entire area of kinesthesis (and/or haptaesthesis, Muller, 1969) to texture, since other terms – such as viscosity and consistency – were also in general usage, particularly in official grades and standards. The fourth sense involved in quality evaluation of food, namely sound as sensed by the ear, was mentioned occasionally – but usually incidentally– as being only of minor importance in the overall evaluation of food quality.

Another reason for reluctance on the part of some workers to assign a primary role to the term texture in sensory evaluation of food, was it ubiquitous use so that its precise meaning was not immediately evident. Judging by the usual standard dictionary definitions, the term texture was first used in the textile industry in connection with the art of weaving, as "disposition or connection of threads as in a fabric" (Webster's Dictionary). It is only in the last decade or so, with the development of 'textured foods', that this original meaning of texture could be applied directly to the evaluation of food quality. A more broadly applicable definition such as "the disposition or manner of union of particles of a body or substance", could also apply to all natural and processed 'solid' foods, where "the disposition or manner of union" of different types of cells and tissues in the food material could be considered as the 'texture' of the food. This, therefore, is a major attribute of food quality which can be included under a definition of texture and which would be directly related to the internal structure of natural or fabricated foods (Sherman, 1972).

Webster's Dictionary also lists a definition of texture specifically for petrography as "smaller features of a rock – granular" which has been and still is applied in some food areas, as for example, in meats, where the smoothness or coarseness of the muscle fibers is of concern. Thus here again structure is involved, but in this instance it would be difficult to call this characteristic as something pertaining to the sense of feel only since it is obviously noticeable to the sense of sight. Certainly the property of 'marbling', that is the interlacing of fat within the muscle, could well be included under one of the dictionary definitions of texture, but here again it could as well or more readily be classified under the general term of appearance.

Viscosity and consistency, on the other hand, were classified by Kramer and Twigg (1959) as appearance factors, since they were generally used in grades and standards of quality as referring to the flow of liquid and semisolid drinks or slurries which could be readily perceived by the eye of the observer before his kinesthetic sense came into play. Thus, in their book on *Quality Control for the Food Industry*, Kramer and Twigg (1970) treated 'viscosity and consistency' under the general heading of appearance, and texture alone under kinesthetics. Viscosity and consistency were treated in accordance with classic rheological concepts as applying to Newtonian and non-Newtonian liquid and semi-solid materials respectively*, while the term texture was reserved for 'solid' foods whose complex structure caused difficulties in converting sensory responses in terms of classic rheological model systems.

Kramer therefore proposed to confine texture further – from the sensory standpoint to the sense of feel only, and from the physical standpoint to the part of rheology that

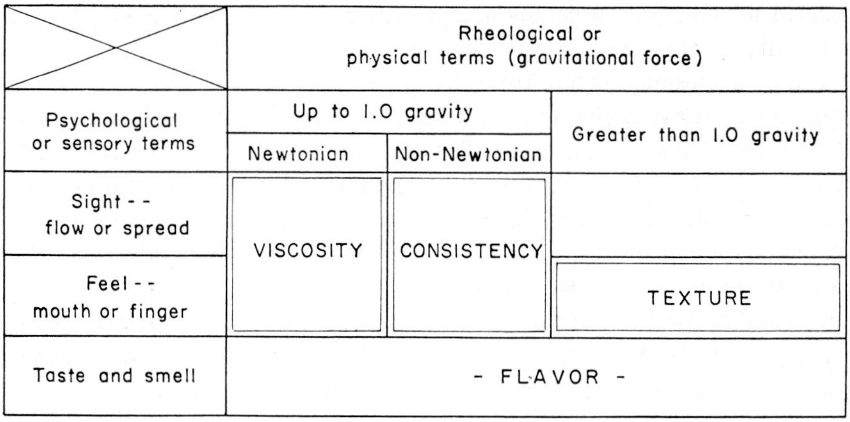

Psychological or sensory terms	Rheological or physical terms (gravitational force)		
	Up to 1.0 gravity		Greater than 1.0 gravity
	Newtonian	Non-Newtonian	
Sight - - flow or spread	VISCOSITY	CONSISTENCY	
Feel - - mouth or finger			TEXTURE
Taste and smell	- FLAVOR -		

Fig. 1.   Classification of texture in relation to force required to initiate flow (Kramer, 1964).

* It should be noted that the British Dairy Industry usage of *consistency* is synonymous to this restricted definition of *texture* (Scott Blair, 1968). Also note the similarity of treatment for fluid systems by Corey and Creswick (1970).

deals with the deformation or flow of matter, but only as a result of the application of forces greater than gravity (Figure 1). Thus 'gel strength' was considered a textural characteristic since a force greater than 1.0 g is required to cause deformation and it is sensed primarily by mouthfeel. On the other hand, consistency of a sauce was not included, since its flow characteristics may be observed by the sense of sight, and flow occurs with the application of gravitational forces alone. He further suggested that which food slurries such as sauces, being invariably non-Newtonian, this characteristic could properly be called 'consistency'. Flow characteristics of an oil or syrup were termed 'viscosity' since they also deformed under the force of gravity alone and, in addition, being quite homogeneous and in many instances true solutions, were more likely to exhibit Newtonian flow.

Separate parameters of texture were further classified in accordance with the nature of the force applied to cause specific types of deformation such as compression, shearing, extrusion, or a combination of these.

During the decade of 1960, a number of additional textural classifications were proposed (Szczesniak, 1963b; Bourne, 1966a, b) which, while dealing largely with 'solid' products, did not specifically exclude liquid or semi-liquid foods, while Sherman (1969) utilized the state (solid, semi-solid, fluid) of the product in his classification of the mechanical (masticatory) properties. Considerable progress was also made, notably by Mohsenin (1970a, b), in defining mechanical properties of solid foods in rheological terms, thereby providing a much more precise understanding of the specific mechanical properties that could be related to very specific human kinesthetic responses.

3. Definition

With the advent of the *Journal of Texture Studies* in 1969, the term texture may be considered to have been generally accepted as a major division of sensory quality covering all kinesthetic responses of foods in whatever state they are in. By the same token, in distinguishing texture from the other major sensory categories, it may be appropriate to relegate characteristics such as visual appearance or particle size (Sherman, 1969) to appearance as well. Other attributes such as sensations of cold, heat, or oiliness would be classified under flavor as well as texture. Our definition of texture may further be limited to sensations of touch or feel by the human hand and mouth parts, since quality attributes of food are not ordinarily sensed by other human organs prior to or during mastication.

Any definition, and one of texture is no exception, must be arbitrary especially in drawing borderlines. It should be recognized that every primary quality attribute (as well as secondary attributes) is not entirely independent and may overlap and certainly be influenced by other attributes. Thus Kramer (1968) portrayed sensory quality in the form of a finite continuum (Figure 2), or circle, with the primary attributes of appearance, texture (kinesthetics), and flavor as sharing the periphery of the ring. At one part of this periphery between appearance and texture there is an overlapping zone where terms such as consistency and viscosity may be placed, since

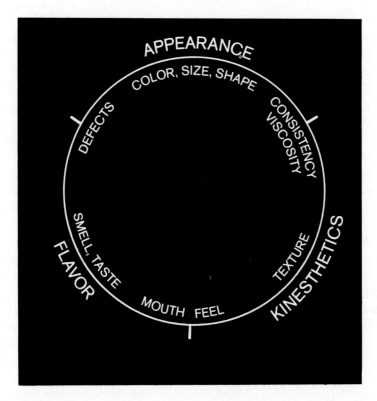

Fig. 2. A schematic presentation of sensory quality of foods as a finite continuum (Kramer, 1968).

these can be classified under both appearance and texture. At the other side of the zone where texture meets flavor, there is a similar overlapping where the term mouthfeel may be placed.

Accepting the inevitability of such overlapping borders, there is nevertheless general agreement that the term texture rather than kinesthesis or haptaesthesis *is* the accepted popular term of *that one of the three primary sensory properties of foods which relates entirely* (or in addition to the other primary properties) *to the sense of touch or feel* and is, therefore, at least potentially capable of precise measurement objectively by mechanical means in units of mass, or force. The equivalent psychological and physical terms are kinesthesis (the muscle sense), or haptaesthesis (the skin sense), and rheology (the deformation or flow of matter), respectively.

4. Measurement

4.1. SENSORY QUALITY AND SENSORY MEASUREMENT

A sensory property is one that is perceived by one of the senses with which the food consumer evaluates the product. Hence, texture is a sensory property of foods. Its measurement may be accomplished directly by the use of the respective sense (in this case the sense of touch) and in such instances the test is subjective (Stevens, 1966b).

Although texture is a sensory property, there is the opportunity of measuring it by physical, more specifically rheological objective methods. In general, objective techniques for measuring sensory properties have the basic disadvantage of measuring sensory properties only indirectly and are, therefore, accurate only to the extent that they are analogous to the human sensory response. At the same time they have the advantage of objectivity in that – at least potentially – they are not as subject to drift, fatigue and are more precisely calibratable than human sensors.

Even less direct, non-rheological, objective methods have been used successfully for measuring textural properties of foods, particularly where the sensory quality is not specifically defined, but is a broad concept based largely on the physiological condition of the plant or animal tissues constituting the food (e.g. maturity). Such non-rheological measurements may be colorimetric, densimetric, or determinations of chemical composition. Success of such measurements depends not so much on the equivalency, but on the coincidental relationship that the test results happen to change consistently with changes in consumer responses for the quality attribute. Thus although such procedures may be successful in many instances, they may suddenly become unsatisfactory. Maturity of lima beans for example was measured adequately for many years by the lightness of the product, since lima beans, as they mature, lose chlorophyll which provides them with green pigmentation. This method was successful until a new variety was developed which retained its green coloration as it matured. Consequently large quantities of hard, mature undesirable lima beans were graded as young, immature succulent lima beans, simply because they retained greenness and did not turn 'white' on maturity (Kramer and Hart, 1954).

The lima bean problem was solved simply by replacing the color (lightness) test with a physical test of hardness, similar to the 'tenderometer' test for raw peas. This new hardness test was satisfactory, since the specific parameter to be measured was hardness, and fresh lima beans and peas became harder as they increased in maturity. For canned peas however, the hardness test was not satisfactory. In fact, tender, less mature raw peas remained firmer upon canning than more mature peas. Furthermore, any lot of pea of any maturity could be 'softened' to any desired level simply by extending the time and temperature of heat processing. For the canned peas, however, the sensory property to be tested was not specifically that of hardness, but a combination of crispness and lack of mushiness, which was more accurately determined by a chemical measurement of the alcohol insoluble solids, largely starch, plus a mechanical test for hardness (Angel et al. 1965).

Szczesniak has recently prepared a comprehensive review of instrumental (1973a) and indirect (1973b) methods of texture measurement. She lists the basic elements of such instruments as being: (1) a probe contacting the food sample; (2) a driving mechanism for imparting motion (and stress); (3) a sensing element for detecting the resistance of the foodstuff (strain); and (4) a readout system. She further classified the types of texture measuring devices as: (1) penetrometers; (2) compressimeters; (3)

shearing devices; (4) cutting devices; (5) masticometers; (6) consistometers; (7) visco-meters; (8) extrusion measurements; and (9) multi-purpose units. The indirect meth-ods are classified as: (1) chemical; (2) enzymatic; (3) microscopic; and (4) physical.

4.2. ACCURACY OF OBJECTIVE METHODS

At the risk of redundancy, it may still be appropriate to point out at this time that objective measurements are unquestionably useful and in many instances preferable to subjective measurements in the evaluation of rheological-mechanical properties of new fabricated food products, but their use in the evaluation of complex natural products may be limited. Liquid foods exhibiting Newtonian characteristics such as some oils, syrups, and juices, may be measured objectively and accurately and reported in generally acceptable rheological terms. Non-Newtonian liquids or semi-solids, and particularly natural 'solid' foods, may be more difficult to measure objectively with sufficient accuracy, and for this reason any such proposed objective method should first undergo thorough examination and validation to establish its accuracy before it is approved for routine use. A generalized procedure for accomplishing this was suggested by Kramer (1956).

(1) Definition of parameters to be measured – the more precisely the individual parameters are defined, both in sensory and in physical terms, the greater the prob-ability of developing a successful objective test. This is essentially a search for psycho-physical analogues (or sensory-rheological analogues).

(2) Statistical analysis for selecting the best objective method or combination of methods for each sensory parameter by use of correlation-regression analyses, where sensory evaluation of texture is always the dependent variable (y) and the objective test value is always the independent variable (x); thus the regression equation assumes the form:

$$y = a + \beta x.$$

Since relationships between subjective and objective values are usually non-linear, a truer and higher correlation can frequently be obtained by converting sensory data to their logarithms (Hopkins, 1950; Stevens, 1966a). Where relationships are more com-plex and such a simple expedient will not suffice, there may be a need for adding higher order functions (e.g. quadratic, cubic, etc.). Thus the regression equation will appear as:

$$y = a + \beta x + \beta^2 x^2 \cdots + \beta^n x^n.$$

Where more than one independent variable (objective test) is involved, the simplest form of the regression equation will be:

$$y = a + \beta_1 x_1 + \beta_2 x_2 \cdots \beta_n x_n.$$

Higher order functions may be added as required.

By such analyses, those measurements that contribute significantly to the predict-ability of the sensory parameter will be retained while others will be dropped. The partial regression coefficients (β) will also indicate the relative importance of each of the objective methods while the coefficient of determination (R^2) will indicate the extent to which the selected method or methods are capable of predicting the sensory quality level of the parameters studies.

(3) Validation – since there is always the chance that the set of samples on which the original studies were made yielded results that may be limited to little more than chance relationships among these samples only, it is necessary to validate the accuracy (or predictability) of the selected procedure. This may be accomplished by repeating the study with additional samples that cover but do not exceed the usual commercial range for each quality.

(4) Integration of selected objective scales for predicting overall textural quality. In the above mentioned search for objective methods for measuring specific parameters, some methods may have been selected that may contribute significantly to the prediction of more than one specific parameter of texture. Thus, by means of multiple regression and other multivariate analyses (Kapsalis et al., 1973) it may be possible to reduce the number of objective tests that will provide adequate predictability for overall textural quality.

It is only after such elaborate studies that an indirect, objective method or a combination of methods can be developed to predict with sufficient accuracy a sensory quality such as texture. A common flaw in the development of these procedures is that the 'validation' step (3) is either omitted or not performed thoroughly, so that in application, under conditions which are always subject to change, gross errors may be made in assigning quality levels to specific lots of given commodities simply because variations in certain variables were not encountered during the studies leading to the procedures used. The reverse may also be true – results obtained from objective tests performed with certain instruments may be taken as the true sensory quality levels when in fact the consumer is indifferent or unaware of the differences indicated.

Another word of caution may be called for, and that is to use an adequate number of samples. With the rapid proliferation in available equipment on one hand, and improved training and precision of taste panelists on the other, large numbers of dependent and independent variables may be defined, measured and analysed statistically. With the availability of high speed computers, the statistical analysis is usually not a serious problem. There is the risk, however, that one may ignore the fact that with the addition of every variable (i.e. parameter, and/or test procedure) we lose one degree of freedom. Thus, if the number of test procedures entered into a multiple regression analysis approaches or exceeds the number of samples, we are left with no degrees of freedom. This inevitably leads to a conclusion that the combination of objective tests predicts perfectly the sensory parameter. This, however, is nothing more than a meaningless statistical exercise, since a negative number of degrees of freedom must result in a perfect fit (Kramer, 1966; Rasekh, 1968).

4.3. PRECISION OF OBJECTIVE MEASUREMENTS

While precision (i.e. reproducibility) of objective measurements is potentially superior to sensory, this cannot be taken for granted. Thus, an imprecisely manufactured instrument can result in as great or greater or intra instrumental error than an error arising as a result of differences in human responses particularly those obtained from trained panelists. It is, therefore, important that instrumentation for objective measurement of sensory properties be produced with the utmost precision, and the tests performed under carefully controlled and specified conditions. Preparation of the food sample and its loading in the instrument must also be rigidly specified and precisely performed.

Size and shape of the test cell as well as the test sample must be rigidly specified and maintained. Thus for example, enclosing the sample in a container so that deformation can occur in only one direction could simplify substantially the nature of the forces employed in the test, thereby substantially reducing the testing error.

In general, the larger the sample units or number of units in the sample to be tested, the more precise and efficient can the expected results be. Thus for example, the average of 100 individual puncture tests on 100 kernels of corn may be less precise and more time-consuming than one compression-shear-extrusion test on one sample consisting of 400 kernels of corn.

Given a thorough understanding of the specific problem involved, following the performance of a thorough and adequate study, there is no reason why the texture of any product cannot be measured accurately and precisely with the appropriate instrument or instruments. In fact, it is frequently found that as few as 3 or 4 instrumental measurements can predict adequately all the textural properties of a food product; in some instances where a specific parameter is dominant (as hardness of peas) only one test may suffice. This is an indication that the objective measurement of textural quality may well be far less complex than the objective measurement of other sensory properties such as odor.

References

Angel, S., Kramer, A., and Yeatman, J. N.: 1965, 'Physical Methods of Measuring Quality of Canned Peas', *Food Technol*, **19**, 1278.

Bourne, M. C.: 1966a, 'A Classification of Objective Methods for Measuring Texture and Consistency of Foods', *J. Food Sci.* **31**, 1011.

Bourne, M. C.: 1966b, 'Measurement of Shear and Compression Components of Puncture Tests', *J. Food Sci.* **31**, 282.

Corey, H. and Creswick, N.: 1970, 'A Versatile Recording Couette-Type Viscometer', *J. Texture Studies* **1**, 155.

Crocker, E. C.: 1945, *Flavor*, McGraw Hill Book Co., Inc., New York, 172 pp.

Hopkins, J. W.: 1950, 'A Procedure for Quantifying Subjective Appraisals of Odor, Flavor and Texture of Foodstuffs', *Biometrics* **6**, 1.

Kapsalis, J. G., Kramer, A., and Szczesniak, A. S.: 1973, 'Quantification of Objective and Sensory Texture Relations', in A. Kramer and A. S. Szczesniak (eds.), *Texture Measurements of Foods*, D. Reidel Publishing Co., Dordrecht, Holland, p. 130.

Kramer, A.: 1955, 'Food Quality and Quality Control', Chapter 23, *Handbook of Food and Agriculture* (ed. by Blanck), Reinhold Publ. Co., New York, p. 733.

Kramer, A.: 1956, 'The Problem of Developing Grades and Standards of Quality', *Food, Drug, Cosmetic Law J.* **7**, 23.

Kramer, A.: 1964, 'Definition of Texture and Its Measurement in Vegetable Products', *Food Technol.* **18**, 304.

Kramer, A.: 1966, 'Sensory Evaluation of Food Flavor. Flavor Chemistry', *Amer. Chem. Soc. Adv. in Chemistry Series* **56**, Washington, D.C., p. 64.

Kramer, A.: 1968, 'The Judging of Food Quality – A Consideration of Uniform Scoring', *Proc. Tech. Mtg. Food and Dairy Ind. Expo.*, p. 79.

Kramer, A. and Hart, W. J., Jr.: 1954, 'Recommendations on Procedures for Determining Grades of Raw, Canned, and Frozen Lima Beans', *Food Technol.* **8**, 55.

Kramer, A. and Twigg, B. A.: 1959, 'Principles and Instrumentation for the Physical Measurements of Food Quality with Special Reference to Fruit and Vegetable Products', *Adv. Food Research* **9**, 153.

Kramer, A. and Twigg, B. A.: 1970, *Fundamentals of Quality Control for the Food Industry*, 2nd ed., Avi Publ. Co., Westport, Conn., ch. 4 and 7.

Lee, F. A.: 1940, 'Determination of the Quality of Vegetables', *Proc. 1st Food Conf.*, Inst. Food Technologists, p. 33.

Mohsenin, N. N.: 1970a, *Physical Properties of Plant and Animal Materials*, Volume I – 'Structure, Physical Characteristics and Mechanical Properties', Gordon and Breach Science Publishers, New York, 734 pp.

Mohsenin, N. N.: 1970b, 'Application of Engineering Techniques to Evaluation of Texture of Solid Food Materials', *J. Texture Studies* **1**, 133.

Muller, H. G.: 1969, 'Mechanical Properties, Rheology, and Haptaesthesis of Foods', *J. Texture Studies* **1**, 38.

Rasekh, J.: 1968, 'The Application of Headspace Gas-Liquid Chromatography for Measuring Quality of Fresh and Processed Vegetables', Ph.D. Thesis, Univ. Maryland.

Scott Blair, G. W.: 1968, Comments made during symposium on sensory evaluation at Swedish Institute for Food Preservation Research, Goteborg, Sweden, September.

Sherman, P.: 1969, 'A Texture Profile of Foodstuffs Based on Well-Defined Rheological Properties', *J. Food Sci.* **34**, 458.

Sherman, P.: 1972, 'Structure and Textural Properties of Foods', *Food Technol.* **26**, 69.

Stevens, S. S.: 1966a, 'Matching Functions Between Loudness and Ten Other Continua', *Perception Psychophysics* **1**, 5.

Stevens, S. S.: 1966b, 'On the Operation Known as Judgement', *American Scientist* **54**, 385.

Szczesniak, Alina S.: 1973a, 'Instrumental Methods of Texture Measurement', in A. Kramer and A. S. Szczesniak (eds.), *Texture Measurements of Foods*, D. Reidel Publishing Co., Dordrecht, Holland, p. 71.

Szczesniak, Alina S.: 1973b, 'Indirect Methods of Objective Texture Measurements', in A. Kramer and A. S. Szczesniak (eds.), *Texture Measurements of Foods*, D. Reidel Publishing Co., Dordrecht, Holland, p. 109.

CHAPTER VI

INSTRUMENTATION FOR DETERMINATION OF MECHANICAL PROPERTIES OF FOODS*

PETER W. VOISEY

*Engineering Research Service, Research Branch,
Canada Department of Agriculture, Ottawa, Ont., Canada*

1. Introduction

In research, product development and quality control for the food industry many parameters must be measured with precision. This has been accomplished successfully in many areas, but the scientific measurement of food texture is still in its infancy. There has been increasing interest during the past decade in developing sensory and objective techniques for evaluating this critical factor in relation to consumer reaction. Sensory analysis is evolving into a sophisticated measuring tool as indicated by recent work (Brandt *et al.*, 1963; Sherman, 1969; Szczesniak and Bourne, 1969 and Szczesniak and Smith, 1969). The technique, however, is time consuming and requires the use of considerable amounts of trained labour. Thus, in recent years interest has accelerated in objective measurement of textural characteristics of foods to provide efficient and precise quantitative descriptions.

Texture of foods is related to the physical properties sensed by the eyes before eating (except color), the sense of touch in handling the food, and the tactile receptors in the mouth during consumption. Thus, the consumer is aware of factors such as size and shape, particle size, moisture content, fat content, structure and the mechanical properties. These properties are interrelated and together create the complex food quality termed texture.

Because food texture is composed of so many variables, it is not possible to obtain an overall index in a single measurement. In general, therefore, only those properties which have the greatest influence on consumer acceptance are measured. The mechanical properties are the most critical for many foods. The purpose of this paper is to review recent advances in the instrumentation used to measure these properties and to point out the advantages of employing more sophisticated apparatus. Bourne (1966a) has proposed a classification of objective methods into force, distance, time, and energy measuring instruments and there have been several recent reviews of instruments in use (e.g.: Finney, 1969 and Szczesniak, 1963, 1969). There are common problems in objective texture measurements which can be solved with readily available equipment. It is hoped that this review will assist in selecting such instru-

---

* This chapter is based on Review Paper entitled, 'Modernization of Texture Instrumentation', *Journal of Texture Studies* **2** (1971) 129–195 with minor editions.

ments and their components from the bewildering array on the market. While this chapter deals mainly with instrumentation, in general Chapter VII, Texture Measurments in Vegetables, discusses texture measurements and results. Some of the instruments discussed in this chapter are also discussed in Chapter VII.

The mechanical properties of food encompass the reaction of the material to applied forces. Since foods are viscoelastic, both time-dependent and time-independent measurements are required. The history of such measurements for engineering materials is extensive. For example, Timoshenko (1953) notes that Leonardo da Vinci (1452–1519) tested the strength of iron wires and proposed a method for measuring the strength of stone beams. Since that time the theory, apparatus and experimental techniques for mechanically testing materials have become a sepcialized field. It is, however, only in the past decade that these developed tools became more widely adopted for testing foods. This has been brought about in part by the increasing interest of engineers in the fields of texture measurements and physical properties of biological materials related to agricultural production (Mohsenin et al., 1963; Mohsenin, 1968, 1970). This has led to intensive studies of the mechanical and rheological aspects of foods, but the relation between this behaviour and the consumer reaction is still not fully understood. It is not the intention to discuss this point here, but it should be noted that, although modern instruments can be used to make precise measurements, the result must be compared with sensory analysis before such techniques can be adopted for routine use (Kramer, 1969; Szczesniak, 1968).

To fully understand all the aspects of mechanical texture properties, the topics of strength of materials, rheology, mechanics (including statics and dynamics) should be considered. Space precludes such a discussion here; therefore, the following general principle is used as a basis for describing the general type of instrumentation employed. To measure the texture of foods in terms of elastic or viscoelastic parameters, the product is deformed in some arbitrarily selected but defined manner, and the force and time required to produce a given deformation or flow are recorded. Such measurements have been used to test many kinds of biological materials for different reasons. Since the techniques and requirements in all of these applications are similar, instruments* developed for other purposes will be cited as examples to make specific points in relation to texture measurements.

2. Texture Measuring Systems

A great variety of instruments have been developed for measuring food texture and several have been widely adopted. Examples such as the Kramer Shear Press (Kramer et al., 1951), the F.M.C. Pea Tenderometer (Martin, 1937) and most recently the General Foods Texturometer (Friedman et al., 1963) have appeared extensively in the literature. These instruments are discussed in Chapter VII. There are also the fields of viscometry and rheometry which encompass a wide range of instruments.

* It should be noted that the use of trade names does not imply an official recommendation by the Canada Department of Agriculture.

A recent review by Finney (1969) indicates the ingenuity that has been used to develop such apparatus. However, all these systems have a common concept and common requirements. Samples of food are subjected to compression, tension or shear (in combination or as a single operation) and force, deformation and time are recorded. A texture measuring system, thus, consists of five components: (a) a mechanical mechanism for deforming the sample, (b) a means of recording the force, (c) a means of recording the deformation, (d) a method of recording time during deformation, and (e) a test cell to hold the sample. The first four components are readily available since these items have already been developed for other purposes. The test cell should, therefore, be considered as the most critical component of the system. The cell may be a single or a multiblade shearing device such as is used in the Kramer Shear Press, a pair of grips to apply tension, flat surfaces to apply compression, a probe to puncture the sample, or a container and paddle used in a viscometer. The design of the cell determines the manner in which forces are applied to the food, how they are distributed throughout the specimen, and which type (tensile, compressive or shearing) predominates in the resulting reaction.

The purpose here is to show that a vast array of equipment is available that can be adopted to measuring food texture provided a test cell which can produce the proper loading of the sample is available. The development of such a cell has been described elsewhere (Voisey, 1970) and will not be discussed here.

3. Units of Measurement

In measuring the force and deformation of test samples, the data must be expressed in a rational set of units. This is obvious for either fundamental or empirical type tests. Units used in testing engineering and other materials are, therefore, being adopted for textural studies. This can, however, lead to errors.

When a material is subjected to a tensile, compressive or shearing force (F in Figure 1), its original length (L) will deform by an amount (D). The force will distribute over an area (A) depending on the homogeneity of the material and the uniformity of contact between the material and the loading surfaces. The results can be recorded in two ways: (a) force and deformation, and (b) stress and strain, using metric or English systems of measure. If either of these methods are used, it is implied that all the dimensions of the specimen are known throughout the test since, obviously, the area supporting the load will influence the result. However, materials subjected to stress change dimensions both along the axis of the applied force and along the axis perpendicular to the applied force. This is the Poisson effect which has been demonstrated in several biological materials (e.g. in apples by Chappell and Hamann, 1967) and a technique has been developed for measuring this parameter in biological materials (Hammerle and McClure, 1971). These dimensional changes are difficult to measure and the Poisson effect is not often considered in testing foods.

If only force and deformation are recorded, then the samples tested must have identical dimensions to make meaningful comparisons. If the relationship between

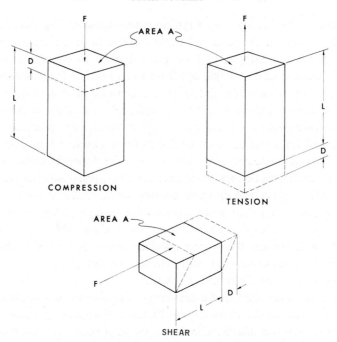

Fig. 1.   Units of measurement. A. area supporting applied force; D. deformation due to force; F. force; L. original length of specimen.

force and deformation is linear during loading, then the ratio of force to deformation F/D, or stiffness, can be used as an index of product properties. The maximum force provides an index of breaking strength which is related to cohesiveness.

If the area (A in Figure 1) over which the force is distributed is known, then the data can be recorded as stress and strain, where stress is the force per unit area, or F/A, and strain is change in length per unit length, or D/L. However, this implies that the stress and the strain are uniformly distributed throughout the sample. This can only be the case if the material is perfectly homogeneous which is most unlikely in biological materials. Also, A is varying by an unknown amount as the specimen is loaded. For example, considerable wasting occurs in a tensile test in the region where the material breaks and the area A is reduced to a when fracture occurs. Thus, F/a should be used to precisely express failure stress.

It is becoming customary to use the elastic moduli to quantify the properties of foods. These moduli are the ratio of stress to strain (tensile, compressive or shear). This concept was developed to describe the behaviour of metals and uses the basic assumption that stress is directly proportional to strain and that all the deformation is recoverable upon the removal of force. Since most foods are viscoelastic, this behaviour is seldom exhibited and the term 'modified elastic constants' has been adopted to reflect this deviation from an ideal state.

Theoretical stress analysis is a highly developed subject and several workers are adopting to foods analyses that have been formulated to calculate the distribution of stresses in engineering structures. This has the advantage that textural properties can be expressed in terms of fundamental physical constants, stress, strain and the elastic moduli. A recent example of this was the determination of the elastic modulus of uncooked potato tuber using a simple puncture text (Finney, 1963; Timbers, *et al.*, 1965). In this case, the dimensions of the test specimen were not critical, which is a particular advantage of the test. There is, however, great danger in adopting the simple classical elastic stress theory to tests of biological materials because such analyses are based on two assumptions: (a) the material is homogeneous, and (b) the deformations are small. Caution must, therefore, be exercised in analyzing stresses and strains in texture measurements where these conditions do not prevail. Sherman (1969) has pointed out that "classical elastic theory deals with very small strains so that when considering the much larger strains involved in mastication the accepted definition of strain may not be valid."

In designing texture testing apparatus, therefore, two approaches can be used to the selection of specimen: (a) make all the specimens the same size and report either force and deformation or stresses and strain; (b) use specimens of a defined shape (but not necessarily the same size), measure the size of each sample and report stress and strain. The method selected depends primarily on practical considerations in preparing samples of each specific product. If, however, the distribution of stress throughout the sample is not known, as for example in the General Foods Texturo-meter, only force and deformation can be reported.

4. Deformation Mechanisms

There are three basic design philosophies which can be considered in designing a mechanical system to deform foods during texture tests: (a) a universal machine that will handle a majority of tests, (b) a general purpose machine that will accommodate a specific range of test cells, and (c) a special purpose machine that will test only one sample type in a specific manner. There are advantages and disadvantages in each, but there are also common requirements to all these approaches.

The speed at which the test material is deformed must be considered as being critical because foods are viscoelastic and their reaction to force is time dependent. Also, the speed of deformation governs the rate of force rise which must not exceed the response rate of the recorder. Two types of deformation control are used in material testing: constant rate of force increase and constant rate of deformation. The latter is used for testing foods because it is simpler and less expensive than the automatic controls required to achieve a constant rate of force increase. The load capacity of the mechanism must also be adequate to handle the resistance forces generated in the deformed samples. Control of deformation speed is affected if this is not the case.

The most reliable and simple method of achieving constant deformation speeds is to use a synchronous electric motor to drive a screw carrying the deforming member of the test cell or the stirring component of a mixer or a viscometer. The deformation speed can be selected by changing the gear (or belt or chain) ratio between the motor and the screw. A reasonable degree of flexibility can be achieved at modest cost and constant reproducible speeds can be made available. This may, however, be inconvenient because a set of gears (pulleys or sprockets) must be kept on hand for each speed required. An alternative approach is to use a variable speed electric motor. These are now available at low cost because the silicone controlled rectifier, a solid state device, is now widely used to construct the electronic motor controls. These provide infinitely variable speeds which are usually selected by turning a knob on the electronic control. Such motors, however, do not provide the same degree of constant speed regulation as synchronous motors. Variations of 1 to 2% are quite common unless expensive motors are used which can provide control up to 0.1%. There is also the problem that there must either be provision to indicate the deformation speed or the speed control knob must be calibrated in terms of the deformation speed. The former involves additional expense, while the latter procedure is time-consuming and must be rechecked frequently if accuracy is required.

There are a number of ways to indicate the deformation speed; the most convenient is to measure the rotational speed of the motor. The deformation speed is then determined by allowing for the velocity ratio (i.e. gears, pulleys and screws) between the motor and the deformation surface. A tachometer (about $ 150) can be used (Voisey et al., 1970b) but then resolution and accuracy become a problem. An accurate measurement can be achieved (e.g. within 0.1 rpm) using an electro-magnetic pulse generator ($ 30) whose output is shown on a digital counter ($ 350) (Voisey et al., 1970c; Voisey and deMan, 1970b).

An inexpensive speed indicator (about $ 50) can be made by attaching a multilobed cam to the motor shaft or to one of the screws to operate a microswitch. An electro-mechanical counter can then be used to record electric pulses from the switch for a preselected time.

The load capacity of motor driven deformation mechanisms is limited by the output torque of the motor at its operating speed (or speeds) and the mechanical advantage of the motor over the deformation surface, i.e. the ratio of the gears, pulley or screw. The maximum allowable force exerted by the mechanism is also limited by the strength of the machine itself including components such as gears, frame and screws.

Hydraulic or pneumatic cylinders can be utilized to provide the motive power for compression, tension and shear machines. The Kramer Shear Press, for example, uses a hydraulic ram. The load capacity in this case depends on the working area of the cylinder piston and the available air or hydraulic pressure. In general, hydraulic systems are used for pressures up to 3000 psi, while pneumatic systems are limited to 100 psi. Machines using cylinders are simple to manufacture because the required components are reduced to a minimum. Accurate control of deformation speeds, however, is difficult unless sophisticated control systems are used to offset the effect of temperature changes in the hydraulic oil or air. A definite advantage exists because the controls operating the up and down motion of the ram consist of simply inexpensive valves which can be operated manually or electrically. Also, different speeds for up and down strokes are easily arranged and rapid return strokes can be used to speed the testing time. One practical problem is the noise generated by the pumps which should preferably be located away from the test area. This is an obvious requirement to anyone that has used a Kramer Shear Press. The author prefers the use of a motor and screwdrive because the speed control is more predictable and, for a given load capacity, the increased costs are not excessive.

Manually operated deformation mechanisms can be constructed inexpensively and there are numerous such examples in the literature. Water or lead shot may be poured into a container (Tyler and Coundon, 1965), a hand powered screw or a weight moved along a balanced beam (Romanoff, 1925) have been used to increase the force. Such devices should be avoided, if possible, because the rate of deformation or the increase of force is not controlled with the accuracy required for precise texture measurements.

4.1. Universal testing machines

A universal testing machine is the most useful apparatus because the same basic mechanism can be employed to conduct almost all of the test methods that have been reported in the literature. All that is required is to manufacture the attachments for holding the specimen or for mounting the test cell, as noted, for example, by Zachringer (1969) in testing apple slices. It can be used for compression, tension and shear tests and can be employed with special test cells.

There are many commercially available universal testing machines which have been developed for different industries such as paper, textiles, plastics and steel. These range from simple hand powered units with a spring scale to indicate force, to complex fully automated test systems. Hydraulic and pneumatic cylinders and electric motors are used for motive power by different manufacturers. Tatnall (1967a, b) recently reviewed the evolution of screwdrive and hydraulic test machines. The major factor to consider in the specification is the load capacity since this determines the maximum test forces which can be achieved while maintaining control of the deformation speed within the manufacturer's specifications. In general, the forces required in texture tests depend on the type and the size of the sample and the nature of the test cell. In the author's experience, forces involved are generally less than 250 kg and do not exceed 500 kg. This would be considered low forces in the material test industry. Also, in many instances the force required can be reduced by changing the size of the test cell or of the specimen until it is within the capacity of the machine. The selection of a machine can, thus, be restricted to those developed for areas such as the textile, paper and similar industries. Taking economic factors into consideration, the choice should be made from the viewpoint of operating flexibility, i.e. the range and number of compression speeds and sensitivities of the force recording system.

One series of machines that has become popular for texture tests is manufactured by Instron Ltd. (Canton, Massachusetts) (Figure 2). Machines manufactured by this company have been used for testing a number of biological materials, e.g., eggshells (Hunt and Voisey, 1966). The earliest reported work with food was by Bourne et al. (1966). This is a screwdrive machine and its principle components were described by

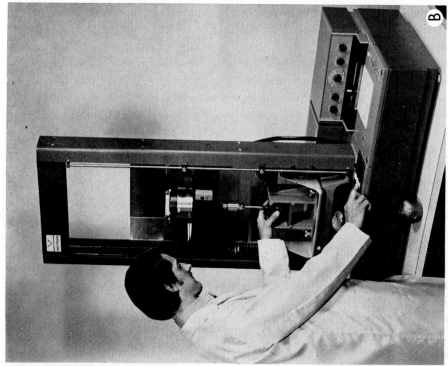

Fig. 2.   Instron Universal Test Machines. A. model TM-M, 250 kg capacity; B. model 1132, 500 kg capacity. (Photo courtesy of Instron Corporation, Canton, Mass.)

Hindman and Burr (1949). A motor drives two vertical screws via a series of gears. The crosshead, or the moving component, is driven up and down the vertical columns by the screws and the length of the down and up strokes and reversal of motion (i.e. deformation) is automatically controlled at preselected positions by switches which are adjusted by calibrated dials. The up and down speeds can be selected independently from an extremely wide range (e.g. 0.05 to 100.00 cm/min) by changing gears in the drive mechanism. A load cell is mounted on the frame of the machine, above or below the crosshead, to detect the force applied to the test specimen. There is a wide range of load cells available having maximum capacities ranging upwards from 100 g, in tension or compression. In addition, the sensitivity of each load cell can be adjusted, within its capacity, by electronic controls in the recording system.

Force is recorded on a potentiometric type strip-chart recorder. The chart is driven by an independent motor at pre-selected speeds ranging from 0.5 to 100 cm/min so that different crosshead-chart speed ratios can be used to record force-time curves. Optional plug-in attachments allow the automatic control or cycling of the crosshead according to either the force on the specimen or its deformation (i.e. extension). When these controls are installed, provision is also made to automatically start, stop and reverse the recorder chart in synchronization with the crosshead, so that either force-deformation or force-time curves can be plotted. The recording system may be built into the loading frame or be in a separate chassis. If desired, other load cells built into different test machines can be conveniently connected to utilize the same recording system.

The most suitable machines for texture tests from the available series are the table models with 250 and 500 kg capacities. The 500 kg capacities come in two price ranges and provide alternative degrees of operating flexibility. Instron Ltd. supplies these machines complete with load cells and a force recording system, and has a range of test accessories that can be added to the basic machine. The most recent machine (Model 1132, Figure 2B) is a specially equipped version of Model 1130, replacing the obsolete Model TM-M.

While the Instron, and other universal testing machines, are intended for up and down linear motion it can be adapted, by attachments, to perform a levered motion as in the General Foods Texturometer or torsional tests. This is not the most convenient, or economic method to accomplish these tests. However, Instron Ltd. do supply a torsion testing attachment for one of the machines pointing out the operational flexibility.

The Instron is discussed here because it has been widely adopted for testing foods following the pioneering work of Bourne et al. (1966). However, many other similar units are available and have been used by different researchers, for example Curtis and Hendrick (1969). Several inexpensive machines are available for testing at loads up to 5 kg (e.g. Model VTM-11 Imass Inc., Accord, Massachusetts 02018). Machines that are simpler or more sophisticated than the Instron and that have different features are also available. The choice may depend on considerations other than economy: availability of service, range of available accessories and so on. Considerable economies can be achieved in some cases by combining components from different manufacturers. The load cell and the electronic recording apparatus may possibly be purchased separately. This depends greatly on the technical support available to the researcher to modify and assemble the equipment.

It should be noted that the table model Instrons are primarily tensile test machines, i.e., force should be applied to the drive screws in the direction of the crosshead pulling downward. This is because the drive screws are spring loaded against the upper thrust bearings. If the crosshead is used to pull upward, the screws will compress the springs when sufficient force is applied and the screws can then move axially. This introduces an error in determining the deformation of the specimen. Thus, tensile testing must be done above the crosshead pulling against a load cell mounted at the top of the frame. Compression testing, particularly at high loads, should be done in the same position using a cage attachment accessory to convert the tensile pull to compression. In general, compression tests can be accomplished below the crosshead against a load cell mounted on the machine base providing the loads used in testing foods are small. Testing in this manner places the long drive screws in compression and under these conditions the strength of the screws is reduced because of the possibility of the screw buckling as a column. Also at high loads – e.g. in using a Kramer Shear cell – accurate force deformation records cannot be obtained for both entry and withdrawal of the blades because the screws may move axially as the force reverses and the position of the blades cannot be determined accurately at this point. This may not be important in practice and the technique might be used to estimate adhesive properties of foods with the Kramer Shear cell (Voisey, 1970; Voisey and Larmond, 1970). The problem does not arise when the test forces are small, such as in testing gels under tension (Henry and Katz, 1969). The limiting force when the crosshead is pulling upward can easily be measured using one of the Instron load cells to record the force required to overcome the springs at the top of the screws. The springs are adjustable, but care should be taken in doing this that the machine is not damaged.

There are several drawbacks associated with the Instron table model test machines when testing foods. The crosshead controls (i.e., extension load cycling and gage length, etc.) are in the base of the machine at the front. They are, therefore, vulnerable to any spilled food or juice pressed out of test samples or cells. A plastic cover is provided but care must be taken and drip trays installed to carry away any liquids. Such trays have to cover all or part of the controls which make their operation difficult. For this reason it is recommended that, in any deformation mechanism designed for testing foods, motors and electrical controls be mounted at the top. It is also a wise precaution to have the drive screws covered with expandable rubber sleeves.

There is also the problem of making fixtures to attach various standard food texture test cells to the machine. Access to a machine shop solves this problem (Figure 3) and various researchers have developed many attachments and fixtures (Jackman, 1960).

The selection of a universal testing machine is primarily dictated by economic considerations or by having access to a machine used for other purposes. For example, Bourne et al., 1966 and Bourne and Mondy, 1967, have used an Instron floor model

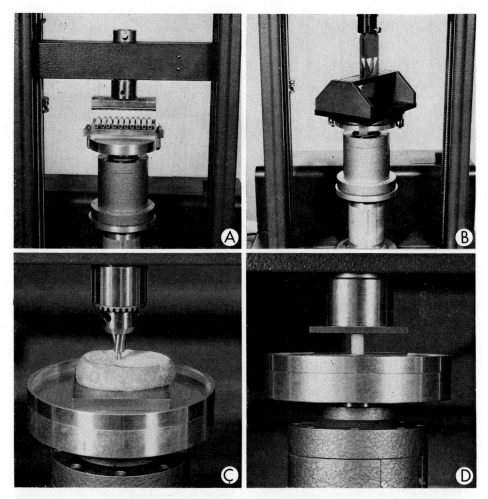

Fig. 3. Attachments installed in an Instron for texture measurements. A. cigarette firmness; B. Warner-Bratzler meat shear; C. puncture test; D. compression test.

with a capacity of 10 000 kg which is more expensive than the table models (e.g. 2500 kg, $ 11 000 or 10 000 kg, $ 16 000). As previously noted, a capacity of 250 kg is probably sufficient for testing most foods. The Kramer Shear Press has a capacity of 2500 kg which is seldom required except when testing materials such as rehydrated dried beans which produce forces of 750 to 1250 kg (Food Technology Corp., Reston, Virginia 22070). Forces up to 500 kg are required for testing meat and 1500 kg for peanuts (Szczesniak *et al.*, 1970). Thus, before making the selection it must be decided which test cells, sample size, and type of products are to be tested. The load capacity required can then be estimated from published data. The lower the capacity required, the lower the cost. However, if the budget allows, it is wise to purchase a greater load capacity machine to handle future unforeseen tests.

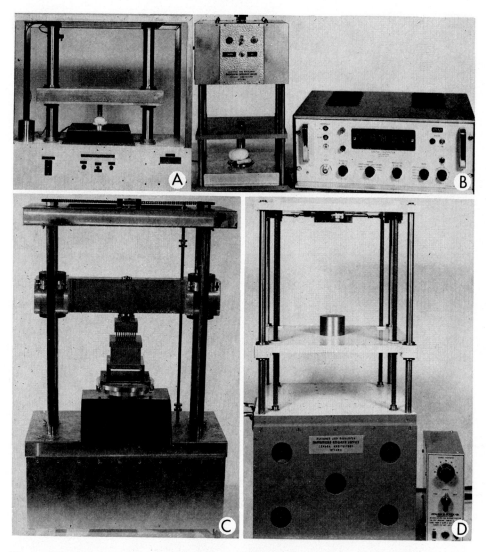

Fig. 4. 'Home-made' universal test machines. A. single speed; B. multispeed – electric drive using a gearbox at top; C. single speed with Kramer shear cell installed; D. multispeed using a variable speed electric drive motor.

It is quite feasible to construct simple universal deformation machines for testing foods; Voisey *et al.* (1967a) have described the principles of screw-type machines and pointed out the critical aspects of design and construction. The design can range from single speed machines with manual or automatic up and down controls (Figure 4A, C) to multispeed units with gearboxes (Figure 4B) or electric variable speed drives (Figure 4D). The examples shown indicate that the deformation mechanism can be constructed quite crudely using readily available components. These can be used to conduct most tests in the same way as commercial machines except that flexibility of operating conditions and convenience are reduced in order to minimize the cost of construction. In each example in Figure 4 the crosshead is moved by two or four screws which are driven by a motor via sprockets and a chain, or a timing belt and pulleys. Other machines have been described in the literature. For example, Mohsenin (1963) constructed a machine based on a pneumatic cylinder for measuring mechanical and rheological properties of agricultural products. Similar designs were used by Finney (1963) and Timbers (1964).

5. Components Available for Modernization of Texture Instruments

Forces and deformations applied to foods to study textural properties have been measured for several decades by simple inexpensive apparatus. Force may be measured by a spring scale, a balance, or directly by weights and recorded on a chart by a pen coupled to the weighing system using a mechanical linkage. A typical example was reported by Romanoff (1925) for testing eggshells. Mechanical recording dynamometers have been used extensively in the measurement of dough quality (Voisey et al., 1966b, e). There are, however, several disadvantages in this type of force recording, particularly when the forces involved are small:

(1) The mechanism may be cumbersome, delicate and subject to errors as moving parts become worn.

(2) The operating techniques may be critical and time-consuming.

(3) The sensitivity of the recording system may be difficult to change.

(4) Movement must take place in the mechanical linkages, springs and pen, etc. This introduces the frictional forces of the recording mechanism into the measurement causing hysteresis between records of increasing and decreasing force.

(5) The calibration is difficult because static methods using weights cannot be used conveniently and do not take frictional errors into account.

(6) The relationship between the recorder pen deflection and the applied force may not be linear, thus introducing errors.

(7) The recording charts generally have curvilinear scales.

Deformation has been measured using graduated scales which must be read directly. This is difficult particularly when deformations are small and a vernier scale must be used to magnify the movement. Deformation can also be recorded by a pen attached to the linkage. A popular method is to use a dial gauge; these are available with graduations as small as 0.0001 in or 0.001 mm. Deformation of a sample can be recorded inexpensively by noting the movement of a beam applying the load (Figure 5A). Distance between compressing surfaces can also be noted in this way or the size of the sample measured directly (Figure 5B) (Voisey and Hunt, 1968a). There are, however, disadvantages to these methods:

(1) The recording linkages or dial gauges are delicate and the sensitivity may be fixed or difficult to adjust. Unless special linkages are used, the records are made on curvilinear paper.

Fig. 5.   Measuring deformation with dial gauge. A. by detecting the position of a balanced beam applying the load; B. by detecting the distance between the deformation surfaces or the change in size of the sample itself.

(2) The deformation detector applies an unrecorded force due to friction within its mechanism. Thus, an error may be introduced into the measurement. For example, a typical dial gauge uses a spring to return the dial to zero. The spring typically exerts an initial force of 40 g which then increases throughout the deformation (Voisey and Robertson, 1969a). These forces cannot be eliminated from the records because to record the force the detector must be connected between the moving deformation surface and the surface supporting the sample.

The problems and errors noted above can be largely eliminated with electronic equipment. This requires the use of transducers converting force or deformation to an electronic signal which is amplified and recorded on a strip chart recorder. The following immediate advantages are obtained:

(1) Friction is practically eliminated from the measuring systems allowing calibrations to be made accurately under static conditions.

(2) Sensitivity is almost infinitely adjustable and zero on the scale can be fixed precisely.

(3) Accurate permanent records can be made on wide rectilinear charts.

(4) Wide selection of apparatus is readily available covering a considerable range of optional features and price.

5.1. MEASUREMENT OF FORCE

Several different principles are used to convert the applied force to an electric signal: the piezoelectric effects of certain crystals, photoelectric detectors, variable reluctance and resistance systems. An excellent review of these devices and sources of supply was prepared by Minnar (1963). Strain gages are currently the most popular elements used to construct force transducers and it is recommended that they be used for texture measurements. They are reliable and available in many shades, load capacities and prices. The electronic equipment for their operation is supplied by numerous manufacturers. These transducers may be designed to measure tension, compression, or both.

Strain gages are based on the principle, discovered by Lord Kelvin in 1856, that when a wire is stretched its resistance changes by a small amount. A strain gage, therefore, consists of a continuous length of wire folded to form a grid and mounted on an insulating material, usually an epoxy resin. To measure force, the gage is bonded to a structural member carrying the load. The change in resistance of the gage is proportional to the amount by which the member is strained. Since stress in the member is directly proportional to strain (within the elastic limit), the stress at the surface can be measured. This can in turn be related to the load on the member by theoretical analysis or direct calibration. Strain gages were initially developed for determining stresses in aircraft structures during the Second World War. Since then, their use has become widespread in a multitude of applications.

Because the resistance change of the gage is small, transducers generally use four or more gages placed on the structural member within the transducer and connected to form a Wheatstone bridge so that the changes add together to quadruple the output

of the transducer. Details of these arrangements are given in several excellent texts
on the subject (Perry and Lissner, 1955; Aronson and Nelson, 1960; Murray and Stein,
1961; Neubert, 1967) which will introduce the reader to the subject.

Within any given load capacity, commercial transducers are available in different
shapes and sizes and can be selected to suit the design of the texture apparatus
(Figure 6A). The accuracy, repeatability and linearity of these devices are generally
more than adequate for texture work. Economies can, however, be achieved by using
'home made' transducers. This offers a definite advantage since the transducer can
then be designed to suit the test apparatus and may even form an integral part of the
test machine structure. Installation of strain gages has become simple and straight
forward. Manufacturers supply complete installation kits and instructions so that
little skill is required. The design of the transducer can be a simple beam clamped at
one end with the force applied at the free end, or clamped at both ends with the
force applied in the centre (Figure 6C). Beams are probably the simplest type of
transducer to construct, but their output depends not only on the force applied but
also on the distance of the force from the fixed ends (Oberg and Jones, 1959). Thus,
beam transducers should be used only where the position of the test specimen and
the position of the force application are fixed. A useful transducer design that overcomes this is in the form of a cross (Figure 6A, B) with the four arms clamped and
two strain gages on each arm. The force in this case can be applied over a large area
at the centre without introducing errors. The cross design is particularly useful for
measuring small forces (100 to 2000 g).

Discussion of problems caused by elevated temperatures and other hostile environ-
ments have been published. Details can be found in the proceedings of the Instrument
Society of America and the Society for Experimental Stress Analysis (e.g., Anon.,
undated, 1961, 1962). Protective sealing of strain gages against the operating environment has received particular attention (Corkill, 1961) and this point should be
considered critical for texture test equipment because food juices will damage strain
gages. Sealing compounds are available from strain gage supliers.

Strain gages have evolved from simple wire grids to grids etched on metal foil
employing techniques similar to those used in producing electronic 'printed circuits'.
Foil gages are currently the most popular type. The most recent advance is the
development of semiconductor strain gages. The resistance of many semiconducting
materials changes with strain so that thin wafers of these materials can be used instead
of wire or foil. These have the advantage that changes in resistance are much larger
and for many applications the transducer output can be recorded without amplification (Dean and Douglas, 1962). Semiconductor gages are more expensive than the
foil type and installation is more difficult. Many commerical transducers are now
available, however, which utilize these gages. The subject of strain gages has become
highly developed and complex but for texture work it is feasible to construct force
transducers having satisfactory precision for as little as $ 20. The author has observed
several workers master the installation techniques in an hour.

Calibration of force transducers in a texture measuring system is important and

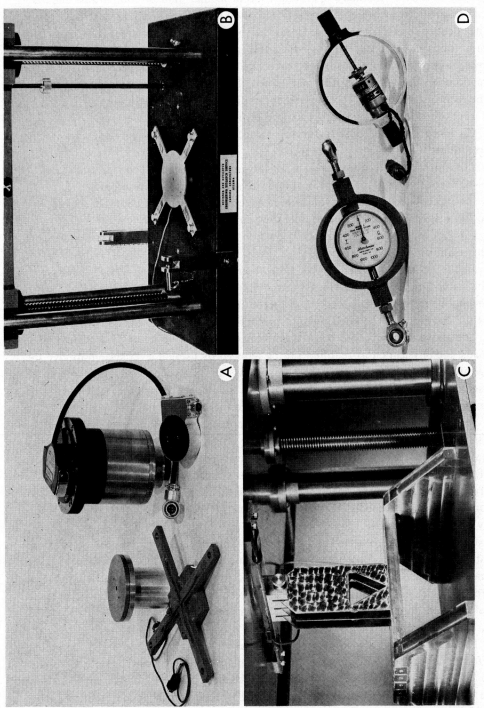

Fig. 6. Examples of force transducers. A. at the left two home-made units, at the right two commercial 500 kg units (top Instron; bottom Strainsert, Bryn Mawr, Pa. 19010); B. a cross-type transducer; C. a beam-type transducer; D. force gauge (left) and hybrid transducer (right).

should be part of the daily operating routine so that data of comparable accuracy can be obtained over long periods. Forces used in testing foods are relatively low so that dead weight calibration, an inexpensive but precise technique, can be used in most cases. The objective is to adjust the transducer output so that a preselected relationship between the applied force and the recorder pen deflection (or other readout device) can be set and maintained. The general procedure used is to adjust the controls of the transducer amplifier-conditioning system as follows:

(1) Adjust the zero control to achieve zero at the pen.

(2) Place a weight equal to the force selected for full scale pen deflection on the transducer.

(3) Adjust the sensitivity (or gain) of the amplifier until the pen is at full scale.

(4) Repeat steps 1 to 3 until zero and full scale conditions are both met.

(5) Place loads on the transducer increasing in increments of about 10% from zero to full scale and back again and note the pen deflection at each load.

(6) Plot a curve of load vs. pen deflection to check the linearity and hysteresis in the relationship.

In the author's experience with both 'home made' and commercial transducers, the maximum deviation from a linear relationship has not exceeded $\pm 0.25\%$ for increasing and decreasing loads. During routine testing, the weight of the sample and of the test cell must be tared by adjusting zero for each specimen.

The above technique can be used conveniently for forces up to 50 kg. For higher loads, the physical effort required can be reduced by using a lever system to give the calibration weight a mechanical advantage over the transducer. This method can be used for forces up to 500 kg. In this case, zero should be adjusted (i.e., tared) with the lever mechanism in place and readjusted when the mechanism is removed.

Another technique is to use a precalibrated force gauge or a proving ring (Figure 6D). The latter are metal rings which are placed on or connected to the transducer to transmit a force exerted by the deformation mechanism. The diametral deformation of the ring is indicated by a dial gauge and is converted to force units using a calibration curve supplied by the manufacturer. In this way, calibrations up to very large forces can be performed (Swindells and Debnam, 1962). The accuracy of calibration by this technique is, however, limited by the resolution of the dial gauge. For this reason, the force gauge used should have a maximum capacity about the same as the full-scale calibration force to obtain a maximum resolution. Other devices such as micrometers are also used to measure diametral deflections of the ring. Ring gauges are sometimes used as force transducers by replacing the dial gauge with an electronic deformation transducer (Figure 6D). This hybrid-type transducer is used in the Kramer Shear Press.

There are other critical points to consider besides accuracy in selecting or designing a force transducer. The ability of the device to withstand accidental overload should be taken into account, particularly with transducers recording small forces. This can be arranged by either making the structural strength much greater than required (which reduces sensitivity) or by installing a mechanical stop to limit the deformation (i.e. strain) of the strain gaged member. The natural vibrational frequency of the transducer is a criterion which governs the maximum rate of force rise that can be used. Errors can be introduced as the rate at which the force is applied increases to near the natural frequency of the transducer. Because foods are relatively soft and low loading rates (up to 100 cm/min) are usually used, the rate of force rise seldom approaches this threshold.

5.2. MEASUREMENT OF DEFORMATION

Deformation of the specimen can be automatically recorded by driving the recorder chart from the deformation mechanism. This is used, for example, in the Kramer Shear Press and in the model 1130 Instron test machine. This may not be satisfactory for many applications because the length of chart used during each test is fixed by the gear ratio between the deformation mechanism and the recorder. This can be changed, but the flexibility of operation is restricted. It is often desired to expand or to contract the deformation scale on the chart during testing to accentuate specific features on the force-deformation curve for easier or more accurate observation. Driving the chart from the deformation mechanism is a definite advantage since the recorded deformation is directly proportional to the movement of the deformation mechanism and is independent of the deformation speed.

The most common method of recording deformation in the testing of foods is to use the relationship between deformation and recorder chart speeds. Providing both of these are constant, the recorder chart can be calibrated in both time and deformation by using a conversion factor. This has been demonstrated with an Instron by Bourne (1967a) who used this approach to develop simple spring models to describe the rheological behaviour of several types of food (1967b). The principle point in the method is that the higher the chart speed, the greater the resolution of the deformation measurement. The accuracy depends on how accurately the chart and the deformation speeds can be measured with a ruler and a stop watch. Also, there will be large errors at the times when the chart and the deformation mechanism motors are accelerating up to speed. The test must, therefore, be arranged so that sample deformation starts after this time. Initiation of sample deformation can then be detected by observing the point at which the force increases above zero.

There is another problem in this technique due to the fact that all force transducers must deflect in order to measure load. In testing foods where sample deformations are large, these errors may be insignificant but this cannot be assumed. The deformation of the transducer under the maximum applied load should be measured with a dial guage. The deformation is generally linearly related to the applied force and the errors involved can be calculated and allowed for. Voisey and Hunt (1967c) demonstrated this in testing the strength of eggshells. One method of reducing the error is to use a force transducer with a much higher capacity than the maximum forces to be recorded, and to use electronic amplification to increase the senstivity. The transducer structure must be stiffer at higher capacity and, therefore, deflects less at a given load.

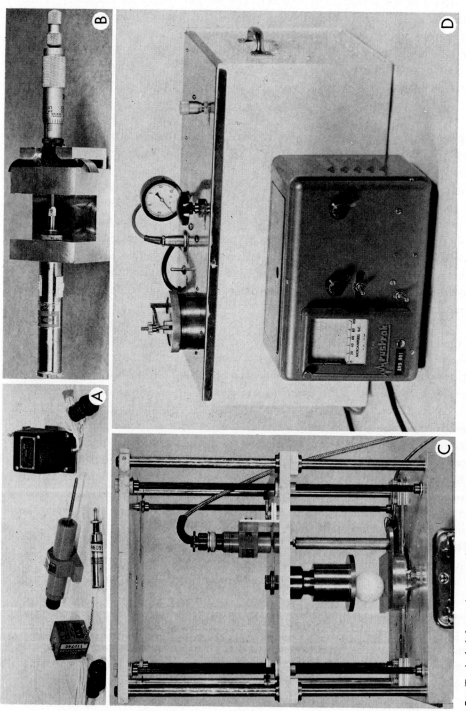

Fig. 7. Typical deformation transducers. A. cable driven potentiometers (left and right), free core (centre top) and restrained core (centre bottom) differential transformer type; B. calibration with a micrometer; C. typical installation in a universal deformation mechanism; D. pressure transducer to test the strength of fruit skins.

To record deformation precisely, particularly at small deflections (<0.02 cm), a separate transducer can be used to record deformation vs. time (or force) directly. This implies that either a two channel or an X−Y recorder must be used to record both force and deformation. This obviously increases costs. Deformation transducers are available in many types of which some can be 'home built'. One of the simplest types is to use a variable resistance or potentiometer. A constant voltage supply (e.g. mercury cell or electronic) is connected across the ends of the potentiometer resistance and a voltage proportional to the rotation of the potentiometer shaft can be obtained from the centre (i.e. wiper) connection. Ten turn potentiometers can be used to give a greater resolution of measurement than ordinary radio type potentiometers. A means of driving such units is to wrap a cable around a drum to drive the potentiometer shaft, using a spring to prevent backlash and rewind the cable (Figure 7A). This type is most useful for large deformations and their resolution is only limited by the coarseness of the resistance wire over which the moving wiper must pass. Linear type potentiometers are also available and have been used by various workers (Rediske, 1962).

One of the most popular types is the differential transformer where the primary and secondary windings are wound on a common spool with a moving core (Chass, 1962a, b). The output of the secondary winding is proportional to the position of the core. These devices are generally called 'Linear Velocity Displacement Transducers' (LVDT) and are available in two types: restrained and free core. Restrained core units have a spring which returns the core to its outermost position and the core is guided by a bearing. Free core units do not have these features and the core does not touch any part of the transducer body. It must, therefore, be attached to and guided by the deformation mechanism. LVDT's have been used for a wide range of applications including measuring the diameter of growing fruit (Tukey, 1964), growth of plants (Klueter et al., 1966) and the diameter of plant stems (Splinter, 1969).

Other instrumental techniques including optical (McClure et al., 1970), ultrasonic, nuclear and audio detection systems could be used but are not convenient for texture measurements.

A simple arrangement for measuring deformation digitally is to connect a mechanical counter directly to the deformation drive mechanism such as described by Summer (1968). If only strain gage recording equipment is available, deformation can be recorded by using a long flexible beam which is deformed with the sample. The output of the strain gages is then calibrated in terms of distance instead of force.

The calibration of deformation transducers is accomplished in the same manner as described above for force transducers except that a precision scale, dial gauge, vernier or micrometer is used as the calibration standard (Figure 7B). In this way, precise results can be obtained for large or very small deformations.

A critical point to consider is the force required to move any transducer during deformation in order to overcome internal friction, mass, return springs, etc. Depending upon how the deformation transducer is installed, this may be sensed by the force transducer and introduce an error into the records. These forces can be measured and plotted against deformation and – if significant – a correction should be made. This is inconvenient and can be eliminated by using a free core LVDT connected between the force transducer and the deformation surface (Figure 7C) to measure sample deformation precisely without introducing additional forces into the system.

5.3. PRESSURE TRANSDUCERS

Pressure transducers are available which use either an LVDT to detect the movement of bellows or similar devices as they are pressurized, or strain gages attached to a diaphragm to detect the strains induced by pressurization. Such transducers are available to detect pressures ranging from a few mm of water to very high pressures (Minnar, 1963). The application of these devices to textural instruments has been limited, although pressure measurements are used in the Alveograph to study bread dough rheology (Kent-Jones and Amos, 1957).

The apparatus designed by Voisey and Lyall (1965b) (Figure 7D), for testing tomato skin toughness demonstrates the possibilities of pressure recording. A cylinder is closed at one end by a strain gaged diaphragm and a sample of skin covers a small hole at the other end. Air is admitted to the cylinder at a slow controlled rate until the skin bursts. Because the rate of pressure increase is very slow, a low frequency response recording system could be used to obtain accurate data with economy.

Pressure transducers are also used to measure forces applied by deformation machines where the motive power is a hydraulic cylinder. This is easy to arrange by connecting a standard pressure transducer into the hydraulic oil supply line. This could be done, for example, with the Kramer Shear Press. There is the problem, however, of friction between the piston and the cylinder walls which is then included in the measurement and must be subtracted. It becomes serious at small forces.

5.4. MEASUREMENT OF TIME

Time can obviously be recorded by standard laboratory clocks and stop watches operated manually or automatically by electric switches on the deformation mechanism at selected points in the test. For example, micro-switches might be used to start an electronic stop watch and then stop timing after a preselected deformation. This principle can be used to construct simple apparatus where the force deformation relationships are known and chart records are not required.

The conventional method is to rely on the speed of the recorder chart which, in general, is driven at a constant preselected speed by a synchronous electric motor. It is, however, recommended that the chart speed by checked with a stop watch to ensure that the correct gears, etc., have been installed in the drive mechanism.

5.5. AMPLIFYING AND TRANSDUCER CONDITIONING EQUIPMENT

The electronic equipment required to operate force and deformation transducers is generally available from the same source as the transducers. In some respect, it is advisable to use this source since the interconnecting cables can then be ordered with the equipment. These take considerable time to make properly and are a common source of operating problems. However, there is often an advantage to buying transducers and amplifiers from different manufacturers since each has different designs offering features and advantages for different applications. The best guide to follow in this respect is to purchase units which have the greatest operating flexibility to handle different applications. This is usually achieved by having a basic electronic unit with plug-in modules. For example, Daytronic Inc. (Dayton, Ohio) supply a series of units (Model 300D) with interchangeable input modules to handle strain gage or LVDT transducers and output modules to drive different types of recorders.

Signal conditioning (Frank, 1965) is an important aspect of any measurement since it controls accuracy. Most of the equipment currently available is satisfactory for texture measurements having a specified accuracy of at least 0.25%. A major point to consider is that the frequency response of the system should not be exceeded by the rate of force rise or the deformation signal.

There are two basic types of electronic systems used to operate transducers:

(1) d–c, where the power supplied to the transducer and the transducer output are d–c, amplified by either a d–c or an a–c amplifier and then produced as a d–c output for recording;

(2) a–c, where the power supplied to the transducer and the transducer output are a–c, amplified by an a–c amplifier and then rectified to give a d–c output for recording.

It is difficult to generalize about the advantages of either type unless cost limits are included. The best choice is probably the a–c type because even inexpensive units are quite stable. The frequency response of these units is limited to relatively low levels (e.g. 400 Hz) by the frequency of the a–c power supplied to the transducer. They will handle both strain gage and LVDT transducers. The d–c systems will handle strain gage transducers only but are available to record signals at high frequencies (e.g. 20 kHz).

The electronic unit performs several functions. It cancels out the effect of connecting leads, balances resistive and capacitive components which are unmatched electrically within the transducer, provides zero adjustment and variation of sensitivity (or gain) and finally converts the signal to a form suitable for recording. An additional useful feature is a meter to provide visual display of the output during initial adjustments. Zero suppression is also available in some units.

Zero suppression is a useful feature which may be included in either the amplifier or the recorder to allow zero on the chart to be any selected load. This is accomplished by generating a voltage within the amplifier (or recorder) which can be connected to add or subtract from the amplifier output (or recorder input). The voltage can be selected by turning a knob, thus moving the recorder pen up or downscale. In some devices (e.g. Model 93 input module Daytronic Inc.) the knob controlling the voltage is calibrated and a separate sensitivity control allows this dial to be calibrated in the same units as the recorder chart. This allows the calibration of the system at much higher sensitivities for the same maximum load or deformation. For example, the system can be calibrated to record 10 kg (or 3500 to 3510 kg) over 100% of the recorder scale. Thus, if forces up to 3500 kg are not of interest, they can be eliminated (suppressed) from the recorded and the force range of interest can be expanded over the full scale.

5.6. ANALOGUE RECORDERS

There are two basic recorder types, potentiometric (Cameron, 1965a; Anon., 1968b) and galvanometric (Cameron, 1965b; Anon., 1967b). The former are used for recording slowly varying signals (Anon., 1969b), while the latter are available for both high and low speed signals.

Recorders are familiar laboratory instruments and are available in a variety of sub-types to suit different purposes. An indication of their diversity is given by the periodic listings in the trade press (Nelson, 1962; Aronson, 1964; Anon., 1967a, b; 1968c; 1969b, c).

In general, textural measurements produce signals varying at relatively slow rates for which the potentiometric-type recorder is most suitable. This uses a motor to drive the pen until the input signal is balanced by a reference voltage proportional to the pen deflection. Instruments of this type are available with charts up to about 25 cm wide which provides excellent resolution of measurement with accuracy usually in the order of 0.25%. Each manufacturer can supply units with different chart speeds and input sensitivities ranging from about 1 mV to high voltages. The ideal choices for texture work are multi chart speed, multi input range models with infinitely variable zero and sensitivity adjustments, usually called 'general purpose laboratory recorders'. These give maximum flexibility of operating conditions and will handle any texture test. The most critical performance aspect is the rate at which the recorder pen responds to the input signal. For potentiometric recorders, this is given as the time required for the pen to travel from zero to full-scale deflection for a full-scale step input signal and currently ranges from 0.2 to 30.0 sec depending on the manufacturer.

Serious errors can be introduced if the rate of rise of the input signal exceeds the rate at which the recorder pen can move. The following example will illustrate the point. Eggs selected to be equivalent in strength by the non-destructive technique of Schoorl and Boersma (1962) were compressed at speeds ranging from 0.5 to 30.0 cm/min in an Instron equipped with a 1.0 sec response time recorder. Theoretically, the eggs should have fractured at about the same force. The results, however (Figure 8) indicated that the eggs became weaker as the compression speed increased. This is incorrect, since it has already been shown that the eggshell strength increases with the compression speed (Voisey and Hunt, 1969). The reason for the discrepancy is that the force tracings were attenuated by the response of the recorder pen. These errors occur not only at the maximum force, but also at all points along the recorded curve. This may explain why some researchers have observed a bioyield point in a foodstuff while others have not. For all practical purposes the rate of rise of the input signal should be limited to about one-third of the theoretical maximum rate of the recorder. This can be accomplished by reducing the speed at which the test samples are deformed. An alternative method is to reduce the sensitivity of the recording system so that

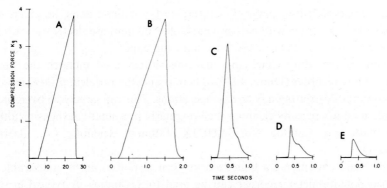

Fig. 8.   Records of eggs compressed at: A. 0.5; B. 5.0; C. 10.0; D. 20.0 and E. 30.0 cm/min.

the pen does not travel as far for a given load. This, however, degrades the accuracy and the resolution of measurement.

The effect of recorder response time can be evaluated by recording curves for typical samples. A line is then drawn from 0 to 100% on the chart at an angle which simulates the theoretical minimum response time of the pen. The angle of the sample record should be less than this line. As it approaches the theoretical maximum angle, the accuracy of the record decreases. This is because the sample signal is dynamic and is being followed by the pen. The pen lags the actual signal by an amount which increases gradually as the pen response time is approached and then increases greatly as this time is exceeded.

The response time of the recorder and amplifier can also be checked directly by placing a load (or deformation) instantaneously on the transducer. The resulting curve then shows the minimum full-scale response time. There is, however, a danger that impact loads will be introduced into the transducer causing it to oscillate. The resulting signal will not, therefore, be a true step input. Care should, thus, be exercised in applying this technique. An alternative method is to add a calibration resistance to the Wheatstone bridge to simulate a full-scale load (Perry and Lissner, 1955). Provision is made for this in most strain gage and LVDT amplifiers because the same technique is used as an indirect method of calibrating transducers that are inaccessible or where the forces are too great to allow a direct calibration with weights (Murray and Stein, 1961).

Another method of checking the response of the recorder is to connect a full-scale voltage directly to its input. This can be accomplished by using a battery with an appropriate resistance network to reduce the voltage or a function generator (e.g. Model F1000 Esterline Angus, Indianapolis, Indiana). The voltage is switched on and off and the resulting record gives an accurate estimate of pen response.

For accurate recording of high speed signals, galvanometer type recorders must be used (Voisey and Hunt, 1969). These are available to record signals at frequencies up to 10 kHz. A light beam type or oscillograph provides the widest charts (Hoadley 1967; Parsons, 1960) because the galvanometer movement is amplified optically and recorded on photosensitive paper. Operating costs for these units, however, is high. Other recorders (generally with narrow charts, 40–100 mm) use electrostatic discharge, heat of even ink sprays (Sweet, 1965) to mark the paper.

The type of recording chart and marking system used dictates the operating economy of any recorder (Dunn, 1970). There is also the problem of keeping the pen writing continuously during day to day operations. This can largely be overcome with commonly used ink pens by cleaning them regularly in a small inexpensive ultrasonic cleaning bath (e.g. Catalog No. 71,003JX, Edmund Scientific Co., Barrington, New Jersey).

Under certain conditions, a high recorder pen response rate is undesirable and a slow speed potentiometer recorder can be used to advantage. A typical example is the electronic recording mixer (Voisey et al., 1966f) where minor fluctuations in the mixing torque are not of interest. This is demonstrated by mixing dough in a 30 g National Mixograph modified to record torque electronically. Three samples mixed under identical conditions using different recorders show that the slower the recorder response, the closer is the record to the average torque during mixing (Figure 9).

Other analogue recording techniques that could be utilized for texture tests include analogue computers and magnetic tape recorders (Vandenberg, 1966). Their high initial cost has so far prevented their use. Event recording, where the instant at which operations occur is recorded against time (Freilich, 1966), is one of the least expensive recording methods. While this type of instrument is not applicable to force-deformation recording, it can be used to indicate points where particular events take place, such as the completion of the deformation stroke in texture test cells. A typical application was described by Voisey and Kloek (1964). Many recorders have available an optional event marker channel that can be added at one side of the recorder chart.

Every recorder introduces some errors into the measurement of quasistatic and dynamic data. In general, however, providing the precautions outlined above are taken, the force-deformation records of adequate precision, accuracy and resolution (Deutsch, 1965) can be obtained. Greater accuracies can be achieved if errors inherent in different components of the system are determined (Irwin and Korsch, 1964; Aaker, 1965; Salvador, 1966) and appropriate corrections made.

5.7. DIGITAL RECORDING

It is feasible to record the force deformation curve directly in digital form on magnetic or punched tape for computer analysis. This technique is currently too expensive for texture tests because rapid sampling of the data (e.g. 100 points per sec) is required to record the transient peaks that occur when specimen fail. Slower rates would introduce errors. It may, however, be economic to invest in such equipment if large numbers of samples must be tested where the labour saved in interpreting strip-charts can offset the cost of the equipment (Boyd, 1964). Recent developments (Buckley, 1969) have integrated a small computer into such data acquisition systems. Thus, the data sampled can be both controlled and analyzed during the test. Sophisticated data processing systems have been developed for this purpose in engineering appli-

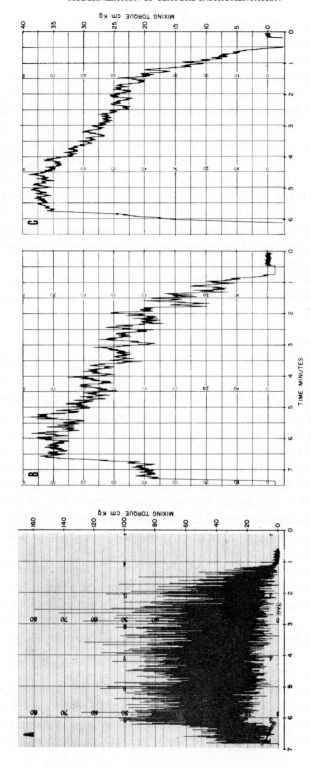

Fig. 9. Torque required to mix bread dough recorded with a full-scale pen response time of: A. 0.2; B. 3.0 and C. 12.0 sec.

cations (Bement and Irvin, 1964; Smith, 1970). However, use of such methods implies that the relationship between force, deformation and texture in the test is completely understood and that the method has become routine.

The digital voltmeter (DVM) (Weitz, 1969a, b; Anon., 1969a) provides a means of displaying static signals in digital form and can be selected from the wide range available commercially (Bakey, 1968a, b). The cost of these units has been reduced considerably in recent years. They can be used to indicate maximum points on the force-deformation curve by using peak detection techniques. Peak detection is accomplished electronically, the circuit follows the input signal as it increases and, as the slope of the signal changes when the specimen breaks, the circuit holds and maintains the maximum at its output. Several DVM's are available with the required electronic circuits built in. A typical application was shown in testing fruits and vegetables (Voisey et al., 1964b) where the maximum force during compression was recorded. An inexpensive method of accomplishing this is to purchase a strain gage amplifying and conditioning system that has the peak detection system built in (e.g. Daytronic Model 300D-93-M). The output can then be recorded by one of the many inexpensive ($ 100) digital panel meters that have become available during the past five years (Walton, 1969).

5.8. INTEGRATION OF ENERGY MEASUREMENTS

The area under the force-deformation curve represents the work done or the energy used to deform (or otherwise affect) the sample. Traditionally, this has been measured manually using planimeters, e.g. in determining fibrousness of asparagus with the Kramer Shear Press (Backinger et al., 1957). Area measurements have been employed by many workers using the Kramer Shear Press, the General Foods Texturometer and other texture measuring devices. Manual measurement, however, is tedious and subject to human errors. This can be eliminated by means of integrators which are readily available and use several operating principles ranging from chemical, mechanical, electromechanical to electronic. Electronic types (Tobey, 1969) are the most useful for texture work because the normal digital readout provides a high degree of resolution and accuracy. Many of these types, however, are designed to integrate slowly varying signals (Avery and Yoerger, 1958; Funk and Dowe, 1965; Brach and Mack, 1967; Brach and St. Amour, 1969; Brach et al., 1969; Lomask, 1969; Brach et al., 1970a, b) and are not suitable for texture measurements unless very slow loading rates are used or the test spans a long time period such as in a mixing test. Other electronic integrators are designed to record precisely rapidly varying signals by converting the analog input to pulses by means of a voltage-to-frequency converter and then counting the pulses. As the frequency used at this conversion increases, so the frequency of the input signal can increase and still maintain accuracy.

The advantages of integrating techniques are that the data are easily presented in digital form eliminating the human interpretation of strip charts and the error which this introduces. Also, the energy used is due to the sum of all the operations on the test sample. For example, in the Kramer Shear Press, the food is compressed, sheared and then extruded (Szczesniak *et al.*, 1970). The energy used can thus serve as a gross textural index related to viscoelastic and strength properties. It should be noted that because of this, control of sample size is critical for meaningful comparisons. Obviously, the larger the sample, the more energy is required to deform and break the food. This has been demonstrated by Voisey (1970) with baked beans and peas, and by Szczesniak *et al.* (1970) with a variety of foods.

Integrators suitable for texture work are expensive, and should be used for routine measurements only after it has been determined that energy is a satisfactory index of the textural parameter under study. For this reason, it is desirable that the integrator be capable of operating in parallel with a strip chart recorder so that the results can be compared with analog data. A typical method for accomplishing this is employed by Instron Corp. A gear is added to the recorder pen drive mechanism to rotate a potentiometer which controls the speed of an electric motor. The speed of the motor is proportional to the pen deflection and mechanical counters are used to record the number of revolutions of the instrument. At full-scale, the motor produces 5000 revolutions or counts per minute. This is converted to energy units by an appropriate factor.

A readily available solution in many laboratories is to use the electronic integrator installed in gas chromatography equipment. These instruments are quite sophisticated but they can be connected in parallel with the force deformation recorder (Voisey *et al.*, 1967b). Calibration is then performed in the usual way with weights and a curve is plotted of counts per minute vs force (or deformation). The conversion factor is then determined in the following manner: if $F=$ applied force g; $c=$ resulting number of counts/min; $V=$ rate of deformation cm/min; work done per min $= FV$ cmg; $\therefore c = FV$, or 1 count $= FV/c$ cmg.

5.9. GENERAL CONSIDERATIONS

In any recording system, the frequency response is limited by that component having the minimum frequency capability. Thus, before using a system, the recorded output should be analyzed to ensure that the input signal frequency is well within the limits imposed by the transducer, amplifier or recorder.

Modern electronic force and deformation recording systems are capable of giving accurate and reproducible data only if accurate calibration methods are used. The cost of the electronic instrumentation required to do this has decreased rapidly to the point where the major cost is in the deformation mechanism. For example, it is now feasible, using a 'home made' force transducer based on semiconductor strain gages, to assemble a complete electronic system for recording force on a 25 cm wide chart for about $ 500. This investment provides a limited purpose facility that can be used with almost any texture test method or mechanism currently in use.

It is a worthwhile investment to assemble all the electronic equipment in a single cabinet. This can be done inexpensively by using standard 19 in. cabinets available from electronic supply houses. All the instruments should then be ordered 'to fit standard 19 in relay racks'.

The terminology used by manufacturers to specify the performance of different

instruments can be confusing. It is, therefore, recommended that it be clearly understood before selecting a particular instrument. Glossaries of instrument terms (Anon.,
1966) are available from various sources.

6. Examples of Instruments

Many instruments for texture measurements have been recently redesigned, or new
ones constructed taking advantage of the above described modern components.
An early example was the addition of a force transducer and a recording system to
the Kramer Shear Press by Decker *et al.* (1957). The following examples are given
to indicate the way transducers and recording systems are used to improve accuracy
and flexibility of texture tests. They should also serve to indicate how the existing
apparatus can be modified to take advantage of the developments in the field of
electronics.

6.1. THE PUNCTURE TEST

The puncture test was one of the first measurements used to detect textural characteristics. One of the pioneering applications was by Willard and Shaw (1909). Early
instruments were manually operated and used the deflection of a spring to indicate
the force applied to the probe. These instruments known as fruit 'pressure testers'
have been widely adopted for field tests to determine the maturity of fruits (Magness
and Taylor, 1925; Blake, 1929; Mohsenin *et al.*, 1965). There are several variations
in the designs available (Haller, 1941) including the 'mechanical thumb' which controls
the probe penetration to a preselected amount by a switch operating a light (Figure 10).
The use of these instruments has been investigated (Nichols, 1960) and their operating
limitations are well known. Fruit pressure testers have one major undesirable feature
in that the force is applied manually and the rate of application is not controlled.
Additional discussion on puncture testing is referred to in Chapter VII. The mechanical
thumb has been modernized to indicate force and penetration by electronic apparatus
(Textprobe Agricultural Specialty Co., Beltsville, Md. 20705).

The puncture test has the advantage that specimen can be tested with a minimum
of preparation and that under certain conditions physical constants of the test material
can be derived from the data (Timbers *et al.*, 1965). The test cell, i.e., the punch, can
be easily reproduced at a very low cost. The technique has, therefore, been adopted
by many workers to mechanical deformation mechanisms in order to apply force at
controlled rates in conjunction with electronic recording systems. For example,
Mohsenin *et al.* (1963) used an apparatus specifically developed for testing fruits and
vegetables and later investigated the effect of rapid penetration rates (Fletcher *et al.*,
1965). The Instron has been used extensively for the same purpose, e.g. in testing
potato tubers (Voisey *et al.*, 1969c) (Figure 11A) in an attempt to find a more reliable
quality index than specific gravity (Voisey *et al.*, 1964a). Bourne (1965) has reported on
fundamental studies of the effect of penetration rate and compared the Magness-
Taylor pressure tester with results from the same probe mounted in an Instron.

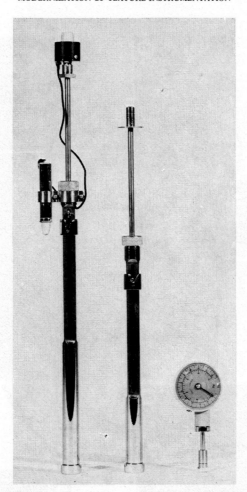

Fig. 10. Fruit pressure testers. A. mechanical thumb; B. pressure tester; C. compact instrument manufactured in Italy (Effi. gi, Alfonsine). (See also Figure 2a, Chapter VII.)

Other work has shown the effect of probe shape (Bourne, 1966a) and that the modulus of elasticity can be estimated from probe force and deformation (Finney, 1963; Timbers *et al.*, 1965).

The puncture test is widely used and since the test machine requirements are relatively simple it is worthwhile constructing a special purpose unit so that more expensive equipment is not tied up. This approach has been used by different workers and several designs were proposed for the purpose (e.g. Jacobson and Armbruster, 1968).

A typical machine was developed for estimating tomato skin strength (Voisey *et al.*, 1964c; Voisey and MacDonald, 1964). It is based on a cross-type force transducer mounted under a cast aluminum base (Figure 11B). The test specimen or puncture probe can be aligned anywhere within a 12 cm diameter platform supported at the

PETER W. VOISEY

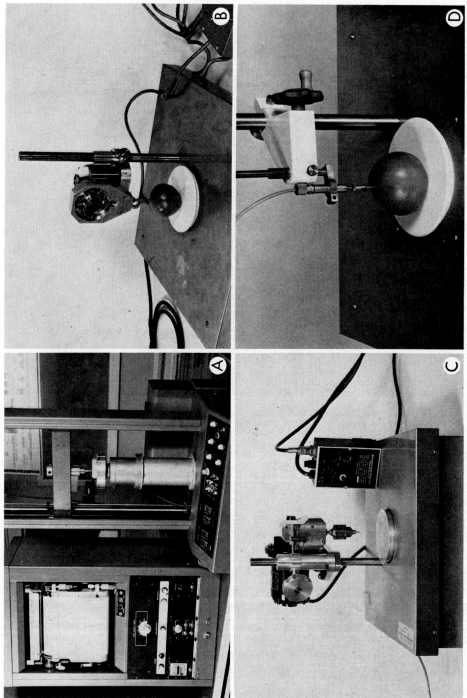

Fig. 11.   Puncture tests. A. using an Instron; B. cam driven at fixed speed; C. variable speed; D. pneumatic loading cylinder.

centre of the cross. Different probe drive mechanisms can be mounted on a column and moved up and down to accommodate specimens of different thicknesses. The original design (Figure 11B) used a synchronous motor rotating a multilobed cam to drive the probe down at constant velocity. This was later replaced by a variable speed motor (Figure 11C) so that the constant probe speed could be selected from a wide range. The later design is now manufactured commercially for about $ 450 (Queensboro Instruments, Ottawa). Other loading devices, such as a pneumatic cylinder (Figure 11D), can be used. The output of the force transducer can be connected to any suitable electronic recording apparatus.

The usefulness of this simple instrument is indicated by its applications. Extensive work with tomatoes (Voisey and Lyall, 1965a, b, 1966; Voisey et al., 1970a) has demonstrated that it can measure skin toughness in certain fruits and that it operates efficiently. Other applications include the measurement of pericarp tenderness in sweet corn (Voisey and Nuttall, 1965), the measurement of potato texture (deMan, 1969a), the study of the behaviour of fats (deMan, 1969b) and the firmness of heat induced milk gels (Kalab et al., 1970a, b), the measurement of French fry (Kenkars et al., 1967) and strawberry texture (Blatt et al., 1969). A recent innovation by Tanaka et al. (1971) was to use cone penetrometers in place of puncture probes to test dairy products. Instead of using a constant weight cone, the cone was driven into the product at a constant velocity by the puncture tester. This allowed the calculation of yield value and apparent viscosity for several foods.

An even simpler design than above was based on a cantilever beam transducer using

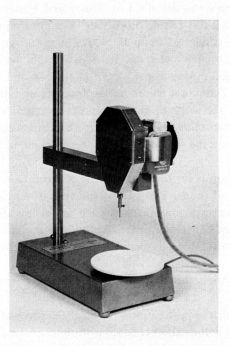

Fig. 12. Simplified puncture tester using a semi-conductor strain-gage force transducer.

semi conductor strain gages (Voisey and MacDonald, 1966). This machine (Figure 12) is suitable only for measuring the penetration force because the beam deflects a relatively large amount introducing an error into deformation measurements. This error may be excessive for certain foods. However, it can be calculated and allowed for. The force transducer is powered by a battery and its output fed directly to a strip chart recorder without amplification, thus reducing the cost of the apparatus considerably.

A new instrument recently developed by Armour and Co. to measure meat carcass tenderness is another example of a puncture test using modern instrumentation. The force required to manually insert 10 pointed probes into the meat is detected by strain gages and used to classify each carcass.

The maximum force during penetration is commonly the point of greatest interest and peak detection techniques can be used to advantage. When the force and the deformation are recorded on a chart, however, the stiffness of the sample can be estimated from the initial slope of the curve.

6.2. DEFORMATION AND CREEP TESTING

It is often more convenient to record deformation only, particularly when using a constant test force. For example, in an elastic material where the force is proportional to the deformation, elasticity or stiffness can be determined by a non-destructive technique as in the case of eggshells (Hunt and Voisey, 1966). There is a distinct economic advantage to the method because a simple special-purpose deformation mechanism will suffice.

A unit for testing eggs was described by Voisey and Foster (1970) where a 500 g weight is lowered onto a rod resting on the egg. The position of the rod is detected by an LVDT which is connected to an amplifier (Figure 13A). The amplifier output is indicated by a digital panel meter and provides a resolution of 0.00025 mm in the measurement. Deformation due to the weight of the rod and then the rod plus the weight are recorded. Thus, two points on the force deformation line exactly 500 g apart are known and the slope or stiffness can be precisely calculated. Pneumatic cylinders are arranged to raise and lower the rod and the weight at constant speed. Foot pedal controls leave the operator's hands free to manipulate the sample making for rapid operation.

A different application is in the measurement of creep or relaxation to study the rheological behaviour of foods. This is a special type of tensile or compressive test to study flow occurring over long periods when forces less than those required to cause failure are applied. There are several methods of doing this:

(1) applying a constant force and recording the resulting deformation changes with time;

(2) applying a constant deformation and recording the resulting force changes with time;

(3) using a universal testing machine to control the force at a constant level and recording the resulting deformation changes with time.

The first two methods are used most frequently because they are easily arranged

Fig. 13. Deformation recording machines. A. recording movement of a rod applying force to sample; B. recording position of balanced beam pulling (or compressing) the sample; C. tensile test clamps that can be replaced with compression surfaces.

for both tensile and compressive testing. A typical example of the constant force technique is to use a balanced beam to apply a preselected force to the specimen, e.g. fruit skin (Figure 13B). The movement of the beam is detected by a free core LVDT which does not introduce friction into the measurement and provides an output for recording. For initial calibration, a dial gauge is installed as shown so that the beam can be moved a preselected amount and the amplifier adjusted until the recorder pen deflects a corresponding amount. The instrument can be modified for compression testing by replacing the clamps (Figure 13C) with compression surfaces. A definite problem in this type of a test is the drift of the recording system. This should be determined and the necessary corrections made to the data.

The deformation test machine (Figure 13A) can also be used for the same purpose as the balanced beam.

6.3. Tensile and Compression Testing

Special purpose machines can be designed for tensile or compressive tests on specific products. Figure 14A shows a typical arrangement for tensile testing of fruit skin. A small synchronous motor rotates a screw to move a carriage carrying a clamp (or compression surface) mounted in the centre. A second clamp is mounted in the centre of a strain-gaged beam. Thus, the force exerted on a specimen between the two clamps can be detected. To facilitate calibration of the force transducer, a spring scale is mounted behind the beam and is pressed against it by a screw until the pre-selected calibration force is applied.

Obviously, such tests can be accomplished more conveniently in 'homemade' (Figure 14B) or commercial universal testing machines and many such applications have been reported. There are, however, several points which must be considered. In tensile tests, the clamps used to grip the specimen always present a problem particularly with biological materials. Any damage due to the pressure of the clamps may cause premature failure of the sample. This can be overcome by careful clamping arrangements. A more satisfactory method is to reduce the area of the specimen over the mid portion along the strain axis. The specimen should then fail in this zone and the effect of the clamps is reduced to negligible proportions. In compression testing, the finish of the compression surfaces can influence the results under certain conditions since it has a great bearing on contact stresses developed. This is particularly important in testing hard brittle materials. It is, therefore, the best policy to select a material such as commercially available stainless steel ground to a specified surface finish and make all compression surfaces of this material (Voisey and Hunt, 1967c).

The action of teeth in repeated compression of food can be simulated mechanically and the forces recorded during the process. An example of this imitative type machine is the General Foods Texturometer which is now in commercial production (Zenken Co., Tokyo, distributed in U.S. and Canada by C. W. Brabender Co., Hackensack, N.J.). Again General Foods Texturometer is discussed further in Chapter VII. A simple version of this type of tester can be made from a motor driven crank driving a plunger up and down (Figure 14C). The deformation can be varied by adjusting

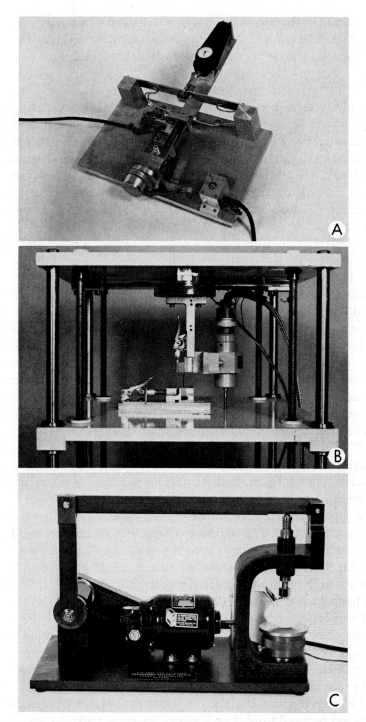

Fig. 14. A. simple tensile test apparatus; B. tensile testing in a home-made universal test machine; C. repeated load tester.

the crank radius and the force is detected by a transducer between the plunger and the specimen. A variable speed motor allows testing at different deformation rates and the forces vary sinusoidally with time (Voisey and Hunt, 1968a).

6.4. SHEAR TESTS

The most widely used shear measurement is based on the principle suggested by Warner (1928) and Bratzler (1932) and finds popular application in testing of meat. Cores are inserted into a triangular hole in a thin blade which is then pulled through a slot to cut the specimen. In the original apparatus, the maximum force was indicated by a spring scale. The technique has been adopted widely for testing other foods and the shearing blade has been made to fit the Kramer Shear Press (Hartman et al., 1963) and the Instron (Howard and Heinz, 1970) so that force deformation records can be obtained.

The routine Warner-Bratzler shear test does not require a sophisticated deformation mechanism because the test conditions are practically constant. This is another case where it is logical to use a machine specifically designed for the purpose and supply modern attachments. Such a conversion has been described (Voisey et al., 1965; Voisey and Hansen, 1967). A synchronous motor is arranged to rotate two screws which drive a bar up and down a fixed stroke (Figure 15A). A slot in the bar passes over the blade to shear the sample. The blade is supported by a cantilever beam projecting from the top of the frame which has strain gages mounted near the fixed end to produce a signal for recording the shearing force. This is an example of how the transducer can be integrated into the structure of the deformation mechanism. This apparatus is now available commercially for about $ 450 (Queensboro Instruments). The versatility of the machine can be increased by using additional blades (Figure 15B). This can increase the precision of measurement under certain conditions since a better estimate of the average shear force is obtained. The instrument has been adopted for testing meat (Larmond et al., 1970) and other products such as strawberries, French fries and cheese.

Another popular shear test employs a circular cutter or wire to cut through soft foods such a scheese curd. The original apparatus used by Emmons and Price (1959) to estimate curd firmness in cottage cheese was based on a Cherry-Burrell curd tension meter which indicated the cutting force on a dietetic scale (Figure 16A). The apparatus was recently modified with a strain-gaged cantilever beam to support the slotted curd container to detect shearing force (Voisey and Emmons, 1966; Voisey et al., 1966g). A special purpose deformation mechanism was also constructed (Figure 16B). A synchronous motor rotates a screw to drive a frame carrying the cutting wire through the sample in a container. A mechanism is incorporated so that the frame could be disconnected from the screw at the end of the test and returned manually to the initial position to save time. With the added sensitivity of electronic recording and the additional information then available it was possible to show that reliable estimates of firmness are obtained using 2 or 3 wires. The same apparatus was also used to estimate firmness of milk puddings and custards using the Cherry-Burrell circular

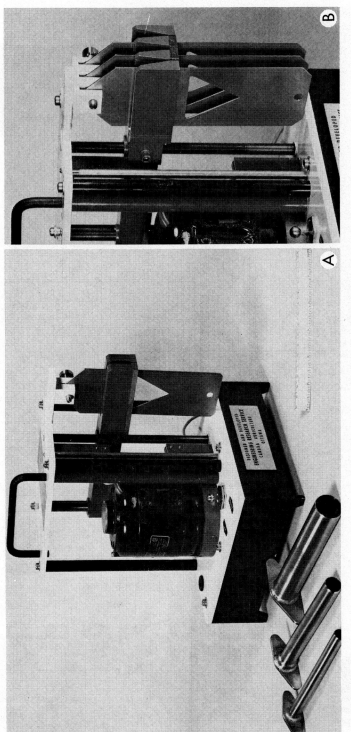

Fig. 15. A. a modern version of the Warner-Bratzler shear apparatus; B. two or three blades can be used instead of one.

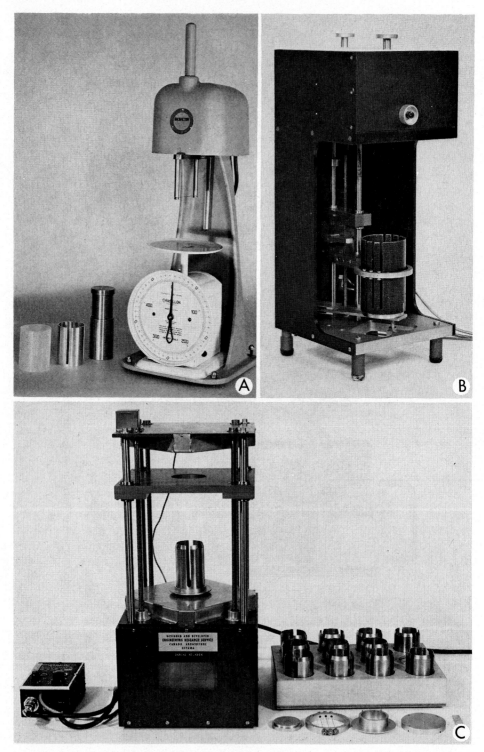

Fig. 16.   Wire shearing devices. A. adaption of a Cherry-Burrell curd firmness meter; B. a modernized version using a force transducer; C. a universal test machine arranged to shear a wire through butter.

cutter instead of the wires (Emmons *et al.*, 1970) and working at full-scale recording sensitivities of 5 to 100 g.

Another example of a wire apparatus is a small universal deformation machine adapted to estimate the spreadability of butter (Figure 16C). A single wire is sheared through a container full of butter conditioned at a preselected temperature. The container has 4 slots in the sides so that it can be rotated 90° and a second measurement obtained on the same sample. The error introduced by the first passage of the wire is negligible and the technique was found to reflect the consistency of butter (Voisey, 1965, unpublished data).

The wire shear apparatus is convenient and economic and has the advantage that the most critical component, the wire, is inexpensive and easily replaced. This was recently demonstrated in extending the technique to canned fruits and vegetables using 10 parallel wires (Voisey, 1970).

The most critical point in any shear apparatus is the shape of the cutting edge since it has a pronounced effect on the shearing force. This is true for any device including the Kramer Shear Press, the Pea Tenderometer and the Warner-Bratzler shear. Warner (1928) and Bratzler (1932) used radiused edges on their blade, whereas other workers have used flat sharp edges and the data can no longer be compared. The shape of the cutting edge determines the distribution of stresses particularly at the points of contact where compression, tension or shear may be generated depending on the shape of the cutting edge.

For a blade type test cell, a sharp edge is easier to manufacture but it is more prone to wear and damage than a radiused edge. For this reason, the wire shear test technique has a distinct advantage because wire is manufactured to close tolerances and yet is inexpensive. The choice between blade or wire type test cells is, however, largely dictated by practical considerations. It would, for example, be unreasonable to use a wire cutter to test meat because the force required to do this would be very high. Under these conditions, the strength of the wire would be inadequate and it would bend. On the other hand the cutting edge in a blade has almost unlimited strength.

Recently, attention has been paid to the effect of sample size on force readings in shear-type instruments. Pool and Klose (1969) and Davey and Gilbert (1969) showed a very significant effect of small differences in sample diameter on forces recorded by the Warner-Bratzler shear. This point has been overlooked by many workers and, undoubtedly, contributed to much inaccuracies. Samples have been prepared by using coring tools which makes difficult a precise control of the diameter, particularly in meat. Recently, Kastner and Henrickson (1970) developed a method of producing uniform cores.

Szczesniak *et al.* (1970) have shown that when the compression-shear cell of the Kramer Shear Press is filled beyond a certain level, the maximum force is independent of sample weight for some foods whereas there is a pronounced relationship for other foods as discussed in Chapter VII. With some foods under certain conditions shearing by wires may take place with an almost constant force (Voisey and Emmons, 1966). In such cases, the sample size may not be critical with wire type cells providing sufficient material is used to achieve this 'steady state' condition.

6.5. Recording Mixers, Consistometers and Viscometers

Up until this point, the application of electronic instrumentation has been shown in relation to linear compression, tension and shear tests. The same basic amplifying and recording apparatus, and in some cases transducers, can be used to modernize mixers, consistometers and viscometers. This presents a further justification for the expenditure on the electronic equipment. The food mixer can be considered as a special purpose deformation mechanism and the viscometer as a special purpose mixer with special mixing paddles.

The cereal chemist has been the primary user of recording mixers to evaluate mixing properties and baking potential of wheat flours. The two most popular mixers are the Mixograph (Swanson and Working, 1933), a pin-type mixer, and the Farinograph, a paddle-type mixer introduced by Hanckoczy (Brabender, 1965). These machines are based on the principle patented by Hogarth in 1889 (Kent-Jones and Amos, 1957). There is, thus, a long history of research based on the use of these instruments. The technique is to combine flour, water and other ingredients and to record the torque required to mix the resulting dough. This record provides a quantitative measure related to the rheological properties of the dough.

Different methods have been used to record the mixing torque including wattmeters which register the electric power used by the motor, and mechanical dynamometers which record the reaction of the mixing bowl (Mixograph) or the reaction of the motor (Farinograph). A mechanical dynamometer consists of an arm restrained from rotating about a pivot by a spring or balance. A pen attached to the arm records its movement which is proportional to the torque applied. There are problems inherent in using wattmeters or mechanical dynamometers which introduce errors into the measurement. These become serious when dealing with small samples and are caused by friction and other variables discussed elsewhere (Voisey et al., 1966, b, e).

Two methods are used to measure torque with an electronic transducer: transmission and reaction. In a transmission type, the torque is detected at a rotating shaft driving the mechanism. The shaft twists under the applied torque. The angle of twist may be measured electromagnetically, optically or, most often, by strain gages applied directly to the shaft to detect shear strains (Foskett, 1968). Since the shaft is rotating, provisions must be made to transmit the signal to the amplifier. This may be accomplished by electromagnetic coupling or, most frequently, by slip rings. Reaction type transducers measure, at a preselected radius, the force required to hold part of the mechanism stationary. This may be done with a lever or a beam type force transducer (Guthrie, 1964) or a stationary shaft with strain gages in shear. Slip rings or similar devices are not required in these cases. Transmission type transducers are expensive (e.g. $ 1500 to $ 2500) because of the addition of slip rings. Reaction type transducers are less expensive and home-made transducers are feasible as will be demonstrated.

At this point, a note on the units of torque may be useful. Torque is the tendency of a force to produce rotation about an axis (Minnar, 1963). Thus, if a force Fg is applied to a body at R cm from the centre of rotation (O in Figure 17):

Torque $(T) = FR$ cmg.

If the body rotates one revolution, then the work done or energy used (E) is:

$E = 2\pi RF$ or $2\pi T$.

If the body is rotating at N rpm, the work done per minute (E^1) is:

$E^1 = 2\pi NT$

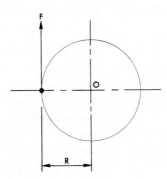

Fig. 17. Torque definition of terms.

Calibration of transmission type transducers is best done dynamically where a preselected torque is applied to the rotating shaft of the transducer using a Prony brake (Jagger, 1952). This applies torque by friction which is counterbalanced by a weight at a known radius from the centre of rotation. This technique is inconvenient, particularly with small instruments, and static calibration is preferred with both reaction and transmission type transducers. A weight is applied at a known radius, often by means of a pulley and a cord, and the response of the recorder is noted. Thus, the techniques for determining the relationship between the torque and the recorder pen deflection are essentially the same as for force transducers.

In calibration for energy measurements using an electronic digital integrator, a preselected torque T is applied for a preselected time (one min) which produces a number of counts (C). The integrator output is then converted to energy units: T cmg applied for 1 min $= C$ counts; work done/revolution $= 2\pi T$ cmg; work done/min $= 2\pi NT$; or, $C = 2\pi NT$; 1 count $= 2\pi NT/C$ cmg.

Using electronic transducers, it is possible to add to any mixer, consistometer or rotating viscometer a system providing a continuous record of torque. This is feasible with practically any mixer size, laboratory, pilot plant or full-scale production units. The conversion becomes more difficult as the mixer or viscometer size is reduced because the effects of friction in any of the working parts become more pronounced.

The most reliable method is to measure the reaction of the mixing bowl or of the viscometer cup because this is generally stationary and the frictional effects in the mechanism driving the mixing paddle or spindle are eliminated. This may be reversed in some instruments where the paddle or the spindle are fixed and the bowl rotates. The input torque to the system can be detected by measuring the reaction of the motor or the torque transmitted to the paddle, spindle, bowl or cup, depending on which part of the mechanism is rotating. However, under these conditions, it is often impossible to eliminate friction in the gears and bearings from the measurement. Nevertheless, these conversions can provide the food technologist with powerful tools for fundamental rheological research and product development as demonstrated by the following examples.

6.6. MEASUREMENT OF MIXING BOWL REACTION

Faced with the need to test small samples for plant breeders, the cereal chemist requires a mixer for studying dough development with a minimum of flour. The Mixograph (National Manufacturing Ltd., Lincoln, Nebraska) has been used for this purpose, but requires 30–35 g of flour. This machine was first modified by installing a beam type transducer to lock the moving arm of the mechanical dynamometer so that the torque reaction of the mixing bowl could be recorded electronically (Figure 18A, B) (Voisey et al., 1966c). The next stage in the development was to remove the mechanical dynamometer entirely and replace it with a reaction type electronic transducer. This demonstrates how much simpler the mechanism can become (Figure 18C). Different recording techniques were investigated and the usefulness of digital integration was demonstrated (Voisey et al., 1966d, 1967b). The Mixograph was, thus, converted from a crude machine into a precise instrument (Figure 18D). Using the same techniques, it was found quite practical to scale down the size of the mixer to handle 5 and 10 g samples (Figure 18E) (Voisey et al., 1966b) and then further to 1 to 5 g samples (Voisey et al., 1969b) (Figure 18F). This was possible because the sensitivity of the electronic dynamometer system could be made to be as high as required. Repeatable results were feasible even at these small sample sizes (Voisey and Miller, 1970) and the cereal quality laboratories in Ottawa and Winnipeg now work routinely with 3.5, 5 and 10 g flour quantities (Voisey, 1966). The modified 30–35 g mixer has also been used in other applications, e.g. in recording mixing torque for margarine (deMan, 1970). This pin type mixer is now applicable to a range of food products.

Larger mixers have been modified using the same techniques, e.g. the Hobart-Swanson (Model C-100, Hobart Manufacturing Co., Troy, Ohio) another pin-type mixer (Voisey et al., 1970b) – so that dough development curves could be recorded with baking test size samples (Figure 19A, B). Similarly, a household mixer (Model K5A, Hobart Manufacturing Co.) was converted to a 'Food Technologist's Recording Viscometer' by mounting the bowl on a spindle prevented from rotating by a beam type transducer (Figure 19C, D). This unit was used to record the consistency of whipped toppings, egg albumen, cake mixes, and other food products (Voisey and deMan, 1970b). It has been demonstrated that instant potatoes can be graded with this instrument (Voisey and Dean, 1970).

6.7. MEASUREMENT OF MIXING PADDLE REACTION

Almost the same principles can be used to measure the reaction of a mixing paddle

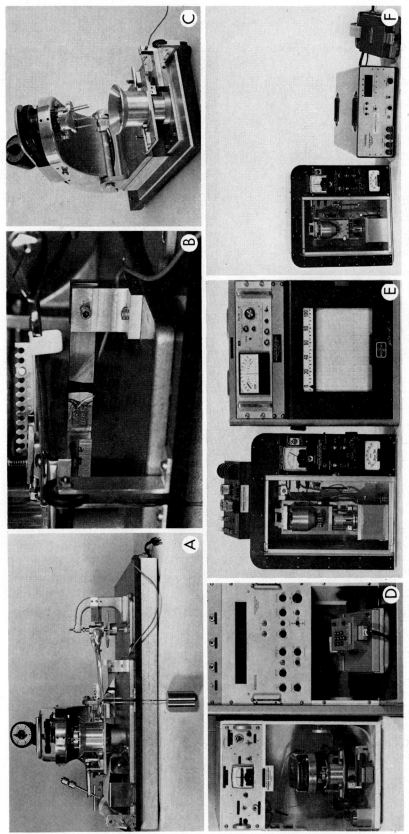

Fig. 18. Electronic recording Mixograph type mixers. A. National 30–35 g converted; B. close-up of beam-type transducer in A.; C. the final simplified version; D. same connected to a digital integrator; E. a 10 g version with a strip-chart recorder; F. a 5 g model with a digital integrator.

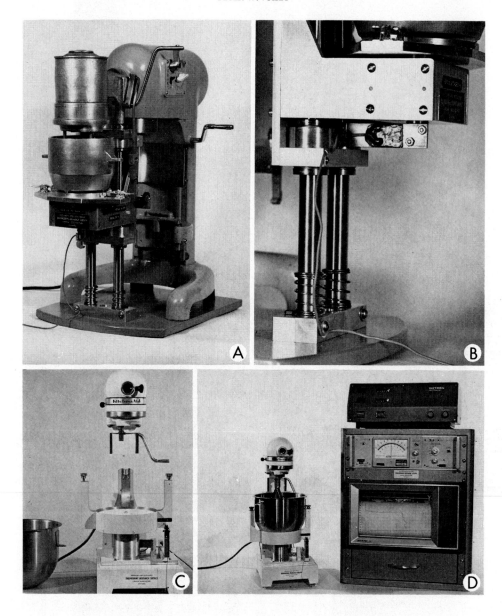

Fig. 19.   Other mixer conversions. A. Hobart pin-type; B. close-up of beam-type transducer; C. a Food Technologist's Recording Viscometer based on a household-type mixer; D. same with a recorder and digital integrator.

or a viscometer spindle. This has been demonstrated with the Amylograph (Type DC2 Brabender Corp., Rochelle Park, New Jersey) by replacing the interchangeable coil springs used in the mechanical dynamometer with an electronic beam-type transducer (Voisey and Nunes, 1968) (Figure 20A). The number of components required was minimal and the instrument could be quickly reverted to its original design (Figure 20B). This latter feature can be arranged when converting any mixer or viscometer equipped with a mechanical dynamometer.

6.8. MEASUREMENT OF INPUT TORQUE FROM MOTOR

A transmission type transducer can be used to record the torque input to the mixing head (or bowl) by placing it between the drive motor and the mixer drive shaft.

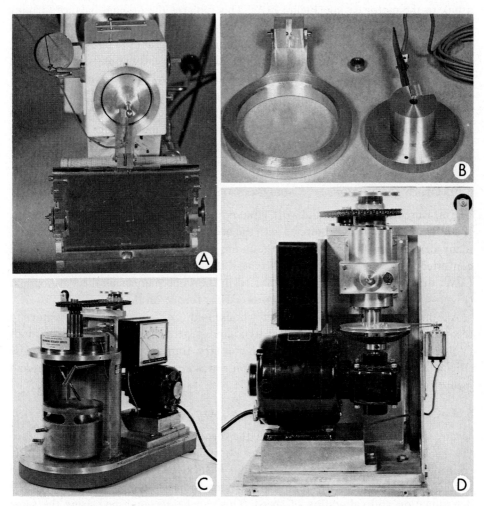

Fig. 20.   A. converted Amylograph; B. components required for A; C and D. transmission-type transducer installed on a GRL mixer.

A typical example is the addition of torque recording to the Grain Research Laboratories (GRL) mixer developed by Hlynka and Anderson (1955) for mixing bread dough in an air-tight bowl. The motor was arranged to drive the shaft of the transducer which in turn rotated the mixing head (Figure 20C, D) (Voisey *et al.*, 1970b). In this case, however, the high friction in the complex mechanism driving the mixing pins was recorded and had to be corrected for taking certain data from the resulting curves.

6.9. MEASUREMENT OF MOTOR REACTION TORQUE

The Farinograph is one of the most popular mixer-viscometers and is used in many applications beyond its original one in cereal chemistry. It can be quickly modified and incidently worn out machines rescued by installing an electronic transducer. The simplest and most inexpensive method is to use a beam-type unit connected by a link to the motor casing (Figure 21A). This replaces the original balance forming the dynamometer and considerably simplifies the apparatus (Figure 21B). The modification was first suggested for standard 50 g bowl Farinographs (Voisey *et al.*, 1966e). This method was found not suitable for scaling down the sample size because of friction in the motor reduction gear. A 10 g Farinograph (Figure 21C) was, therefore, constructed using a transmission-type transducer which eliminated all frictional effects except that of the paddle drive gears built into the mixing bowl. This friction was minimized by careful reconstruction of the gears and of the bearings supporting the paddle shafts (Voisey *et al.*, 1971). The apparatus gave reproducible results with 10 g flour samples.

6.10. OTHER APPLICATIONS OF TORQUE TRANSDUCERS

The measurement of torque and work input is not restricted to mixers and viscometers. If a laboratory is already equipped with the necessary electronic recording apparatus, many devices can be purchased without the normal torque recording system and converted to record electronically. A typical illustration is an extruder (Prep Center C. W. Brabender) which can be converted in the same manner as the Farinograph (Figure 22A, B) using a transducer to record the torque on the motor.

Another application is in the use of a household (or commercial) food grinder to record the torque and energy needed to compress, stretch, shear and extrude foods (Voisey and deMan, 1970a). A transmission type transducer was used in this case (Figure 22C and D). This particular apparatus was applied to testing peas, beans and meat (Voisey, 1970, unpublished data).

6.11. GENERAL CONSIDERATIONS IN RECORDING TORQUE

In any mixer, grinder or extruder, etc., the torque required will be proportional to the rotational speed of the mechanism. For this reason speed must be carefully controlled. This is best achieved by using a synchronous electric motor. If the added flexibility of variable speed is required, then provisions should be made for a precise determination and calibration of speed. This can be accomplished using the same methods as described for universal testing machines.

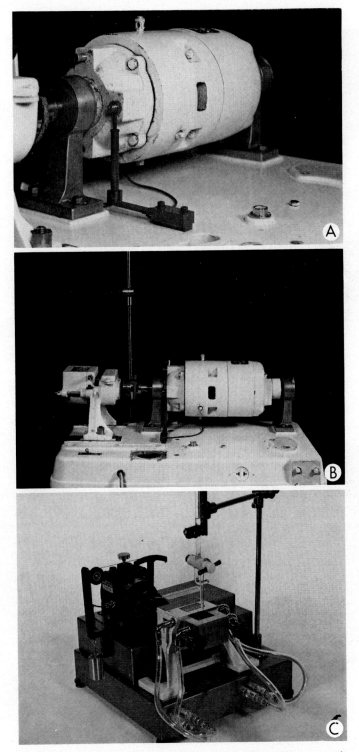

Fig. 21. The Farinograph. A and B. standard machine converted by adding a transducer to measure reaction of motor; C. a 10 g Farinograph using a transmission-type transducer.

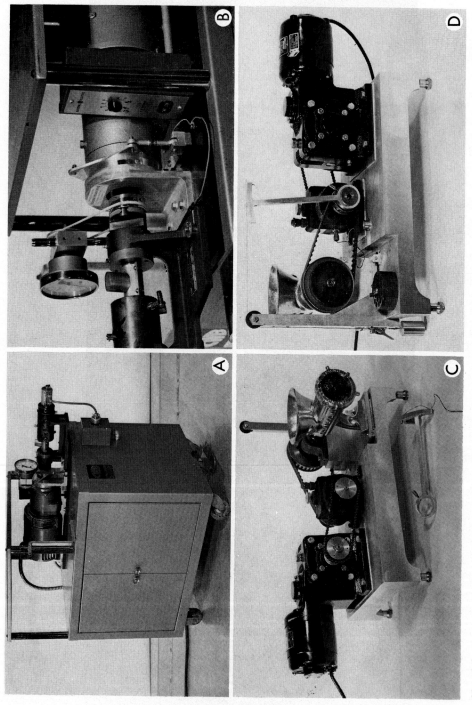

Fig. 22. Other torque measurements. A. applied to a laboratory extruder; B. close-up of transducer detecting torque reaction of drive motor; C and D. a transmission type transducer installed to record torque input to a domestic food grinder. Note the method of calibration with a lever shown in D.

Recorders used in the cited applications have different requirements than those normally used for texture measurements. During mixing, extruding, or grinding, the torque may fluctuate widely and randomly about the average operating torque. The frequency and amplitude of these changes depend on the mixing speed, the type of mixing action and the product being tested. This has been demonstrated in several of the previously discussed applications. The average torque is usually of major interest and it is desirable to damp out the fluctuations. This can be done by including a viscous damper in the dynamometer mechanism, as in the Farinograph, or an inertia damper, as in the Mixograph. Under these conditions, however, the effect of the damper is unknown, particularly when static calibration techniques are used. A better solution is to use a recorder with a slow pen response. This was demonstrated in mixing dough in a pin-type mixer using different recorders (Figure 9) and has been reported for the Mixograph (Voisey et al., 1966b, c, f, 1967b, c; Voisey and Miller, 1970), the Farinograph (Voisey et al., 1966e, 1970c), other mixers (Voisey and Nunes, 1968; Voisey and deMan, 1970b; Voisey et al., 1970b) and a food grinder (Voisey and deMan, 1970a).

If the torque amplitude about the mean is considered important, as for example by some cereal chemists in testing dough, then the frequency response of the recorder must be adequate. In recording a signal from a pin-type mixer fluctuating at 4 Hz, it was found that a recorder capable of full-scale response of at least 20 Hz was required (Voisey et al., 1966b).

The bowl, paddles or motor are rigidly fixed when electronic torque transducers are used. Thus, the relative speed of the paddles and bowl in a mixer, or stirrer and cup in a viscometer is constant. This is not the case with a mechanical dynamometer which must rotate to record torque; the mixing affects the measurement and the measurement affects the mixing. Modern instrumentation gives a major advantage of allowing constant shear rates. This is important in measuring viscosity, particularly in products where consistency is developed mechanically.

In mixers and similar devices, the food is worked continuously in a repeated pattern which may involve compressive, tensile and shearing actions. The torque and the energy used are indexes of multiple deformations performed on the test sample. The energy used, the integral of torque and time, should thus provide an index extremely sensitive to textural changes. It can be easily recorded with a high degree of accuracy by digital integration even when the torque is fluctuating rapidly (Voisey and Brach, 1968). The future of these types of energy measurements can only be guessed at this time, but their influence in the bread-making industry has been significant. For example, the Chorleywood process for continuous dough making relies in part on controlling the energy input to the dough (Chamberlin et al., 1962).

7. A Case History in The Development of Modern Texture Instruments

Modern texture testing instruments offer other advantages besides increasing accuracy and resolution of measurement. Often considerable labour can be saved by updating a widely accepted technique. The additional information available from a force-deformation curve may provide a clearer picture of the product's behaviour leading to a better understanding of the characteristic being measured. This, in turn, may lead to improved techniques of measurement. Once the reaction of the food to applied force has been precisely measured, then the specification for an economic, efficient test instrument can be formulated. This is particularly important in developing instruments for production quality control where the apparatus may be widely adopted. This ultimate goal is often overlooked by many workers. The following example deals with tobacco, not food, but it indicates the possibilities of a planned attack to improve a mechanical test on a biological product.

An important quality factor in shredded tobacco is its 'filling value', i.e. its ability to expand and fill a cigarette as it emerges from a cigarette making machine. An accepted technique is to place 57 g of tobacco in a cylinder and press on it for 10 min with a constant weight, i.e. a piston. The penetration of the piston is the filling value (f.v.) (Voisey and Walker, 1969). This is a measure of the time dependent properties of the tobacco, i.e. a relaxation or creep text. The method was first investigated using an Instron test machine to drive the piston into the cylinder at a constant speed of 20 cm/min (Figure 23A). It was found that the force to compress the tobacco to a preselected volume was highly correlated with the standard method for a wide range of tobacco types (Voisey and Walker, 1969). It was also found that the sample size could be reduced to 10 g by using a smaller cylinder, thus, saving preparation time. In addition, 9 min were saved in executing each test. The force to compress to a preselected volume was also more sensitive to changes in f.v. than the original method. A simplified special purpose deformation apparatus (Figure 23B) was, therefore, designed specifically to test 10 g samples. This included a strip-chart recorder and a peak force detector with digital readout since the maximum force coincided with the force needed to compress the tobacco to the selected volume (Voisey and Walker, 1970).

The new apparatus was considered to be too sophisticated for industrial production and quality control applications. A second apparatus was, therefore, developed (Figure 23C) in which a pneumatic cylinder forced the piston into the sample holder and a spring scale with a 'hold at maximum' indicator displayed the peak force (Voisey and Walker, 1970). Errors in compressed volume due to the deflection of the spring scale were reduced to acceptable limits by using a stop on the piston to control the pneumatic cylinder. Thus, this technique evolved from an existing time consuming measurement, through sophisticated instrumentation, back to a relatively simple and inexpensive apparatus that gave more precise results than before with less labour.

8. Improvement of Existing Instruments

Techniques described in the preceding examples can be used to convert any existing texture testing instrument by using some ingenuity. It is logical to do this once the required electronic apparatus has been purchased for one application.

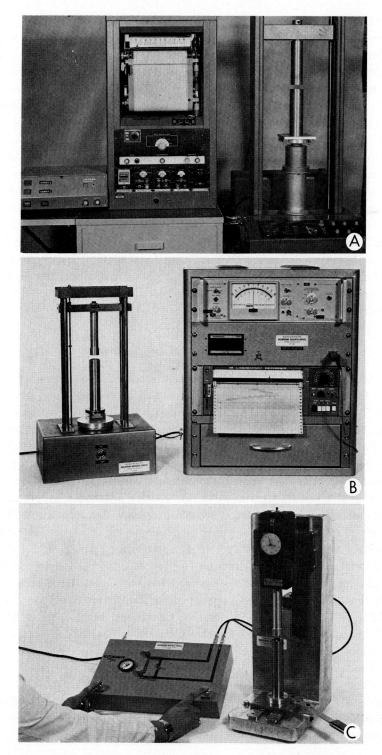

Fig. 23. Evolution of a mechanical test apparatus. A. investigation of force-deformation relationships in an Instron; B. a special purpose apparatus based on results; C. final simplified apparatus.

For example, the Pea Tenderometer can be modified by removing the existing pendulum and scale used to measure force. The counterbalanced grid is then fixed rigidly between the pendulum socket and the frame by a tensile force transducer (Figure 24A). The shearing torque can then be recorded on a strip chart. The rate of shearing becomes constant because only the driven set of blades move. The increased sensitivity allows the testing of many products other than peas, e.g. baked beans (Voisey and Larmond, 1970), still obtaining high accuracy and resolution of measurement. In the cited case, calibration with weights was not practical so force was applied to the transducer via a force gauge by a hydraulic jack built into the machine.

The operating flexibility of the Kramer Shear Press, particularly older models, can be improved in a similar manner. In the example shown (Figure 24B) the hybrid transducer is replaced by a home-made compression transducer connected to an appropriate amplifier. The recorder used has a built-in chart drive motor so that the time (i.e. deformation) scale can be expanded if required. Accuracy force detection is improved because a 250 mm wide chart is used and a simple lever system installed to allow frequent calibration with weights (Figure 24C). In older models, the hydraulic controls operating the press must be improved by adding a flow valve to achieve constant deforming speeds.

9. Other Instruments and Test Techniques

Many devices have been developed for the measurement of food texture but none have advanced the art of mechanical testing as practised by engineers and physicists for centuries. The field of stress analysis has evolved far beyond what is being applied to foods. There are numerous test techniques as reviewed by several workers (e.g. Durelli et al., 1958; Hetényi, 1960; Dove and Adams, 1964; Dally and Riley, 1965; Tuppeny and Kobayashi, 1965; and Durelli, 1967) which have not been applied to foods or have yet to be fully investigated in food technology applications.

A logical goal in texture measurements is to develop instantaneous non-destructive techniques preferably based on inexpensive instruments. The tools available for this purpose cover a wide range. The entire electro-magnetic spectrum is available for investigation. Currently, only the audible portion is being investigated for measuring elastic properties of foods by vibration (Abbott et al., 1968; Finney, 1969, 1970) using techniques similar to those developed for other materials (Anon., 1968a). It is only recently that suitable equipment has become commercially available (e.g. Model PPM-5R – H. M. Morgan Co. Inc., Cambridge, Massachusetts; Model VI, Nametre Acoustic Spectrometer, Abbott et al., 1968). Several companies are now specializing in the manufacture of such equipment (e.g. Stisen, 1970). Nuclear techniques is another area which may find more widespread future applications. It is used to estimate eggshell thickness and density (Voisey et al., 1969a).

It would also be a distinct advantage if the texture of a product could be monitored continuously during production processes. Viscosity can be measured in this manner, but techniques for solid and semi solid products have yet to be achieved. An example of

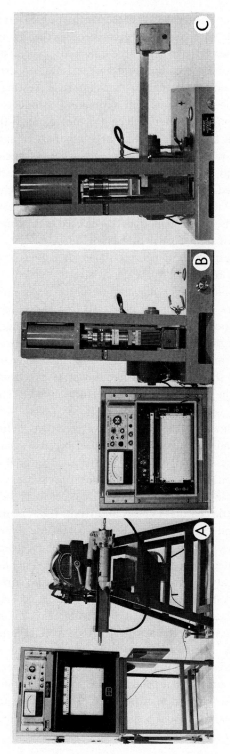

Fig. 24.   Conversion of existing machines. A. pea tenderometer; B. Kramer shear press; C. calibration technique.

existing possibilities was given by Voisey *et al.* (1966a) in the measurement of bread dough resilience while being sheeted. Force transducers recorded the rolling force on the dough at two stages to estimate the recovery of the dough after the first stage (Figure 25A).

Impact testing has not been used for estimating textural properties but considerable data have been published in relation to mechanical harvesting (Mohsenin, 1968). The maximum allowable impact force on agricultural products must be known in order to design harvesting machinery which will not damage the produce. Impact testing offers a practical advantage in that large numbers of samples can be tested rapidly. The equipment required is relatively simple, e.g. a force transducer attached to a rod can be dropped directly on the sample (Figure 25B). Special piezoelectric force transducers are used for this purpose which have a very high frequency response (Voisey and Hunt, 1967a). An oscillograph equipped with a camera is required to obtain permanent records. These techniques can be used to determine elastic properties which are independent of time (Voisey and Hunt, 1967d).

The cost of impact testing equipment is higher than that of the quasi-static test apparatus. However, in recent years the costs have decreased considerably and the techniques may be economically feasible and useful for texture testing. The measurement of impact, however, is a complex science (Harris and Crede, 1961) and should be approached with caution. Impact stresses travel through materials as waves (Stein, 1964) generated in both the rest specimen and the striking object. To obtain a clean signal of force and deformation in testing eggs, it was necessary to use a rod 2m long (Figure 25C) to prevent shock waves reflected down the rod from being recorded by the transducer during the impact (Voisey and Hunt, 1968b). This would not be as critical with soft foods.

Before the above methods can be utilized, however, extensive fundamental research is required to give an understanding of the relationships between force, deformation and texture and how stresses and strains are distributed throughout the food during mastication and in empirical test cells. Modelling provides a valuable tool for the latter problem and the use of techniques such as photoelastic modelling (Coker *et al.,* 1957; Frocht, 1965), brittle coatings (Voisey and Hunt, 1967b), and high speed motion photography (Voisey and Hunt, 1964) have yet to be fully appreciated for textural research. Such techniques have been widely used in medical research.

To-date, there is limited evidence to support the use of nondestructive test techniques for measuring texture. Finney (1971) found that the correlation between the Magness Taylor pressure tester (a destructive test) correlated better with taste panel evaluations of apple texture than acoustic measurements (a nondestructive test). Nondestructive techniques and other methods are used to determine physical properties of engineering materials. The determination is, however, indirect and is based on theoretical analysis. There is no doubt that successful measurements are accomplished in this way for a range of materials and there is, thus, a distinct possibility that it can be done with foods. The major problem in applying the techniques to foods, however are the complex shape and structure involved and the high water content of many products.

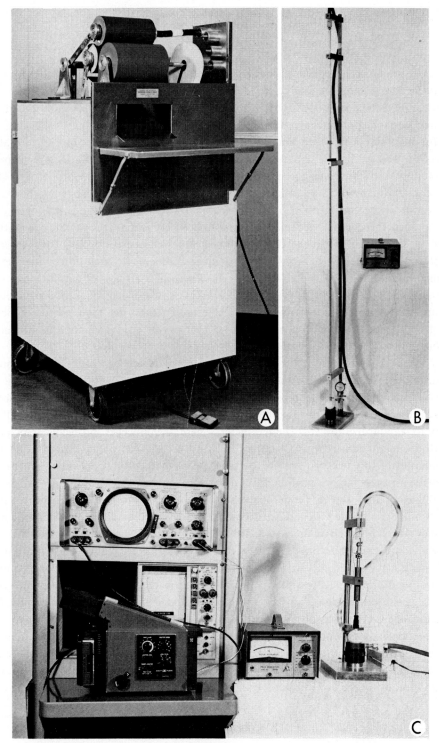

Fig. 25.   A. a recording dough sheeter; B. impact apparatus; C. length of impact rod required to test brittle materials.

10. Standards

Sensory standards have been developed so that taste panels can be trained to recognize and evaluate different textural characteristics (Szczesniak *et al.*, 1963). This is accomplished by using a series of foods of different types, which allow the establishment of scales for each specific characteristic.

In objective tests, force and deformation recording systems can be calibrated mechanically so that these data are recorded precisely. The question is often raised, however, as to how accurately the instrument measures textural properties. This depends on many factors including the accuracy of the recording system (which can be verified), the control of sample size, the environmental conditions, the operating techniques, the sampling techniques, the sample heterogeneity, the sample orientation and the condition of the test cell. All these factors can be controlled within experimental limits. The test cell presents the greatest problem. Its condition can be determined by measuring its dimensions and checking the sharpness of cutting edges etc. This is difficult, time consuming and requires expensive meteorological instruments. Thus, the only reliable information available is that the force required to compress pull or shear the sample under the test conditions at any given time is measured accurately.

There is a definite need for a standard material which can be used to monitor changes taking place within the test cell. Relatively limited work has been done in this respect and to-date the most successful material has been various waxes. Warner (1928) used beeswax to test the meat shear and Staley (1970) used a microcrystalline wax to standardize the Pea Tenderometer. There are, however, problems associated with the manufacture of wax samples of uniform size, weight, and density and the properties of this material are greatly affected by temperature. Providing allowance is made for these factors, meaningful comparisons can be made. Voisey and Nonnecke (1971) compared 31 Pea Tenderometers used in Ontario to grade green peas. Tests with microcrystalline wax indicated that there were serious differences between machines. Much research on these standardizing techniques is, however, still required.

Providing careful techniques are used, it is feasible to reduce the variation of mechanical measurements within a given food sample to a low order. For example, in testing dough it was feasible to reduce variation to less than 5% and, since this includes natural biological variation, it was concluded that the variation attributable to instrument error was of a lower order (Voisey *et al.*, 1969b, 1970c).

Sensory analysis is still the final yardstick of how well an instrumental measurement reflects textural parameters. However, this can only be considered as a crude means of calibrating the instrument which, in theory, should have greater accuracy and resolution of measurement than a trained taste panel.

11. Concluding Remarks

This chapter attempted to show the advantages and to indicate some of the pitfalls in modernizing instruments for measuring texture. It is fairly obvious that forces and deformations can be detected with precision and high resolution. The major stumbling block in measuring texture precisely is establishing the relationships between these measurements and factors governing consumer reaction. While an equipment manufacturer (Food Technology Corp., Reston, Virginia) claims to measure texture precisely in relation to 25 descriptive terms, this is probably an optimistic viewpoint. Research tools required to achieve objective tests of physical properties are, however, readily available and sophisticated systems can be assembled to investigate the textural parameters from 'off the shelf' items.

Deformation mechanisms can be designed to suit any purpose and can be made almost universal. Automatic features such as rapid return, synchronization with recorder chart, etc. can be added to the basic design which is only limited by ingenuity and economic considerations. The greater the degree of automation, the more samples can be tested in a given time (Voisey and Robertson, 1969b). The most critical part of the equipment however, still remains the test cell holding the sample during deformation.

Electronic transducers are easily calibrated by inexpensive weights. The techniques required are simple enough that semi-skilled labour can be used for the purpose with a minimum of training. Thus, it becomes practical to control accuracy of measurement in research or production tests.

The cost of electronic apparatus is greater than that of crude mechanical devices. However, it is more reliable, accurate and generally simplifies construction of the apparatus. Under the latter conditions, costs can often be reduced. There have been significant price reductions for transducers, amplifiers and recorders during the past decade and economics of modernization are, therefore, improving. There is also the added attraction that electronic systems are universally applicable to all deformation devices. Thus, a single system can be utilized not only for texture measurements but also for other measurement problems in food technology. For example, the shrinkage of turkey rolls during cooking was recorded with the same recording apparatus used for measuring texture (Voisey and Aref, 1967). The potential applications are probably unlimited.

While the great majority of objective texture tests can be considered as only empirical, it is realistic and desirable to take any measurement as accurately as possible. Electronic instrumentation is the most efficient way of accomplishing this.

References

Aaker, D. A.: 1965, *How Recorder Response Affects Analysis*, Conf. Anal. Chem. Appl. Spectroscopy, Pittsburg (March).
Abbott, J. A., Bachman, G. S., Childers, N. F., Fitzgerald, J. V., and Matusik, F. J.: 1968, 'Sonic Techniques for Measuring Texture of Fruits and Vegetables', *Food Technol.* **22**, 635.
Anon.: *Strain Gage Instrumentation*, Reprint J. Inst. Soc. Amer. (Undated).
Anon.: 1961, *Strain Gage Instrumentation*, Proc. Inst. Soc. Amer. Fall Conf., Los Angeles, Calif.
Anon.: 1962, *Strain Gage Instrumentation*, Proc. Inst. Soc. Amer. 17th Ann. Conf., New York.
Anon.: 1966, *Glossary of Instrument Terms*, Clevite Corp., Cleveland, Ohio 44114.

Anon.: 1967a, 'Recorders, Analog-direct Writing and Oscillographic', *Electron. Instr. Dig.* **3** (12) 88
Anon.: 1967b, 'E.I.D. Surveys the Scope of Direct Writing Oscillographs', *Electron. Instr. Dig.* **3** (8) 38.
Anon.: 1968a , *Standard Method of Test for Modulus of a Flexible Barrier Material by Sonic Method*, Standard F89–68. Amer. Soc. for Testing and Materials.
Anon.: 1968b, 'The Potentiometric Recorder, A Re-examination', *Electron. Instr. Dig.* **4** (2) 25.
Anon.: 1968c, 'Chart Recorders', *Electromechanical Design* **12** (7) 111.
Anon.: 1969a, 'Digital Multimeters; Who Makes What Types', *Electron. Instr. Dig.* **5** (8) 45.
Anon.: 1969b, 'Chart Recorders', *Electromechanical Design* **13** (9) 34.
Anon.: 1969c, 'Survey of Recorders', *Can. Controls Instr.* **8** (12) 35.
Aronson, M. and Nelson, R. C.: 1960, *Strain Gage Instrumentation*, Instruments Pub. Co., Pittsburg, Pa.
Aronson, M.: 1964, *Recorder Manual*, Instruments Publishing Co. Inc.
Avery, B. W. and Yoerger, R. R.: 1958, 'A Data Integrator for Use With a Strain Gage Recorder', *Trans. ASAE* **1**, 36.
Backinger, G. T., Kramer, A., Decker, R. W., and Sidwell, A. P.: 1957, 'Application of Work Measurement to the Determination of Fibrousness in Asparagus', *Food Technol.* **11**, 583.
Bakey, T.: 1968a, 'Digital Voltmeter Considerations', *Instr. Control Systems* **41**, 95.
Bakey, T.: 1968b, 'Simplify your D.V.M. Selection', *Electronic Design* **16** (4) 76.
Bement, A. L. and Irvin, J. E.: 1964, 'Automatic Processing of Mechanical Properties Data', *Metal Engrs. Quart.* **4**, 11.
Blake, M. A.: 1929, *A Device for Determining the Texture of Peach Fruits for Shipping and Marketing*, Circ. 212, New Jersey Agr. Expt. Sta.
Blatt, C. R., Simpson, W. G., Dean, P. R., and Stark, R.: 1969, *The Effect of Several Fertilizer Sources on Strawberry Color, Texture, Size and Yield*, Ann. Rept. Res. Sta. Can Dept. Agr., Kentville, Nova Scotia.
Bourne, M. C.: 1965, 'Studies on Punch Testing of Apples', *Food Technol.* **19**, 113.
Bourne, M. C.: 1966a, 'Measure of Shear and Compression Components of Puncture Tests', *J. Food Sci.* **31**, 282.
Bourne, M. C.: 1966b, 'A Classification of Objective Methods for Measuring Texture and Consistency of Foods', *J. Food Sci.* **31**, 1011.
Bourne, M. C.: 1967a, 'Deformation Testing of Foods. 1. A Precise Technique for Performing the Deformation Test', *J. Food Sci.* **32**, 601.
Bourne, M. C.: 1967b, 'Deformation Testing of Foods. 2. A Simple Spring Model of Deformation', *J. Food Sci.* **32**, 605.
Bourne, M. C. and Mondy, N.: 1967, 'Measurement of Whole Potato Firmness With a Universal Testing Machine', *Food Technol.* **21**, 97.
Bourne, M. C., Moyer, J. C., and Hand, D. B.: 1966, 'Measurement of Food Texture by a Universal Testing Machine', *Food Technol.* **20**, 170.
Boyd, L. L.: 1964, 'Digital Recording Methods', *Trans. ASAE* **7**, 92.
Brabender, C. W.: 1965, 'Physical Dough Testing', *Cereal Sci. Today* **10**, 291.
Brach, E. J. and Mack, A. R.: 1967, 'A Radiant Energy Meter and Integrator for Plant Growth Studies', *Can. J. Botany* **45**, 2081.
Brach, E. J. and St. Amour, G.: 1969, *Integrating Spectrophotometers Study the Effect of Radiant Energy Spectra on Plant Growth*, Proc. 8th Int. Conf. Med. Biol. Eng.
Brach, E. J., Mack, A. R., and St. Amour, G.: 1969, 'An Integrator Resetting Switch Using a Combination of Unijunction Transistor and Thyristor', *Electron. Engrs.* **41** (491) 94.
Brach, E. J., St. Amour, G., and Mason, W. J.: 1970a, 'A Three Channel Integrating Spectrophotometer', *Can. J. Agr. Engrs.* **12**, 52.
Brach, E. J., Reid, W. S., and McAdam, W. E.: 1970b, 'A Continuous Integrating Spectrophotometer', *Trans. Inst. Elec. Electronic Engrs.* **IM 19**, 92.
Brandt, M. A., Skinner, E. Z., and Coleman, J. A.: 1963, 'Texture Profile Method', *J. Food Sci.* **28**, 404.
Bratzler, L. J.: 1932, *Measuring the Tenderness of Meat by Mechanical Shear*, M.Sc. Thesis, Kansas State Coll.
Buckley, D. J.: 1969, *Digital Data Acquisition Systems*, Eng. Specif. 6901, Eng. Res. Service, Ottawa.
Cameron, E. A.: 1965a, 'A Basic Guide to Potentiometric Recorders', *Can. Controls Instr.* **4** (4) 38.

Cameron, E. A.: 1965b, 'A Basic Guide to Galvanometric Recorders', *Can. Controls Instr.* **4** (3) 66.

Chamberlin, N., Collins, T. H., Elton, G. A. H., and Cornford, S. J.: 1962, *The Chorleywood Bread Process; Commercial Applications. 3. Methods of Measuring Work*, Brit. Baking Ind. Res. Assoc. Rept. 62.

Chappell, T. W. and Hamann, D. D.: 1967, *Poisson's Ratio and Young's Modulus for Apple Flesh Under Compressive Loadings*, Pap. 67–810, Ann. Conf. Amer. Soc. Agr. Eng., Detroit.

Chass, J.: 1962a, 'The Differential Transformer, Its Main Characteristics', *J. Instr. Soc. Am.* **9**, 48.

Chass, J.: 1962b, 'The Differential Transformer. Its Applications and Circuits', *J. Instr. Soc. Am.* **9**, 37.

Coker, E. G., Filon, L. N. G., and Jessop, H. T.: 1957, *A Treatise on Photoelasticity*, Cambridge University Press.

Corkill, D.: 1961, *Sealing of Strain Gages and Electrical Circuitry*, Proc. Fall Conf. Inst. Soc. Amer., Los Angles, Paper 41-LA-61.

Curtis, L. M. and Hendrick, J. G.: 1969, 'A Study of Bending-strength Properties of Cotton Stalks', *Trans. Am. Soc. Agr. Eng.* **12**, 39.

Dally, J. W. and Riley, W. F.: 1965, *Experimental Stress Analysis*, McGraw-Hill.

Davey, C. L. and Gilbert, K. V.: 1969, 'The Effect of Sample Dimensions on the Cleaving of Meat in the Objective Assessment of Tenderness', *Food Technol.* **4**, 7.

Dean, M. and Douglas, R. D.: 1962, *Semiconductor and Conventional Strain Gages*, Academic Press, New York.

Decker, R. W., Yeatman, J. N., Kramer, A., and Sidwell, A. P.: A. P.: 1957, 'Modification of the Shear Press for Electric Indicating and Recording', *Food Technol.* **11**, 313.

deMan, J. M.: 1970, 'Effect of Mechanical Treatment on the Hardness of Margarine and Butter', *J. Texture Studies* **1**, 109.

deMan, J. M.: 1969a, 'Determination of Potato Texture', *J. Can. Inst. Food Technol.* **2**, 76.

deMan, J. M.: 1969b, 'Food Texture Measurements With the Penetration Method', *J. Texture Studies* **1**, 114.

Deutsch, W. G.: 1965, 'Precision, Accuracy and Resolution', *J. Instr. Soc. Am.* **12**, 85.

Dove, R. C. and Adams, P. H.: 1964, *Experimental Stress Analysis and Motion Measurement*, Charles E. Merrill Books Inc., Columbus, Ohio.

Dunn, L. B.: 1970, 'Recordings Charts', *Inst. Control Systems Buyer's Guide* **13**, 15.

Durelli, A. J.: 1967, *Applied Stress Analysis*, Prentice Hall Inc.

Durelli, A. J., Phillips, E. A., and Tsao, C. H.: 1958, *Introduction to the Theoretical and Experimental Analysis of Stress and Strain*, McGraw-Hill.

Emmons, D. B. and Price, W. V.: 1959, 'A Curd Firmness Meter for Cottage Cheese', *J. Dairy Sci.* **42**, 553.

Emmons, D. B., Becket, D. C., and Larmond, E.: 1970, *Physical Properties and Storage Stability of Some Milk-based Puddings and Custards Made with Starches and Stabilizers*, Ann. Conf. Can. Inst. Food Technol., Windsor.

Finney, E. E.: 1963, *The Viscoelastic Behaviour of the Potato, Solanum Tuberosum; Under Quasi-static Loading*, Ph.D Thesis, Michigan State University.

Finney, E. E.: 1969, 'Objective Measurements for Texture in Foods', *J. Texture Studies* **1**, 19.

Finney, E. E.: 1970, 'Mechanical Resonance Within Red Delicious Apples and Its Relation to Fruit Texture', *Trans. Am. Soc. Agr. Eng.* **13**, 177.

Finney, E. E.: 1971, 'Dynamic Elastic Properties and Sensory Quality of Apple Fruit', *J. Texture Studies* **2**, 62.

Fletcher, S. W., Mohsenin, N. N., Hammerle, J. R., and Turkey, L. D.: 1965, 'Mechanical Behaviour of Selected Fruits and Vegetables Under Fast Rates of Loading', *Trans. Am. Soc. Agr. Eng.* **8**, 324.

Foskett, R.: 1968, 'Torque Measuring Transducers', *Instr. Control Systems* **41** (11) 75.

Frank, E.: 1965, *Transducer Conditioning*, 1965 Spring Conf. Soc. Expt. Stress Anal., Colorado.

Freilich, A.: 1966, 'Use Time Event Marking to Increase the Accuracy and Utility of Analog Recording', *Electron. Instr. Dig.* **2** (13) 18.

Friedman, H. H., Whitney, J. E., and Szczesniak, A. S.: 1963, 'The Texturometer – A New Instrument for Objective Texture Measurement', *J. Food Sci.* **28**, 390.

Frocht, M. M.: 1965, *Photoelasticity*, Vol. 1 and 2. John Wiley and Sons, New York.

Funk, J. P. and Dowe, D. G.: 1965, 'An Integrator for Multichannel Potentiometric Recorders', *J. Sci. Instr.* **42**, 615.

Guthrie, J.: 1964, 'Lever-shaft Torque Measurement', *Instr. Control Systems* **37**, 116.

Haller, M. H.: 1941, *Fruit Pressure Testers and Their Practical Applications*, Circ. 627, U.S. Dept. Agr.

Hammerle, J. R. and McClure, W. F.: 1971, 'The Determination of Poisson's Ratio by Compression Tests of Cylindrical Specimens', *J. Texture Studies* 2, 31.

Harris, C. M. and Crede, C. E.: 1961, *Shock and Vibration Handbook* (3 volumes) McGraw-Hill.

Hartman, J. D., Isenberg, F. M., and Ang., J. K.: 1963, 'New Applications of the Shear Press in Measuring Texture in Vegetables and Vegetable Products. 1. Modifications and Attachments to Increase Versatility and Accuracy of the Press', *Proc. Am. Soc. Hort. Sci.* 82, 465.

Henry, W. F. and Katz, M. H.: 1969, 'New Dimensions Relating to the Textural Quality of Semisolid Foods and Ingredient Systems', *Food Technol.* 23, 822.

Hetényi, M.: 1960, *Handbook of Experimental Stress Analysis*, John Wiley.

Hindman, H. and Burr, G. S.: 1949, 'The Instron Tensile Tester', *Trans. Am. Soc. Mech. Eng.* 71, 789.

Hlynka, I. and Anderson, J. A.: 1955, 'Laboratory Dough Mixer with an Airtight Bowl', *Cereal Chem.* 42, 432.

Hoadley, H. W.: 1967, 'Photo-Oscillographic Recording', *Electron. Instr. Dig.* 3 (8) 44.

Howard, P. L. and Heinz, D. E.: 1970, 'Texture of Carrots', *J. Texture Studies* 1, 185.

Hunt, J. R. and Voisey, P. W.: 1966, 'Physical Properties of Eggshells, 1. Relationship of Resistance to Compression and Force at Failure of Eggshells', *Poultry Sci.* 45, 1398.

Irwin, W. J. and Korsch, D. E.: 1964, 'Determining Data System Accuracy', *J. Instr. Soc. Am.* 11 (9) 55.

Jackman, V.: 1960, *Special Techniques Used in the Study of Material Properties*, Symp. Anal. Methods in the Study of Stress-strain Behaviour. Massachusetts.

Jacobson, M. and Armbruster, G.: 1968, 'A Recording Micropenetrometer, Design and Application', *Food Technol.* 22, 1007.

Jagger, J. G.: 1952, *A Textbook of Mechanics*, Blackie and Sons Ltd., London, p. 480.

Kalab, M., Voisey, P. W., and Emmons, D. B.: 1971a, 'Heat Induced Milk Gels. 2. Measuring the Firmness', *J. Dairy Sci.* 54, 178.

Kalab, M., Emmons, D. B., and Voisey, P. W.: 1971b, 'Heat Induced Milk Gels. 3. Physical Effects on Firmness', *J. Dairy Sci.* 54, 638–642.

Kastner, C. L. and Henrickson, R. L.: 1970, 'Providing Uniform Meat Cores for Mechanical Shear Force Measurement', *J. Food Sci.* 34, 603.

Kenkars, E., Tape, N. W., Brunton, G., and Larmond, M. E.: 1967, *Effect of Storage and Growing Area on French Fry Texture*, Rept. Workplanning Committee on Potato Texture. Food Res. Inst., Ottawa.

Kent-Jones, D. W. and Amos, A. J.: 1957, *Modern Cereal Chemistry*, Northern Publ. Co., Liverpool, England.

Klueter, H. H., Downs, R. J., Bailey, W. A., and Krizek, D. T.: 1966, *Motion Detector*, Ann. Conf. Amer. Soc. Agr. Eng. Chicago, Paper 66–836.

Kramer, A.: 1969, 'The Relevance of Correlating Objective and Subjective Data', *Food Technol.* 23, 66.

Kramer, A., Burkhardt, G. J., and Rogers, H. P.: 1951, 'The Shear-press, A Device for Measuring Food Quality', *Canner* 112, 34.

Larmond, E., Moran, E. T., and Kim, C. I.: 1970, 'Eating Quality of Two Basic Breeds of Broiler Chickens and Their Crosses', *J. Can. Inst. Food Technol.* 3, 63.

Lomask, M. R.: 1969, 'The Volt-second Meter and Similar Integration Devices', *Electron Instr. Dig.* 5 (1) 38.

Magness, J. R. and Taylor, G. F.: 1925, *An Improved Type of Pressure Tester for the Determination of Fruit Maturity*, Circ. 350, U.S. Dept. Agr.

Martin, W. M.: 1937, 'The Tenderometer. An Apparatus for Evaluating the Tenderness in Peas', *Canning Trade* 59, 7.

McClure, W. F., Wysong, C. J., Hammerle, J. R., and Wright, F. S.: 1970, 'Photomicrometric Measurements of Dimensions', *J. Am. Soc. Agr. Engrs.* 51, 26.

Minnar, E. J.: 1963, *I.S.A. Transducer Compendium*, Plenum Press, New York.

Mohsenin, N. N.: 1963, *A Testing Machine for Determination of Mechanical and Rheological Properties of Agricultural Products*, Bull. 701, Pennsylvania State Univ.

Mohsenin, N. N.: 1968, *Physical Properties of Plant and Animal Materials*, Vol. 1, Gordon and Breach, Sci. Pub., New York.

Mohsenin, N. N.: 1970, 'Application of Engineering Techniques to Evaluation of Texture of Solid Food Materials', *J. Texture Studies* 1, 133.

Mohsenin, N. N., Cooper, H. E., and Tukey, L. D.: 1963, 'Engineering Approach to Evaluating Textural Factors in Fruits and Vegetables', *Trans. Am. Soc. Agr. Eng.* **6**, 85.

Mohsenin, N. N., Cooper, H. E., Hammerle, J. R., Fletcher, S. W., and Tukey, L. D.: 1965, *Readiness for Harvest of Apples as Affected by Physical and Mechanical Properties of the Fruit*, Bull. 721, Pennsylvania State Agr. Expt. Sta.

Murray, W. M. and Stein, P. K.: 1961, *Strain Gage Techniques*, Soc. Expt. Stress Anal.

Nelson, R. C.: 1962, 'Recorder Survey', *Instr. Control Systems* **35** (7) 75.

Neubert, H. K. P.: 1967, *Strain Gages Kinds and Uses*, Macmillan, New York.

Nichols, R. C.: 1960, 'Some Observations on the Use of Fruit Pressure Testers', *Quart. Bull. Michigan Agr. Expt. Sta.* **43**, 312.

Oberg, E. and Jones, F. D.: 1959, *Machineries Handbook*, Industrial Press, pages 388–398.

Parsons, A. R.: 1960, 'A Giant Step Forward in Oscillography', *Ann. N.Y. Acad. Sci.* **84**, 543.

Perry, C. C. and Lissner, H. R.: 1955, *The Strain Gage Primer*, McGraw-Hill, New York.

Pool, M. F. and Klose, A. A.: 1969, 'The Relation of Force to Sample Dimensions in Objective Measurements of Tenderness of Poultry Meat', *J. Food Sci.* **34**, 524.

Rediske, A. C.: 1962, 'Linear Motion Transducer', *J. Instr. Soc. Am.* **9**, 83.

Romanoff, A. L.: 1925, 'Study of the Hen's Eggshell in Relation to the Function of Shell-secretory Glands', *Biol. Bull.* **56**, 351.

Salvador, J. G.: 1966, 'Position Errors in X–Y Recorders', *Test Eng.* **15** (1) 37.

Schoorl, P. and Boersma, H. Y.: 1962, *Research on the Quality of the Eggshell, A New Method of Determination*, Proc. 12th World's Poul. Cong. Sydney, Australia 432.

Sherman, P.: 1969, 'A Texture Profile of Foodstuffs Based Upon Well-defined Rheological Properties', *J. Food Sci.* **34**, 458.

Smith, J. H.: 1970, 'Computers in Mechanical Testing', *Instr. Control Systems* **43** (6) 125.

Splinter, W. E.: 1969, 'Electronic Micrometer Monitors Plant Stem Diameter', *J. Am. Soc. Agr. Engrs.* **50**, 220.

Staley, L. M.: 1970, 'Materials for Standardizing the FMC Tenderometer', *J. Can. Inst. Food Technol.* **3**, 116.

Stein, P. K.: 1964, *Measurement Engineering, Vol. 1. Basic Principles*, Stein Eng. Services, Phoenix, Arizona.

Stisen, B.: 1970, *Measurement of the Complex Modulus of Elasticity of Fibres and Folios*, Technical Review No. 2, Brüel and Kjaer, Naerum, Denmark.

Summer, C. J.: 1968, 'Digital Recording Mechanism for Linear Displacements', *J. Sci. Instr.* **1**, 1237.

Swanson, C. O. and Working, E. B.: 1933, 'Testing the Quality of Flour by the Recording Dough Mixer', *Cereal Chem.* **10**, 1.

Sweet, R. G.: 1965, 'High Frequency Recording with Electrostatically Deflected Ink Jets', *Rev. Sci. Instr.* **36**, 131.

Swindells, B. and Debnam, R. C.: 1962, 'Methods of Load Verification of Tension and Compression Testing Machines', *Proc. Inst. Mech. Engrs.* **176**, 911.

Szczesniak, A. S.: 1963, 'Objective Measurements of Food Texture', *J. Food Sci.* **28**, 410.

Szczesniak, A. S.: 1968, 'Correlations Between Objective and Sensory Texture Measurements', *Food Technol.* **22**, 981.

Szczesniak, A. S.: 1969, 'The Whys and Whats of Objective Texture Measurements', *J. Can. Inst. Food Technol.* **2**, 150.

Szczesniak, A. S. and Bourne, M. C.: 1969, 'Sensory Evaluation of Food Firmness', *J. Texture Studies* **1**, 52.

Szczesniak, A. S. and Smith, B. J.: 1969, 'Observations on Strawberry Texture, a Three-pronged Approach', *J. Texture Studies* **1**, 65.

Szczesniak, A. S., Brandt, M. A., and Friedman, H. H.: 1963, 'Development of Standard Rating Scales for Mechanical Parameters of Texture and Correlation Between the Objective and the Sensory Methods of Texture Evaluation', *J. Food Sci.* **28**, 397.

Szczesniak, A. S., Humbaugh, P. R., and Block, H. W.: 1970, 'Behavior of Different Foods in the Standard Shear Compression Cell of the Shear Press and the Effect of Sample Weight on Peak Area and Maximum Force', *J. Texture Studies* **1**, 356.

Tanaka, M., deMan, J. M., and Voisey, P. W.: 1971, 'Measurement of Food Texture with the Cone Penetrometer', *J. Texture Studies* **2**, in print.

Tatnall, F. G.: 1967a, *Tatnall on Testing*, American Society for Metals, Metal Park, Ohio.

Tatnall, F. G.: 1967b, 'Evolution of the Universal Testing Machine', *Instr. Control Systems* **40**, 68.

Timbers, G. E.: 1964, *Some Mechanical and Rheological Properties of the Netted Gem Potato*, Msc. Thes. Univ. British Columbia.

Timbers, G. E., Staley, L. M., and Watson, E. L.: 1965, 'Determining Modulus of Elasticity in Agricultural Products by Loaded Plunges', *J. ASAE* **46**, 274.

Timoshenko, S. P.: 1953, *History of Strength of Materials*, McGraw-Hill Co.

Tobey, G.: 1969, 'Analog Integration', *Instr. Control Systems* **42** (1) 49.

Tukey, L. D.: 1964, 'A Linear Electronic Device for Continuous Measurement and Recording of Fruit Enlargement and Contraction', *Proc. Am. Soc. Hort. Sci.* **84**, 653.

Tuppeny, W. H. and Kobayashi, A. S.: 1965, *Manual on Experimental Stress Analysis*, Society for Experimental Stress Analysis, Westport, Connecticut.

Tyler, C. and Coundon, J. R.: 1965, 'Apparatus for Measuring Shell Strength by Crushing, Piercing or Snapping', *Brit. Poultry Sci* **6**, 327.

Vandenberg, G. E.: 1966, 'Continuous Analog Techniques and Instrumentation for Experimental Research', *Trans. ASAE* **9**, 661.

Voisey, P. W.: 1966, 'Electronic Recording Dough Mixers', *Cereal News* **11**, 11.

Voisey, P. W.: 1970, 'Test Cells for Objective Textural Measurements', *J. Can. Inst. Food Technol.* **3**, 93.

Voisey, P. W. and Aref, M. M.: 1967, 'Measurement of Applied Mechanical Pressure During Processing of Turkey Rolls', *Food Technol.* **21**, 169.

Voisey, P. W. and Brach, E. J.: 1968, 'A d–c Signal Generator for Calibrating Electronic Integrators', *Lab. Pract.* **17**, 1349.

Voisey, P. W. and Dean, P. R.: 1971, 'Measurement of Consistency of Reconstituted Instant Potato Flakes', *Am. Potato J.* (in press).

Voisey, P. W. and deMan, J. M.: 1970a, 'A Recording Food Grinder', *J. Can. Inst. Food Technol.* **3**, 14.

Voisey, P. W. and deMan, J. M.: 1970b, 'An Electronic Recording Viscometer for Food Products', *J. Can. Inst. Food Technol.* **3**, 130.

Voisey, P. W. and Emmons, D. B.: 1966, 'Modification of the Curd Firmness Test for Cottage Cheese', *J. Dairy Sci.* **49**, 93.

Voisey, P. W. and Foster, W. F.: 1970, 'A Non-destructive Eggshell Strength Tester', *Can. J. Animal Sci.* **50**, 390.

Voisey, P. W. and Hansen, H.: 1967, 'A Shear Apparatus for Meat Tenderness Evaluation', *Food Technol.* **21**, 355.

Voisey, P. W. and Hunt, J. R.: 1964, 'A Technique for Determining Approximate Fracture Propogation Rates of Eggshells', *Can. J. Animal Sci.* **44**, 347.

Voisey, P. W. and Hunt, J. R.: 1967a, 'Physical Properties of Eggshells. 3. An Apparatus for Estimating the Impact Resistance of the Shell', *Brit. Poultry Sci.* **8**, 259.

Voisey, P. W. and Hunt, J. R.: 1967b, 'Physical Properties of Eggshells. 4. Stress Distribution in the Shell', *Brit. Poultry Sci.* **8**, 263.

Voisey, P. W. and Hunt, J. R.: 1967c, 'Relationship Between Applied Force, Deformation of Eggshells and Fracture Force', *J. Agr. Engrs. Res.* **12**, 1.

Voisey, P. W. and Hunt, J. R.: 1967d, 'Behaviour of Eggshells Under Impact', *J. Agr. Engrs. Res.* **12**, 128.

Voisey, P. W. and Hunt, J. R.: 1968a, 'Elastic Properties of Eggshells', *J. Agr. Engrs. Res.* **13**, 295.

Voisey, P. W. and Hunt, J. R.: 1968b, 'Behaviour of Eggshells Under Impact. Design Considerations', *J. Agr. Engrs. Res.* **13**, 301.

Voisey, P. W. and Hunt, J. R.: 1969, 'Effect of Compression Speed on the Behaviour of Eggshells', *J. Agr. Engrs. Res.* **14**, 40.

Voisey, P. W. and Kloek, M.: 1964, 'Apparatus for Determining the Gassing Power of Dough', *Cereal Sci. Today* **9**, 393.

Voisey, P. W. and Larmond, E.: 1971, 'Texture of Baked Beans–A Comparison of Several Methods of Measurement', *J. Texture Studies* **2**, 96.

Voisey, P. W. and Lyall, L. H.: 1965a, 'Puncture Resistance in Relation to Tomato Fruit Cracking', *Can. J. Plant Sci.* **45**, 602.

Voisey, P. W. and Lyall, L. H.: 1965b, 'Methods of Determining the Strength of Tomato Skins in Relation to Fruit Cracking', *Proc. Am. Soc. Hort. Sci.* **86**, 597.

Voisey, P. W. and Lyall, L. H.: 1966, 'Measuring Tomato Cracking', *Can. Agr.* **11**, 22.

Voisey, P. W. and MacDonald, D. C.: 1964, 'An Instrument for Measuring the Puncture Resistance of Fruits and Vegetables', *Proc. Am. Soc. Hort. Sci.* **84**, 557.

Voisey, P. W. and MacDonald, D. C.: 1966, 'A Puncturing Device for Estimating the Physical Properties of Fruits and Vegetables', *Can. J. Plant Sci.* **46**, 698.

Voisey, P. W. and Miller, H.: 1970, 'The Ottawa Electronic Recording Dough Mixer. 7. Factors Affecting Performance', *Cereal Chem.* **47**, 207.

Voisey, P. W. and Nonnecke, I. L.: 1971, *Report on the Performance of the F.M.C. Pea Tenderometer with Particular Reference to Its Accuracy of Measurement in the Grading of Peas to Establish the Price Paid to the Grower*, Ontario Food Processors Fieldman's Conf., Guelph, Jan. 6–7.

Voisey, P. W. and Nunes, A.: 1968, 'An Electronic Recording Amylograph', *J. Can. Inst. Food Technol.* **1**, 128.

Voisey, P. W. and Nuttall, V. M.: 1965, 'A Comparison Between Mechanical and Sensory Evaluation of Pericarp Tenderness in Sweet Corn', *Can. J. Plant Sci.* **45**, 303.

Voisey, P. W. and Robertson, G. D.: 1969a, 'Errors Associated with An Eggshell Strength Tester', *Can. J. Animal Sci.* **49**, 231.

Voisey, P. W. and Robertson, G. D.: 1969b, 'The Rapid Measurement of Eggshell Strength', *J. Can. Soc. Agr. Engrs.* **11**, 6.

Voisey, P. W. and Walker, E. K.: 1969, 'A New Research Technique for Determining the Filling Value of Tobacco', *Tobacco Sci.* **169**, 91.

Voisey, P. W. and Walker, E. K.: 1970, 'Apparatus for the Measurement of Tobacco Filling Value and Cigarette Firmness', *Tobacco Sci.* **14**, 40.

Voisey, P. W., Hansen, H., and Thomlison, A. W.: 1964a, *A Specific Gravity Calculator for Potatoes*, Eng. Specif. 6243. Eng. Res. Service.

Voisey, P. W., Kloek, M., and MacDonald, D. C.: 1965b, 'A Rapid Method of Determining the Mechanical Properties of Fruits and Vegetables', *Proc. Am. Soc. Hort. Sci.* **84**, 557.

Voisey, P. W., MacDonald, D. C., and Thomlison, A. W.: 1964c, *A Tomato Skin Strength Tester*, *Eng. Specif.* 6206. Eng. Res. Service.

Voisey, P. W., Hansen, H., and Thomlison, A. W.: 1965, *A Recording Shear Apparatus for Evaluating Meat Tenderness*, Eng. Specif. 6511. Eng. Res. Service.

Voisey, P. W., Miller, H., Foster, W., and Kloek, M.: 1966a, 'A Sheeter for Estimating Dough Quality', *Cereal Sci. Today* **11**, 434.

Voisey, P. W., Miller, H., and Kloek, M.: 1966b, 'An Electronic Recording Dough Mixer. 1. The Apparatus', *Cereal Chem.* **43**, 408.

Voisey, P. W., Miller, H., and Kloek, M.: 1966c, 'An Electronic Recording Dough Mixer. 2. An Experimental Evaluation', *Cereal Chem.* **43**, 420.

Voisey, P. W., Miller, H., and Kloek, M.: 1966d, 'An Electronic Recording Dough Mixer. 3. Additional Methods of Recording', *Cereal Chem.* **43**, 433.

Voisey, P. W., Miller, H., and Kloek, M.: 1966e, 'An Electronic Recording Dough Mixer. 4. Applications in Farinography', *Cereal Chem.* **43**, 438.

Voisey, P. W., Kloek, M., and Thomlison, A. W.: 1966f, *Electronic Recording Dough Mixers*, Eng. Specif. 6208. Eng. Res. Service.

Voisey, P. W., Kloek, M., and Thomlison, A. W.: 1966g, *An Apparatus for Comparing the Textural Quality of Foods*, Eng. Specif. 6262. Eng. Res. Service.

Voisey, P. W., MacDonald, D. C., and Foster, W.: 1967a, 'An Apparatus for Measuring the Mechanical Properties of Food Products', *Food Technol.* **21**, 43A.

Voisey, P. W., Miller, H., and Kloek, M.: 1967b, 'An Electronic Recording Dough Mixer. 5. Measurement of Energy Used in a Mixograph-type Mixer', *Cereal Chem.* **44**, 359.

Voisey, P. W., Miller, H., and Kloek, M.: 1967c, *A Technique for Studying Dough Development During Mixing to Evaluate Flours*, Eng. Specif. 6510. Eng. Res. Service.

Voisey, P. W., Hunt, J. R., and James, P. E.: 1969a, 'A Cooparison of the Beta Backscatter and Quasi-static Compression Methods of Measuring Eggshell Strength', *Can. J. Animal Sci.* **49**, 157.

Voisey, P. W., Miller, H., and Kloek, M.: 1969b, 'The Ottawa Electronic Recording Dough Mixer. 6. Differences Between Mixing Bowls', *Cereal Chem.* **46**, 196.

Voisey, P. W., Tape, N. W., and Kloek, M.: 1969c, 'Physical Properties of the Potato Tuber', *J. Can. Inst. Food Technol.* **2**, 98.

Voisey, P. W., Lyall, L. H., and Kloek, M.: 1970a, 'Tomato Skin Strength. Its Measurement and Relation to Cracking Susceptibility', *J. Am. Soc. Hort. Sci.* **95**, 485.

Voisey, P. W., Bendelow, V. M., and Miller, H.: 1970b, 'Electronic Recording Mixers for the Baking Test', *Cereal Sci. Today* **15**, 341.

Voisey, P. W., Miller, H., and Byrne, P. L.: 1971, 'The Ottawa Electronic Recording Farinograph', *Cereal Sci. Today* **16**, 124–131.

Walton, R. W.: 1969, 'Digital Panel Meters', *Electron. Instr. Dig.* **5** (2) 82.

Warner, K. F.: 1928, 'Progress Report of the Mechanical Test for Tenderness of Meat', *Ann. Proc. Am. Soc. Animal Prod.* 114.

Weitz, B. A.: 1969a, 'Digital Multimeters, Types, Techniques, and Trade Offs', *Electron. Instr. Dig.* **5** (8) 34.

Weitz, B. A.: 1969b, 'Design Concepts for Digital Voltmeters', *Instr. Control Systems* **42**, 71.

Willard, J. T. and Shaw, R. H.: 1909, *Analyses of Eggs*, Bull. 159, Kansas State Agr. Coll. Expt. Sta. 143–177.

Zachringer, M.: 1969, *Home Economics Research Studies Apple Texture*, Quart. Bull. Univ. Idaho Coll. of Agr. 53.

CHAPTER VII

TEXTURE MEASUREMENTS IN VEGETABLES

MALCOLM C. BOURNE

New York State Agricultural Experiment Station,
Cornell University, Geneva, N.Y., U.S.A.

1. Introduction

Texture is one of the principal quality characteristics for most vegetables. In some vegetables (such as sweet corn and celery) it is the most important quality characteristic while in other vegetables it takes second place behind color, (as in tomatoes and beets) or flavor, but even in these cases it is still a very important quality characteristic. In this chapter the measurement of texture of fresh vegetables will be discussed, but texture measurement of processed vegetable products will not be included. Chapter VI discusses the instrumentation for the determination of mechanical properties of food and complement and supplement this chapter.

Szczesniak (1963) classifies texture measuring instruments in three groups: (a) fundamental tests in which properties such as Young's Modulus of Elasticity and viscosity are measured; (b) empirical tests that measure properties that are usually poorly defined but which have been shown by practical experience to be related to textural quality in some way; (c) imitative tests that measure various properties under conditions similar to those to which the food is subjected during mastication.

The best known instrument for imitative tests is the General Foods Texturometer (Friedman *et al.*, 1963). This instrument imitates the reciprocating chewing action of the teeth. It evolved from the M.I.T. Denture Tenderometer (which has never been commercially available). The General Foods Texturometer is manufactured by the Zenken Company Ltd. Kyodo Building, No. 52-Chome Honcho, Nihonbashi Chuw-ku, Tokyo, 103, Japan and it is distributed in the U.S. by C. W. Brabender Instruments, Inc., 50 East Wesley, South Hackensack, New Jersey 07606. This type of instrument gives a substantial quantity of information, but the test itself is rather lengthy to perform and abstracting the data from the recorder chart takes a considerable amount of time. It is unlikely that this procedure will be used for routine testing purposes because of the time involved, although it is a very useful instrument for research purposes and for establishing a basis on which simple repetitive tests can be conducted.

The fundamental tests have the advantage that the parameters that are measured are well defined physically in known units and the tests can be reproduced quite well. Mohsenin (1970a) has reviewed this field. However it should be borne in mind that these tests have been developed by engineers and people who are interested in the

ChoKyun Rha (ed.), Theory, Determination and Control of Physical Properties of Food Materials, 131–162.

theory and practice of materials of construction and that these tests may not be very useful in measuring what is sensed in the mouth when food is masticated. The elegant mathematical theoretical basis of materials science is based on the use of small strains, and on the assumption that the test material is continuous, homogeneous and isotropic. A great part of the theory and mathematics of materials science is not applicable to foods because they are not continuous, homogeneous and isotropic and most texture tests use large strains. Mohsenin (1970b) gives an excellent review of the mathematical treatments available for determining the contact stress between two bodies that are pressed together, but he points out that this can only be done if the level of deformation is kept very low.

The outlook of the engineer and the food rheologist are opposite. The engineer wants to measure the strength of a material in order to design a structure that will withstand the forces applied to it under normal use without breaking. He may want to construct a bridge, a chair or a ball point pen, but his intention is to measure the strength and design a structure such that the product will not break under normal use. The food technologist, on the other hand, has a completely different problem. He wishes to measure the strength of the food but with the intention usually of deliberately weakening the strength by some means so that it will be easily broken down to a fine state suitable for swallowing when it is subjected to crushing forces between the teeth. The food rheologist is interested in knowing how well something will break up into many pieces, in contrast to the engineer who is interested in knowing how well a substance will hold together. When a test piece is broken into two pieces the engineer normally stops his test because at that point it is broken and of no further interest to him. In contrast, when the food product has broken into two pieces, the food rheologist is just beginning to become interested in it and he wishes to continue the test until the product is broken into a great number of pieces. Therefore, food texture measurement might be considered more as a study of the *weakness* of materials rather than of the strength of materials. This important difference places limitations on the direct application of classical engineering theory and techniques to texture measurement of foods. The classical engineering theory and testing techniques can be of substantial help to food rheologists, particularly in the rigorous use of fundamental units and strict methods of standardization of test instruments. However, a large part of engineering theory and practice will be of small use to food rheologists because of the fundamentally different nature of the problem. Eventually the food rheologists will be using fundamental tests, but many of these will be properties that have not yet been defined and that are different from the present fundamental tests that are used by the engineers.

Finney (1971) studied the dynamic elastic properties of apples using an acoustic method, and he correlated this with a taste panel and also with a Magness Taylor pressure test. He found that the dynamic resonance test – a well defined property – gave good correlations with the taste panel, but that the Magness Taylor test, which is empirical and poorly defined gave even better correlations with the taste panel. He concluded, "the dynamic resonance test is not as reliable an index of sensory firmness as the pressure test, but it has the advantage of being nondestructive."

When we survey the objective methods that are used to measure the texture of vegetables we find that the empirical methods are used almost exclusively on a routine basis; therefore, most of this chapter will be devoted to a discussion of the empirical methods. Figure 1 is an attempt to show schematically the relationships among empirical, fundamental and imitative tests. Table I lists the advantages and disadvantages of each type. The ideal texture measuring apparatus should combine the best features

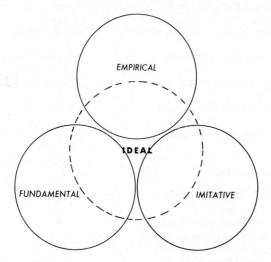

Fig. 1. Schematic representation of the ideal texture measuring apparatus and its derivation from empirical, fundamental and imitative instruments.

of the fundamental, empirical and imitative methods and eliminate the undesirable features of each of these. At the present time there is no ideal texture measuring equipment or system. Empirical methods are used almost entirely. The future direction of the research should be the move from the empirical into the ideal by including more of the fundamental aspects and more of the imitative aspects in our empirical tests. The ideal texture measuring technique will probably be some kind of combination of the empirical, fundamental and imitative methods.

Table II lists and classifies objective methods that are used for measuring the texture of vegetables. This table has been developed from Bourne's (1966b) classification of objective methods for measuring texture and consistency of foods, but with the selection of only those kinds of instruments that are used with vegetables. Under force measuring instruments, are listed puncture, extrusion and crushing tests. Tests that involve pure shear, tension and torque are not listed here because they are rarely used on fresh vegetables. Under distance measuring instruments are listed penetrometers, deformation tests and acoustic spectrometers, and area and volume tests are not included. Instruments that use time and energy and chemical analysis as the basis of measurement are also excluded in this table.

TABLE I

Comparison of different systems of texture measurement of food

System	Advantages	Disadvantages
Empirical	Simple to perform	No fundamental understanding of the test
	Rapid	Incomplete textural measurement
	Suitable for routine quality control	Arbitrary procedure
	Good correlation with sensory methods	Cannot convert data to another system
	Large samples give averaging effect	Usually 'one point' measurement
Imitative	Closely duplicates mastication	Unknown physical equivalent of measurement
	Good correlation with sensory methods	Evaluation of graphs slow
	Complete texture measurement	Not suitable for routine work
		Restricted to 'bite size' units
Fundamental	Know exactly what is measured	Poor correlation with sensory methods
		Incomplete texture measurement
Ideal	Simple to perform	None
	Rapid	
	Suitable for routine wotk	
	Good correlation	
	Closely duplicates mastication	
	Complete texture measurement	
	Know exactly what is measured	
	Can use large or small size samples	

TABLE II

Objective methods for measuring
texture of vegetables

1. Force-measuring
   a. puncture
   b. extrusion
   c. crushing

2. Distance-measuring
   a. penetrometer
   b. deformation
   c. acoustic spectrometer

3. Multiple-variable
   eg.   Durometer

4. Multiple-measuring
   eg.   Instron
            G. F. Texturometer
            Shear Press with recorder
            Ottawa Texture System

2. Puncture Testing

The best known puncture tester is the Magness Taylor pressure tester that is manufactured by the D. Ballauf Company, 619 H Street, N.W., Washington, D.C. 20001.
Two instruments are available, one with a ten pound force spring and the other with
a thirty pound force scale. These instruments are provided with two punches, one of
5/16 in. in diameter and the other 7/16 in. diameter. The large diameter punch, is used
on softer materials and the smaller diameter is used on the firmer materials. The
choice of which spring scale to use, the ten pound or thirty pound depends on the
firmness of the test material. The Chatillon Company (John Chatillon & Sons,

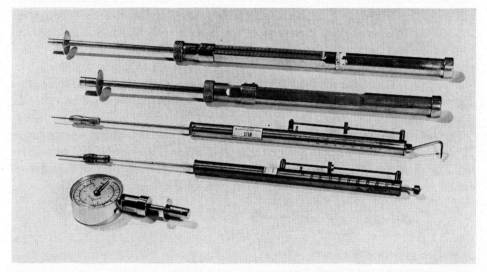

Fig. 2a.   Hand operated puncture testers. Top two are Ballauf instruments with 30 lb and 10 lb
force scales and 5/16″ and 7/16″ diameter tips; center two are Chatillon instruments with 500 g
and 2000 g force scales and 0.032″ and 0.058″ diameter tips; bottom is Effe-Gi instrument with
12 kg force scale and 7/16″ diameter tip.

Division of Aero-Chatillon Corporation, 83–30 Kew Gardens Road, Kew Gardens,
New York 11415) also supplies a line of puncture testers with force scales from 500 g
up to 40 lb full scale force and pressure tips that range from 7/16 in. diameter down
to 0.026 in. diameter. The Effe. Gi Company, C. Garibaldi 102, 48011 Alfonsine,
Ravenna, Italy also make a puncture tester with a 26 lb scale (12 kg) and pressure
tips of 5/16 and 7/16 in. diameter. The Effe. Gi instrument fits comfortably into the
hand, can be easily carried in the pocket and costs less than half the price of the other
two instruments. All three of these instruments are satisfactory. The Effe. Gi instrument
is the most convenient to use and to carry out into the field, while the Chatillon line
provides the widest range of test conditions and the Ballauf instruments are best
known. Figure 2a shows examples of these three puncture testers. Figure 10 in Chapter
VI also shows puncture testers.

Some instruments are available that use more than one punch. The earliest of these was the Christel texture meter (Christel, 1938) which consists of a nest of twenty-five punches of approximately 3/16 in. diameter that are forced down into a metal cup containing the food by means of a rack and pinion hand drive. The maximum force required to do this is recorded on a force gauge above the machine. This instrument has been widely used, however it is no longer commercially available.

The Maturometer (Lynch and Mitchell, 1950, Mitchell *et al.*, 1961) consists of 143 metal punches, each one of 1/8 in. diameter mounted on a plate directly over a matching set of holes in a metal plate that are counter-sunk to hold 143 peas. The maximum force required to punch 143 peas simultaneously is recorded on a gauge. The instrument was developed in Australia and is widely used there for measuring the quality of fresh peas. It is commercially available and can be obtained through the C.S.I.R.O. Division of Food Preservation, North Ryde, N.S.W. Australia. The Maturometer has not been used for any commodity other than peas except on an experimental basis. A recent development in Australia has been to mount a single 1/8 in. diameter punch in an Instron machine for single puncture tests on peas and some other commodities. This test has been called the 'Single Pin Maturometer Reading' (SPMR), (Casimir *et al.*, 1971).

Each of these puncture instruments measures the force necessary to give some degree of penetration of the metal punch into the food. The test is irreversible and destructive in the sense that a hole is left in the food. Bourne (1966a) has shown that the force required to puncture food can be described by the equation:

$$F = K_c A + K_s P + C,$$

where

  F = the puncture force;
  K_c = the compression coefficient, a property of the test material;
  A = the cross-sectional area of the punch;
  K_s = the shear coefficient, another property of the food but different from K_c;
  P = the perimeter of the punch;
  C = constant.

DeMan (1969) confirmed Bourne's work. With processed cheese he found that both perimeter and area were involved while the area only was involved with butter and margarine and the K_s coefficient was 0 thus eliminating the affect of the perimeter of the punch. In studying the force distance curves of a single punch being pressed into foods (Bourne, 1965) it was found that three characteristically different shapes of curves were obtained (Figure 2b). The first type is represented by curve A. There is a rapid rise in force over a short distance as the pressure tip moves onto the food because the whole item is deforming under the load and there is no puncturing of the tissue. This stage ends when the pressure tip begins to penetrate into the flesh and is represented by the sudden change in slope called the 'Bioyield point' which marks the beginning of the second stage. During the second stage, represented by the rising sawtooth line, the pressure tip penetrates the tissue. With type A commodities, the pres-

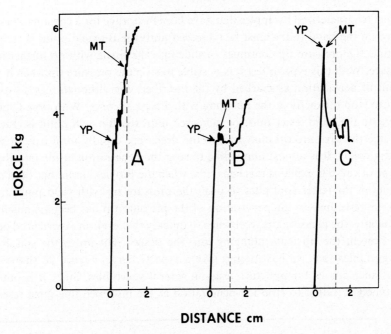

Fig. 2b. Characteristic force-distance curves obtained in puncture tests.

sure test depends upon the depth of penetration of the metal tip into the fruit and it is a higher value than the Bioyield point. A second characteristic type of curve is shown at B. The first stage is the same as type A, but the second stage after the Bioyield point appears as a force plateau which is approximately constant over a considerable distance of penetration of the tip. In this case the pressure test is almost independent of the depth of penetration of the metal tip and the reading that is obtained is approximately equal to the Bioyield point.

The third characteristic shape is shown in curve C. Once again the first stage is the is the same as type A, but in the second stage the force falls back sharply after the Bioyield point, sometimes falling to a plateau at a lower force than the yield point and sometimes continuing to fall but at a slower rate than immediately after the Bioyield point. In a type C curve, the Bioyield point and the pressure test are identical regardless of the depth of penetration of the tip into the tissue.

These three curves are useful in explaining the performance of hand operated puncture testers. With type A products (represented by most varieties of freshly harvested apples) the hand tester must be pushed with increased force to make the pressure tip penetrate to the required depth. After the yield point is passed, each increment in pressure causes a corresponding increment in penetration and there is no further penetration until the pressure is increased by another increment. The steeply sloping second stage of the type A force distance curve traces this behavior. With this type of product it is easy to stop the pressure test exactly when the penetration reaches the required depth as indicated by the inscribed line on the pressure tip. With a type

B product (characterized by apples that have been in storage for a considerable period) the force on the hand tester must be increased until the Bioyield point is reached, at which time the pressure tip continues to slide into the tissue with no further increase in pressure. With this type of test it is possible to stop the pressure tip when it reaches the point of penetration as marked by the inscribed line although it is a little more difficult to stop it exactly at the line than with a type A curve. With type C products the force on the hand tester must be increased until the Bioyield point is reached, at which time the pressure tip plunges into the flesh very rapidly until it is stopped by the splash collar. It is almost impossible to stop the penetration at the inscribed line. The slope of curve C explains this behavior; when the Bioyield point has been reached the spring in the tester continues to push the pressure tip with yield point force although the resistance to the penetration of the pressure tip has become much lower. Consequently, the pressure tip accelerates so quickly that even an experienced operator cannot prevent the tip from plunging into the tissue right up to the splash collar.

All vegetables that we have tested give a type C curve. Figure 2c shows typical force-distance curves for puncture tests on several vegetables. Since it is impossible with a type C product to stop the penetration at the inscribed line most researchers

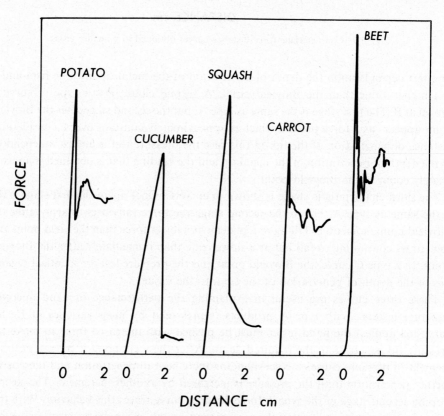

Fig. 2c. Characteristic force-distance curves for puncture tests on vegetables. Note that each is a type C curve.

consider the puncture test to be an unsatisfactory test for vegetables, but this is an erroneous opinion. In a type C curve, the Bioyield point is the maximum force that is encountered during the test and since the hand tester uses a maximum force reading dial, it will read the Bioyield point even though penetration goes beyond the inscribed line. Therefore the hand puncture test can be a very useful test on vegetables even though it appears to be an unsatisfactory test. The pressure test should be used more widely on vegetables because it is a simple instrument that can be carried into the field and it gives data easily.

TABLE III

Comparison of puncture test by
hand tester and Instron

Commodity	Hand tester mean force kg	Instron mean force kg
Irish potatoes	10.76	10.86
Summer squash	9.78	9.50
Beets	12.23	12.59

(5/16″ diameter punch, mean of 25 punches)

Table III shows some data in which the puncture test on some vegetables was performed using the Effe. Gi puncture tester. The tip from this tester was also mounted in the Instron and used to test the same vegetables. In reading off the Instron charts, the Bioyield point was picked off in the normal way while with the hand tester, the dial reading was recorded. It can be seen from this table that the hand tester gives good measurement of the Bioyield point of these vegetables, although the pressure tip plunged right in up to the splash collar in every test.

3. Extrusion Testing

In tests that use the extrusion principle, the food is placed in some receptacle and force is applied until the food extrudes through an outlet in the form of some kind of slot or hole that is provided. During this process the structure of the food is disrupted. Usually the maximum force to accomplish extrusion is measured and used as an index of textural quality. This type of test has been used widely on various vegetables and many other foods. Since extrusion requires that the food flow under pressure, this kind of test is restricted to those commodities that flow fairly readily under an applied force. We have found, for instance, that this type of test is not suitable for measuring the textural quality of raw green beans because they do not flow easily under an applied force.

The oldest commercially available extrusion instrument is the well known Tenderometer (Martin, 1937) which was developed for measuring the texture of fresh green peas. In this instrument approximately 200 g of peas are placed in a container that consists of a grid of metal bars 1/8th of an inch thick with 1/8th in. spacings. A second grid of bars rotates down through the slots and extrudes the peas, some of the peas passing through the stationary grid and some of the peas passing through the moving grid. There is probably an element of shearing in addition to extrusion, but the basic process in this instrument is that of extrusion. This machine has been widely adopted as a means of measuring the quality of fresh green peas and establishing the payment to the grower. Unfortunately, the instrument has a serious problem in calibration which seems to be primarily due to the friction between the moving grid and the stationary grid and the fact that this friction can change if the grids or bars are slightly bent or dented in any way. A number of workers have commented on this problem over the years and a number have attempted to find a standard test material that could be put into the Tenderometer as a means of calibration, but no material has been successful. The most recent and possibly most comprehensive study on the problems of calibrating the Tenderometer is that of Voisey and Nonnecke (1971).

Another very widely used instrument is the Shear Press (Kramer, *et al.*, 1951) which works on the same principle as the Tenderometer in that the food is placed in a metal box with bars across the bottom and another set of bars moves down through the box extruding some of the food through the grid in the bottom and some upward between the bars. The material that extrudes upwards is scraped out from between the blades by the grid across the top of the box and falls back into the cell during the upstroke of the machine. The advantages of the Shear Press are that the box containing the test material is rectangular and that the motion of the moving parts is linear rather than moving through the arc of a circle as in the Tenderometer. Like the Tenderometer the basic Shear Press consists of bars $\frac{1}{8}$ of an inch thick with spaces $\frac{1}{8}$ of an inch wide to allow the moving bars to pass through. The original Shear Press was made by the LEE Company and the box and moving bars were constructed of welded stainless steel. There was quite a problem in this cell with friction between the moving parts. Later, the Allo Company made the Shear Press and now the Food Technology Corporation, 12300 Parklawn Drive, Rockville, Md., 20852, makes the instrument. Although the different manufacturers of the Shear Press built basically the same instrument there are some differences in the design and construction of the test cell (Figure 3a and Chapter VI, Figure 24b). The Food Technology Corp. manufactures a cell in aluminum alloy, the moving blades are bolted, are slightly flexible and have a greater clearance between the stationary bars which reduces the frictional component. The contact surfaces of the moving bars in the Food Technology Corporation model are slightly slanted, whereas in the earlier models the contact surfaces were flat and parallel therefore possibly causing differences in some tests. For example, Ross and Porter (1966, 1968) used the old design of Shear Press cell to measure textural properties of French fries and from the curves that were plotted on the chart they were able to pick off several different textural characteristics. In order to obtain these kinds of curves, it is necessary to use the old design of Shear Press with squared off blades because the typical curves are not obtained with the slanted shear blades that are found on the latest model.

Fig. 3a. Left shows the LEE Shear Press cell constructed of rigid plates of stainless steel and squared-off ends. Right shows the Food Technology Corp. Shear Press cell constructed of aluminium alloy with some flexibility in the plates, and slanted ends.

In using the standard Shear Press cell the most common method is simply to read off the maximum force during the compression-extrusion stroke. Some researchers divide the maximum force by the weight of material that was used in the test to give a result of force per gram. In a study of the effect of the quantity of material in the Shear Press cell, Szczesniak *et al.* (1970) found that foods could be divided into three classes depending on the maximum force and sample weight (Figure 3b).

(1) Linear relationship, found in white bread and sponge cake. (This relationship might become curvilinear if the instrument had sufficient force capacity to allow the shear press to be filled with these commodities.)

(2) Curvilinear relationship, found in raw apples and cooked white beans.

(3) Maximum force is independent of weight once a certain minimum weight has been placed in the cell. Examples are found in canned beets and canned and frozen peas.

In the light of this evidence it seems unwise to use the force per gram in expressing the results of the Shear Press unless preliminary tests show that a linear relationship exists. It would be preferable to use a standard weight of product in the Shear Press and simply express the result as the maximum force instead of force per gram of product.

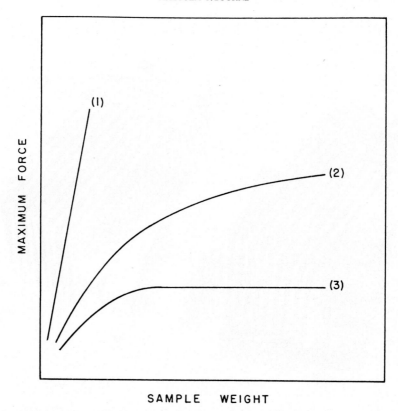

Fig. 3b. Maximum force vs sample weight plots for Shear Press. Behaviour (1) is exemplified by bread and cake, (2) by raw apples and cooked white beans, (3) by canned beets, canned and frozen peas (from Szczesniak, 1970).

Another factor that needs to be considered in the Shear Press is that it is hydraulically driven and as it runs the oil becomes heated and causes the Shear Press to run at a faster speed. In many applications this change in speed would not be important but for very careful work, or in the handling of commodities that show shear-thinning or other rate-dependent effects this should be taken into account. Ang *et al.* (1960) pointed out this feature of the Shear Press and overcame the problem by putting a heater and a thermostat in the oil tank and maintaining the oil at a constant temperature of 170 °F. It should be pointed out, that Ang *et al.* were using very slow speeds to compress whole onions and small differences in speed might have affected their results.

Various other extrusion testing instruments have been described in the literature (Bourne and Moyer, 1968) but most of these are not commercially available and are not widely used.

In order to study the principle of extrusion testing, a simple extrusion cell that is completely free from friction was constructed (Bourne and Moyer, 1968). It consists of a cylindrical aluminum cell of 10.2 cm internal diameter (Figure 4). A set of ½ in. thick circular aluminum disks were prepared and fastened to a solid aluminum bar which forces the discs down into the metal cylinder. The complete cell is mounted in the Instron Universal Testing Machine which provides the driving motion and the force measuring components of the test. The food is placed in the metal box and the plunger is forced down into the box until the food extrudes up through the annulus between the compressing plate and the wall of the metal box. The width of the annulus can be quickly altered by changing the plates. The force required to perform the extrusion is recorded.

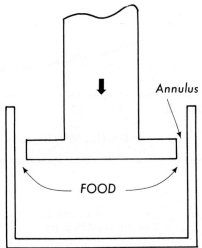

Fig. 4.   Schematic diagram of simple cell for pure extrusion.

One advantage of this kind of test cell is that there is no friction to cause errors in the test or in calibration. During the upstroke there is a tendency for the metal box to follow-up with the pressing plunger and it is sometimes difficult to remove. This objectionable feature could probably be eliminated by placing a small valve in the compressing plate to allow air to flow into the lower part of the box during the return stroke or by physically holding the metal cylinder by means of a clamp. Fresh green peas were used as the test material and the alcohol insoluble solids (AIS) was used as an independent index of the quality of the peas. This test cell is satisfactory and can be used routinely on a wide number of commodities including canned and frozen green beans, sweet corn, beets, peas, cooked dried beans, cherries, raw and processed apple slices and many other commodities.

The width of the annulus that would give the best resolution among samples that have a similar texture was found. In Figure 5, the plateau extrusion force on green peas is plotted against the annulus width; the extrusion force, as expected, is inversely proportional to the width of the annulus. Table IV shows the slope of the force-AIS plot at various annulus widths. The slope increases as the annulus width decreases, but this increasing slope is due to the higher extrusion force that is required for the smaller annulus width. For instance, at 10 mm annulus width the force is in the range of 300 kg, while at 1 mm annulus the force is above 1600 kg. Plotting the logarithm of the

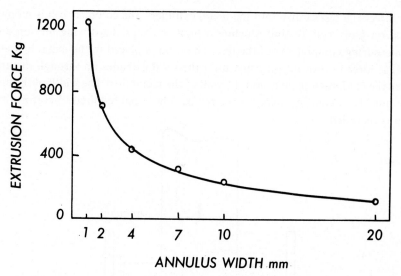

Fig. 5.   Effect of annulus width on extrusion force using fresh green peas in a back-extrusion cell.

TABLE IV

Effect of annulus width on
slope of force-AIS graph

Annulus width mm	Slope Force-AIS	Slope log-force-AIS
1	76.5	0.0186
2	75.3	0.0278
3	65.8	0.0309
4	60.9	0.0364
5	50.0	0.0352
7	37.3	0.0336
10	15.9	0.0186

force against the AIS plot normalizes the effect of the different force ranges and then the slope reaches a definite maximum at 4 mm annulus width. In terms of obtaining the best resolution between samples it becomes clear that with green peas a 4 mm annulus would be the best width to use. A very small annulus width, 1 mm, and a very wide annulus width, 10 mm, give much lower slopes and are not nearly as effective as the 4 mm annulus in resolving differences between samples. From 3 mm to 7 mm annulus width the slope of the log force-AIS plot is close to the maximum. It is interesting to note that the Tenderometer and the standard Shear Press cell each use a gap width of $\frac{1}{8}$ in. (3.2 mm) which is close to the best width for optimum resolution.

The maximum force in this simple extrusion cell is practically independent of the sample size that is used provided that there is enough sample to give some extrusion. It is also practically independent of the amount of water that is mixed with the peas.

Figure 6 shows that the extrusion force is dependent upon the speed of travel at low speeds and practically independent of the speed of travel at higher speeds. Evidently shear thinning occurs at the slower speeds particularly with the smaller annulus widths. From this evidence this test should be conducted at a speed of 30 cm min^{-1} or higher for peas for better results.

A group in Canada (Voisey, 1971), has recently developed the 'Ottawa Texture Measuring System'. The basic machine and accessories can be purchased from Canner's

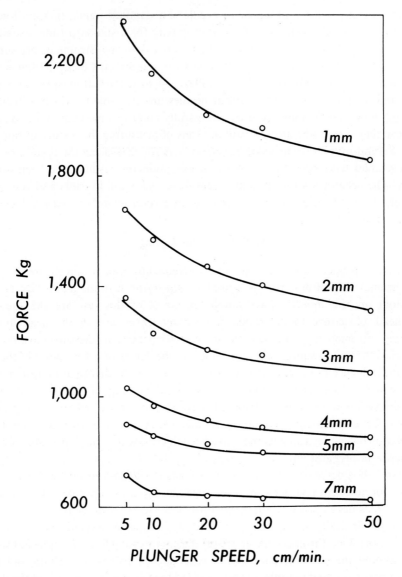

Fig. 6.   Effect of plunger speed on force required to extrude fresh green peas in a back-extrusion cell. Annulus widths are 1, 2, 3, 4, 5, 7 mm.

Machinery, Ltd., Simcoe, Ontario, Canada. The group developed a number of test cells that fit into a basic machine including an extrusion cell in which the material is extruded through slots that are formed by heavy metal wires. These test cells are applied to the measurement of texture of various commodities including peas and other vegetables. Prototype and similar types of 'Ottawa Texture Measuring System' are discussed in Chapter VI.

4. Penetrometers

The Penetrometer consists of a cone or needle of a standard weight that is allowed to sink into a test food under the influence of gravity for a standard time, usually 5 s. The depth of penetration is measured. The two best known instruments are supplied by Precision Scientific Company, 3739 West Cortland Street, Chicago, Illinois 60647 and Lab-Line Instruments, Inc., Lab-Line Plaza, Melrose Park, Illinois 60160. Various angles and weights of cones and various needles are available for these instruments. Although these instruments are used successfully on fats, some puddings and similar products, they seem to be unsuccessful in terms of measuring the texture of raw vegetables. Several investigations using penetrometers for measuring the texture of fruits and vegetables have been found. Without exception the results have been unsatisfactory. The penetration test does not seem to be successful in fruits and vegetables; however, this kind of test may be successful in certain manufactured products such as heavy pastes.

5. Deformation Testing

One physical property of vegetables that is frequently used as a measure of quality is the firmness or softness that is sensed by squeezing it in the hand. Occasionally it is done in the mouth. Two simple examples of this test are the squeezing of a head of lettuce to determine how many leaves are in the head and the squeezing of tomatoes to measure the degree of ripeness. In the subjective estimate of firmness the fingers squeeze the vegetable and the degree of deformation of the food under the influence of the compression force is sensed. The deformation test is one of the most widely used sensory tests for measuring the quality of vegetables. Although many simple laboratory-built instruments have been devised to measure this property there are few commercially available instruments. A review of the great number of deformation testing instruments that have been devised will soon be available (Bourne, manuscript in preparation).

Kattan (1957) described an instrument he calls the Firm-o-meter for performing a deformation test on tomatoes and other vegetables that are approximately spherical in shape. This developed into the Asco Firmness Meter which is available from the Agricultural Specialty Company, Box 705, Hyattsville, Maryland. The interesting features of the Asco Firmness Meter are (a) the compressing force is applied at several points around the circumference of the vegetable by means of a chain and (b) a standard pre-stress load is applied. The standard load is applied for 5 s and the degree of compression of the vegetable is measured. When Kattan and Littrell (1963) used the Asco Firmness Meter on canned sweet potatoes with a pre-stress load of 200 g and a load of 400 g applied for five seconds they obtained a correlation coefficient of $R = -0.98$ between the firmness meter and the AIS of the canned product.

The Marius Company, Ganzenmarkt 4–8 Utrecht, Holland manufactures a deformation tester for eggs. This instrument may be of use for vegetables but no reports of its use on vegetables have been located.

In view of the great use of the deformation principle as a sensory test, it seems surprising that so few instruments for measuring the deformibility of vegetables have been placed on the market. There is a real gap in our instrumentation in this area.

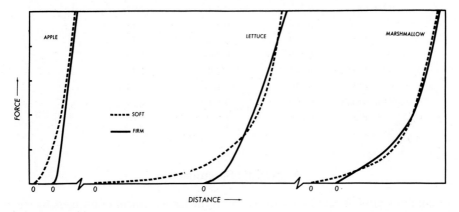

Fig. 7. Deformation curves for three commodities squeezed in the Instron. Note how difference between soft and firm is most pronounced at low force.

Any manufacturer who makes a simple, reliable deformation tester will find a ready market for that instrument.

Bourne (1967a) studied the principle of the deformation test on foods using the Instron Universal Testing Machine. Figure 7 which gives results for three different commodities, shows that the difference between a soft and a firm article of food can be measured with greater resolution under low deforming forces than under high deforming forces because the softer item behaves more and more like the firmer item as the deforming force increases. This is brought out in Table V where the deformation of two potatoes under various forces is listed together with the ratio of the deformations. Under a low deforming force the ratio is much higher, indicating that differences are distinguished more easily than at higher deforming forces. A deformation testing instrument should be designed to use low deforming forces, but since this is accompanied by low deformations in absolute measurements the instrument must be able to measure small differences in distance with a high degree of precision. The fact that a *gentle* squeeze will discriminate between samples much better should be of interest to the supermarket chains because there is less damage to commodities when they are squeezed with a small force. The motto in all deformation testing should be, SQUEEZE IT GENTLY, (Bourne, 1967b).

TABLE V

Deformation of whole potatoes

Force range g	Potato #1 mm	Potato #2 mm	Ratio #2/#1
0–200	0.160	0.532	3.33
200–400	0.142	0.440	3.10
400–600	0.130	0.382	2.93
600–800	0.110	0.302	2.74
800–1000	0.102	0.238	2.34
1000–1500	0.242	0.490	2.02
1500–2000	0.200	0.372	1.86
2000–2500	0.180	0.282	1.57
2500–3000	0·160	0.240	1.50
3000–3500	0.142	0.208	1.47
3500–4000	0.130	0.188	1.44
9500–10000	0.100	0.120	1.20
0–4000	1.698	3.674	2.16

6. Acoustic Spectrometry

A recent innovation in texture measurement is the measurement of firmness by means of acoustic spectrometry (Drake, 1962; Abbott et al., 1968a, b; Finney and Norris, 1968). The Nametre Company, 272 Loring Avenue, Edison, New Jersey 08817 manufactures an instrument for this purpose. The basic principle in this procedure is to subject a vegetable or fruit to sonic vibrations ranging from low frequencies of about 20 Hz up to frequencies of several thousand Hertz. A receiver which is placed on another portion of the vegetable records the amplitude of vibrations within the test piece. A plot of amplitude of vibration of the vegetable against the frequency shows distinct peaks representing the resonance frequencies of the vegetables or fruit (Figure 8). The test may be conducted on intact vegetable or on cylinders of standard shape that have been cut from it. From classical engineering theory it is possible to calculate Young's Modulus of Elasticity, which is a measurement of deformability of the article. Abbott et al. (1968) have used this technique for measuring the textural quality of apples. Finney and Norris (1967) have used it for measuring the texture of Irish potatoes and sweet potatoes, and in other publications they have used this method on a number of fruits. This technique seems to offer considerable promise and it may answer the need for a deformation tester; however, this test is conducted at frequencies of hundreds or thousands of Hertz whereas the normal squeezing performed by the hand is at a frequency of less than one Hertz, and chewing in the mouth is at a frequency of about 1 to 2 Hz. Therefore we cannot be sure that the firmness as measured by the acoustic spectrometer will correlate well with the firmness as measured by sensory methods until it is tested on each commodity.

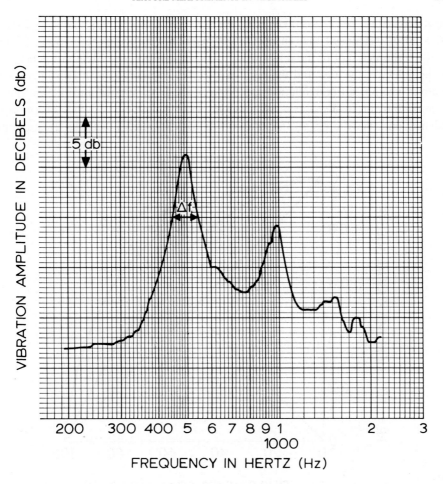

Fig. 8. Recorded amplitude-frequency curve on a cylinder of potato tissue showing resonance at the fundamental longitudinal vibration at 500 Hz and the first harmonic at 1000 Hz (from Finney and Norris, 1967).

7. The Durometer

The Durometer which is manufactured by the Shore Instrument Company, 90–35 Van Wyck Expressway, Jamaica 35, New York, is a small hand instrument that was designed for measuring the softness of rubber, but it has been used for the measurement of firmness of potatoes, onions, and some other vegetables. The instrument consists of a spring loaded hemispherical indentor that protrudes through a metal anvil. In Model A-2, the indentor requires a force of 0.78 kg to make it recede to a point level with the anvil face. A circular dial calibrated from 0 to 100 indicates the force exerted on the spring inside the instrument. In operation the instrument is held in the hand and the indentor is pushed against the vegetable until the anvil face contacts the vegetable and the reading is taken from the dial. The Durometer is a good example of the kind of instrument that should not be used in measuring the texture of foods. Figure 9 shows a plot of the relationship between applied force, depth of penetration of the indentor, the area and the perimeter of the indentor, against Durometer scale reading. There is a linear relationship between the force applied and the Durometer scale reading, an inverse linear relationship between the depth of penetration of the indentor and the reading, and inverse curvilinear relationship between area and per-

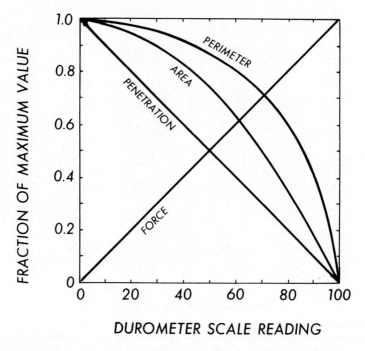

Fig. 9. Relation between force, area, perimeter and depth of penetration of the ball, and the Durometer scale reading.

imeter of the indentor in contact with the food and the Durometer scale reading (Bourne and Mondy, 1967). The manufacturer of the Durometer explains in the literature that readings above 80 on the dial "may be unreliable". Figure 10 plots the force/unit area against the Durometer scale reading for Model A-2. In the lower part of the scale the relationship is approximately linear; it becomes curvilinear above a reading of about 40 and above a reading of 80 it increases steeply. This is probably the reason for the unreliable readings above 80. In using the Durometer the force, the distance, the area and the perimeter all vary independently and the readings cannot be compared with any other instrument. This is the type of instrument that should not be used in texture measurement.

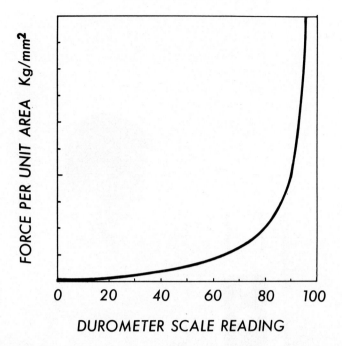

Fig. 10.   Relation between force per unit area on the ball and the
Durometer scale reading.

8. Multiple Measuring Instruments

Over the last decade there has been a tendency to move away from texture measuring instruments that are 'one point' systems to instruments that plot on a recorder a complete force distance or force time curve that records the entire history of changes in force during the test. Instruments with built-in recorders were initially used for research purposes but there is an increasing tendency to use them also for routine testing purposes. The four best known machines of this type that are commercially available are the Food Technology Corporation Texture Test System, which grew out of the Kramer Shear Press, the General Foods Texturometer, the Instron and, quite recently, the Ottawa Texture Measuring System. These four instruments will now be briefly discussed. All these instruments are also discussed in Chapter VI.

8.1. FOOD TECHNOLOGY CORPORATION TEXTURE TEST SYSTEM

Figure 11 shows a picture of the Food Technology Texture Test System. Figure 24b in Chapter VI also shows the same instruments with a recording system. It consists of a ram that is hydraulically driven and it has a number of test cells including the standard shear compression cell, a similar cell with thin blade and narrow slits to handle materials that have a small particle size, a shear blade, a compression plunger, a succulometer cell and several kinds of extrusion cells. The force is sensed by proving rings and forces in the range of a few pounds up to 3000 lb can be measured. The

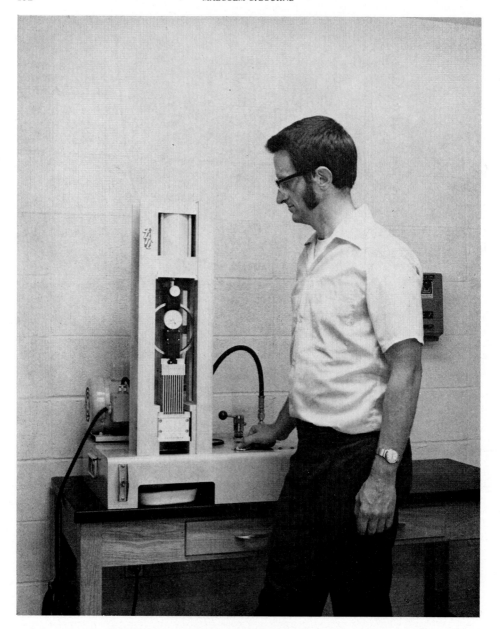

Fig. 11.   The Food Technology Texture Test System (by courtesy of
Food Technology Corp.).

speed of the plunger can range from 0 up to 34 cm a minute, and this can be varied
continuously over the entire range. The speed of travel of the plunger may change as
the oil warms up while the machine is running. The length of stroke is 3-$\frac{1}{2}$ in. The
recorder is driven by a shaft attached to the ram and chart width is 5 in. This

is the most widely used multiple-measuring instrument for measuring texture of vegetables.

8.2. GENERAL FOODS TEXTUROMETER

This instrument consists of a 'tooth' which can come in various shapes but is usually some kind of a flat faced cylinder that compresses bite-size samples of food (Figure 12). The tooth moves through the arc of a circle and its speed is not constant but follows a sine wave pattern because the driving mechanism is an eccentric. The machine has two chewing rates: 12 and 24 chews per minute. The maximum force available at the tooth is approximately 60 kg and forces can be measured to about 100 g. The total stroke length is approximately 1.4 in. By a suitable analysis of the curve obtained from this instrument it is possible to measure the properties of hardness, fracturability, gumminess, chewiness, elasticity, and cohesiveness. This instrument is very good for research purposes but it is restricted to small size pieces of food and reading the data off the chart is time consuming.

8.3. INSTRON UNIVERSAL TESTING MACHINE

The Instron Corporation makes a range of machines for measuring a wide variety of physical properties of many different kinds of materials. It was adapted for use on foods by Bourne *et al.* (1966). In 1972, the Instron Corporation announced a new model that has been especially designed for measuring textural and physical properties of foods (Figure 13).

However, it should be remembered that the other units in the line can be readily adapted to use with foods and can extend the range of force, speed and most other specifications of the machine. This machine is mechanically driven through a synchronous motor system and gives exact control of speeds from 50 cm min^{-1} to 1 cm min^{-1}. One of the strong features of the machine is the constant speed of the moving cross-head and also the constant speed of the chart drive. The force range runs from a few grams to 500 kg, and the stroke length is adjustable up to 36 in. This machine has been built to take extrusion cells, shear cells, puncture tests, deformation tests, the shear press, and the Ottawa test cells and it can also be adapted to perform the General Foods texture profile type of test. Many laboratories now use this machine.

8.4. OTTAWA TEXTURE MEASURING SYSTEM

This instrument was announced in 1971 and results from the work that was done by the Engineering Research Service of the Department of Agriculture in Ottawa, Canada. It consists of a screw operated ram, a load cell with a force capacity of 1000 lb, a stroke length of up to 8 in. and a recorder chart that is 10 in. wide (Figure 14). It can accept a variety of test cells including some new designs of test cell that were developed by the Engineering Research Branch.

Each one of the above four instruments can give far more information from a test than a simple maximum force or other one-point reading instruments. Because so much more information can be obtained from instruments of this type, they will be

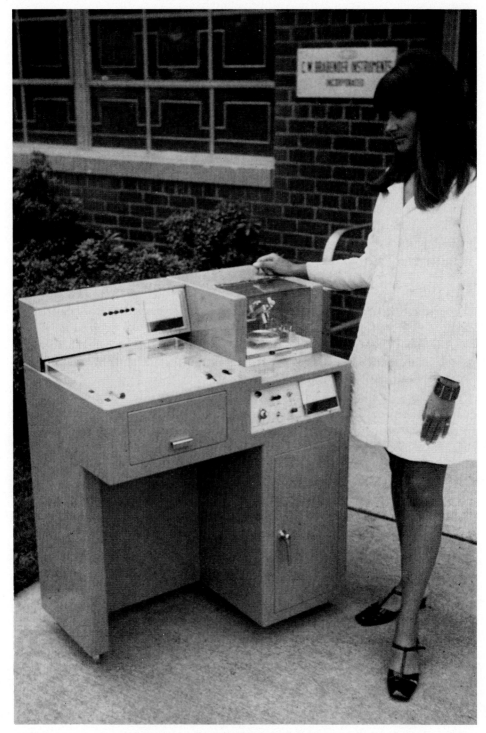

Fig. 12.   The General Foods Texturometer.

Fig. 13. New model Instron designed for use with foods (by courtesy of Instron Corp.).

used more widely in the future. Among the major tasks that lies ahead for food rheologists are the design of the test cells for specific purposes, the understanding of what the tests are doing, how they should be performed, the interpretation of the charts and the reading off of the pertinent data. In many cases, it will be possible on

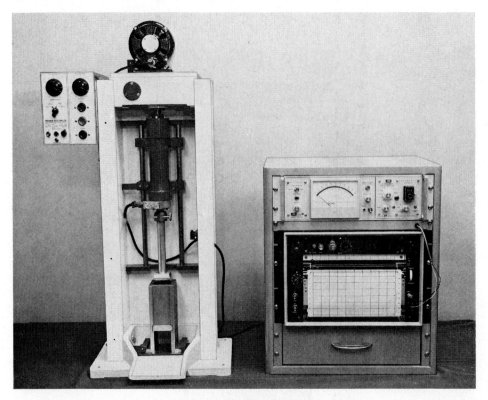

Fig. 14. The Ottawa Texture Measuring System (by courtesy of Canners Machinery Ltd.).

the basis of the work done on instruments of these types to design simple, inexpensive instruments for specific routine tests on particular commodities.

9. Unsolved Problems

In addition to the problems of test cell design and interpretation of charts there are many aspects in the texture measurement of vegetables that are not well understood. One of these is that fact that texture is a multi-faceted property of foods. We should talk about the 'textural properties' of foods rather than 'texture'. There is still a widespread impression that texture is a single property of food that can be measured with one number in a manner similar to the measurement of pH, but this is incorrect. 'Texture' is a combination of different physical properties of foods and instrumental methods measure only a few of them whereas the mouth can detect many more. It is quite common to find 10–30 different textural properties in a single commodity when using the General Foods Sensory Texture Profiling technique (Brandt *et al.*, 1963). It is obviously impossible to measure all of these by means of instruments but the advantage of instrumental measurements is that they are quicker and cheaper and more constant than a sensory panel. There are occasions when the textural properties of a food change in unison; for example, Bourne (1968), has shown that as pears ripen a number of different textural properties decrease in unison. A single textural measurement becomes a good index of the entire texture profile of ripening pears because there is an excellent correlation between the different textural properties as the pears ripen. Ahmed and Dennison (1971) showed that the textural properties of irradiated mangos and peaches decreased in unison, while Bourne (unpublished data, 1966) showed that all the textural properties of freestone peaches and clingstone peaches decrease approximately the same rate during ripening. Bourne (unpublished data, 1967) found that the values for textural qualities of apples during maturation and in storage did not change in unison and this probably explains why the puncture test is an effective test for pears but a less effective test for apples.

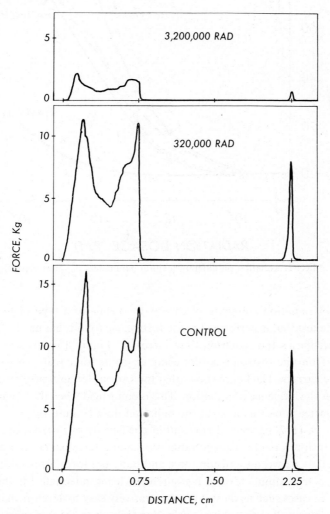

Fig. 15. Force-distance curves of cylinders of irradiated carrot tissue subjected to a G.F. Texture Profile test in the Instron.

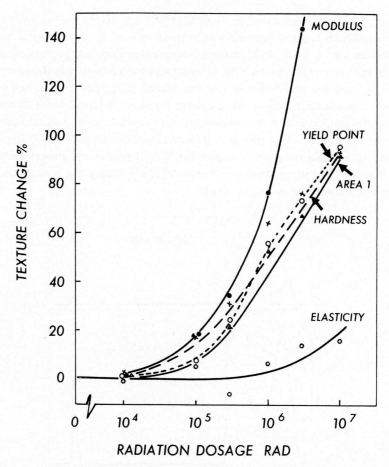

Fig. 16.   Effect of radiation on different textural parameters of core tissue of carrots.

In studies of radiation softening of carrots by means of a type of General Food Texture Profile test that is performed in the Instron we found that most of the textural parameters decline as the radiation dose increases. Figure 15 shows texture profile curves traced from the Instron recorder using the same force scale for irradiated and non-irradiated carrots. The figure shows that the various textural properties of carrots decrease under the influence of radiation. This is confirmed when the % change in each textural parameter is plotted against the radiation dose (Figure 16).

Shannon *et al.* (1971) obtained correlation coefficients of $r = 0.91$ and $r = 0.98$ between the extrusion force for canned table beets and a taste panel; and a correlation coefficient $r = 0.91$ between extrusion force and puncture force. This indicates a good possibility of a 'one point' texture measurement being adequate for the texture of canned beets. In cases such as these it is comparatively easy to develop an instrumental test that correlates well with sensory evaluation, but when the various textural parameters do not change in unison it is more difficult to obtain good correlation with sensory methods. This is a research area that has not been given the attention it deserves.

More work along these lines is needed in order to identify the textural properties of vegetables, to find those vegetables in which the textural properties change in unison and those vegetables in which the textural properties do not change in unison. In those vegetables where textural properties change in unison a simple 'one point' textural measurement will be satisfactory for describing the texture of the product, while for those vegetables in which the texture does not change in unison several measurements will be needed in order to describe the texture and the changes that occur during growth, processing and storage.

A good example of a vegetable that requires more than one textural measurement is sweet corn. The textural properties of sweet corn change rapidly during maturation and since the price paid to growers by processors is correlated with quality, considerable effort has been devoted to establishing instrumental methods for measuring the textural quality of sweet corn. A group from the University of Maryland have been the most active in this field (Kramer and Smith, 1946; Kramer et al., 1949; Kramer, 1952; Kramer, 1958; Kramer, 1959; Kramer and Cooler; 1962). This group developed the test for pericarp; the Succulometer in which the volume of juice expressed from 100 g sweet corn subjected to 3000 lb pressure for 3 min is measured; the trimetric procedure that measures pericarp, moisture content and kernel size; and the bimetric procedure that is a combination of Shear Press and Succulometer readings.

The problem with sweet corn is that the pericarp of the corn which represents about 5% of the weight of the kernel is responsible for about 50% of the textural sensations in the mouth. The various measurements that have been devised for measuring the textural quality of sweet corn measure essentially the changes in quality of the starchy endosperm and they do not measure the quality of the pericarp. There is a test for determining the quantity of pericarp (Kramer et al., 1949) but it does not measure the quality of the pericarp. As the corn matures the pericarp becomes increasingly tough and this is detected easily on the teeth. At the present time there is no instrument to measure the toughness of the pericarp. Because the pericarp is tough, thin and elastic, it is difficult to tear it in some way in which the force required for tearing can be measured. We have placed sweet corn in the back-extrusion cell using a 1 mm annulus and examined the product that has been extruded through this narrow slit. We find that the endosperm is greatly changed in nature while the pericarp can be recovered almost intact with just a few small tears in it. The problem of measuring the quality of pericarp is as yet unsolved, although it is a very pressing problem in the grading of sweet corn.

A related problem is that of determining the toughness of fiber in vegetables such as asparagus and celery. The interaction between the fiber and the matrix of the vegetable in which it lies makes the problem difficult. In raw asparagus and celery the crisp matrix in which the fibers lie makes it possible to cut and shear the fibers comparatively easily, while after cooking the matrix becomes soft and the fibers become more prominent in the mouth, although probably they are no tougher than in the uncooked vegetable. This class of problem of a single vegetable with two distinct textural phases also appears in the skins on tomatoes and other vegetables. The interaction between the skin and the flesh lying beneath the skin is very difficult to separate. The correct measurement of skin toughness free from interference of the texture of the underlying fleshy tissue is still an unsolved problem.

Another class of problem on which insufficient work has been done is the correlation between the type of test used in sensory evaluation of food with the type of instrumental test that is performed. Szczesniak and Bourne (1969) asked a group of people to measure the firmness of a number of commodities and observed the type of tests that were used. They found that with semi-solid foods, such as puddings, people would use some kind of viscosity test to measure the firmness; with foods such as tomatoes and lettuce a deformation test was used; while with apples and some pears, a puncture test was used; and with carrots, a bending or flexure test was used. In this range from very soft to very firm texture the sensory evaluation showed that there were four distinct methods for measuring firmness: viscosity, deformation, puncture, and flexure. In establishing instrumental methods it is important to know what firmness range is involved and to establish the correct kind of test. It would for example be foolish to use a puncture test for measuring the firmness of tomatoes when people use a deformation test. Considerably more work is needed in this area in terms of finding out what kind of tests people are using, the conditions under which they use them and then using this information to design instruments that copy the pattern that people use. The ultimate calibration of all texture measuring instruments must be the senses of the human body.

10. Conclusion

The literature indicates that more progress has been made in the practice of texture measurement of vegetables in the last twenty years than in the entire preceding period. With a greater number of people now working in this field and the increasing sophistication of the instrumentation available, the next twenty years will lead to even greater progress in texture measurement of vegetables.

References

Abbott, J. A., Bachman, G. S., Childers, R. F., Fitzgerald, J. V., and Matusik, F. J.: 1968, 'Sonic Techniques for Measuring Texture of Fruits and Vegetables', *Food Technol.* **22**, 635.

Abbott, J. A., Childers, N. F., Bachman, G. S., Fitzgerald, J. V., and Matusik, F. J.: 1968, 'Acoustic Vibration for Detecting Textural Quality of Apples', *Proc. Am. Soc. Hort. Sci.* **93**, 725.

Ahmed, E. M., and Dennison, R. A.: 1971, 'Texture Profile of Irradiated Mangoes and Peaches', *J. Texture Studies* **2**, 489.

Ang, J. K., Isenberg, F. M., and Hartman, J. D.: 1960, 'Measurement of Firmness of Onion Bulbs with a Shear Press and Potentiometric recorder', *Proc. Am. Soc. Hort. Sci.* **75**, 500.

Bourne, M. C.: 1965, 'Studies on Punch Testing of Apples', *Food Technol.* **619**, 413.

Bourne, M. C.: 1966a, 'Measurement of Shear and Compression Components of Puncture Tests', *J. Food Sci.* **31**, 282.

Bourne M. C.: 1966b, 'A Classification of Objective Methods for Measuring Texture and Consistency of Foods', *J. Food Sci.* **31**, 1011.

Bourne, M. C.: 1967a, 'Deformation Testing of Foods. I: A Precise Technique for Performing the Deformation Test', *J. Food Sci.* **32**, 601.

Bourne, M. C.: 1967b, 'Squeeze it Gently!' *Farm Research* **32**, 8.

Bourne, M. C.: 1968, 'Texture Profile of Ripening Pears', *J. Food Sci.* **33**, 223.

Bourne, M. C. and Mondy, N.: 1967, 'Measurement of whole Potato Firmness with a Universal Testing Machine', *Food Technol.* **21**, 1387.

Bourne, M. C. and Moyer, J. C.: 1968, 'The Extrusion Principle in Texture Measurement of Fresh Peas', *Food Technol.* **22**, 1013.

Bourne, M. C., Moyer, J. C., and Hand, D. B.: 1966, 'Measurement of Food Texture by a Universal Testing Machine', *Food Technol.* **20**, 522.

Brandt, M. A., Skinner, E. Z., and Coleman, J. A.: 1963, 'Texture Profile Method', *J. Food Sci.* **28**, 44.

Casimir, D. J., Coote, G. G., and Moyer, J. C.: 1971, 'Pea Texture Studies Using a Single Puncture Maturometer', *J. Texture Studies* **2**, 419.

Christel, W. F.: 1938, 'Texturemeter, a New Device for Measuring the Texture of Peas', *The Canning Trade* **60**, 10 (March 28).

DeMan, J. M.: 1969, 'Food Texture Measurements with the Penetration Method', *J. Texture Studies* **1**, 114.

Drake, B. K.: 1962, 'Automatic Recording of Vibrational Properties of Foodstuffs', *J. Food Sci.* **27**, 182.

Finney Jr., E. E.: 1971, 'Dynamic Elastic Properties and Sensory Quality of Apple Fruit', *J. Texture Studies* **2**, 62.

Finney Jr., E. E. and Norris, K. H.: 1967, 'Sonic Resonant Methods for Measuring Properties Associated with Texture of Irish and Sweet Potatoes', *Proc. Am. Soc. Hort. Sci.* **90**, 275.

Finney Jr., E. E. and Norris, K. H.: 1968, 'Instrumentation for Investigating Dynamic Mechanical Properties of Fruits and Vegetables', *Trans. Am. Soc. Agr. Eng.* **11**, 94.

Friedman, H. H., Whitney, J. E., and Szczesniak, A. S.: 1963, 'The Testurometer – a New Instrument for Objective Texture Measurement', *J. Food Sci.* **28**, 390.

Kattan, A. A.: 1957, 'Changes in Color and Firmness during Ripening of Detached Tomatoes, and the Use of a New Instrument for Measuring Firmness', *Proc. Am. Soc. Hort. Sci.* **70**, 379.

Kattan, A. A. and Littrell, D. L.: 1963, 'Pre- and Post-harvest Factors Affecting Firmness of Canned Sweet Potatoes', *Proc. Am. Soc. Hort. Sci.* **83**, 641.

Kramer, A.: 1952, 'A Tri-Metric Test for Sweet Corn Quality', *Proc. Am. Soc. Hort. Sci.* **59**, 405.

Kramer, A.: 1958, 'Developing a Schedule for Payment for Raw Materials on a Quality Basis with Special Reference to Sweet Corn', *The Canning Trade* **80**, 6 (April 21).

Kramer, A.: 1959, 'Rapid Measurement of Quality of Sweet Corn for Processing', *Proc. Am. Soc. Hort. Sci.* **74**, 472.

Kramer, A. and Cooler, J.: 1962, 'An Instrumental Method for Measuring Quality of Raw and Canned Sweet Corn', *Proc. Am. Hort. Sci.* **81**, 421.

Kramer, A. and Smith, H. R.: 1946, 'The Succulometer, an Instrument for Measuring the Maturity of Raw and Canned Whole Kernel Corn', *Food Packer* **27**, 56.

Kramer, A., Guyer, R. B., and Ide, L. E.: 1949, 'Factors Affecting the Objective and Organoleptic Evaluation in Sweet Corn', *Proc. Am. Soc. Hort. Sci.* **54**, 342.

Kramer, A., Burkhardt, G. J., and Rogers, H. P.: 1951, 'The Shear Press: a Device for Measuring Food Quality', *The Canner* **112**, 34 (February 3).

Lynch, L. J. and Mitchell, R. S.: 1950, 'The Physical Measurement of Quality in Canning of Peas', *Comm. Sci. Inst. Res. Org. Australia Bull.* **254**, 35 pp.

Martin, W. McK.: 1937, 'The Tenderometer, an Apparatus for Evaluating Tenderness in Peas', *The Canner* **84**, 108.

Mitchell, R. S., Casmir, D. J., and Lynch, L. J.: 1961, 'The Maturometer – Instrumental Test and Redesign', *Food Technol.* **15**, 415.

Mohsenin, N. N.: 1970a, 'Application of Engineering Techniques to Evaluation of Texture of Solid Food Materials', *J. Texture Studies* **1**, 133.

Mohsenin, N. N.: 1970b, *Physical Properties of Plant and Animal Materials*, Gordon and Breach, New York.

Proctor, B. E., Davison, S., Malecki, G. J., and Welch, M.: 1955, 'A Recording Strain-Gage Denture Tenderometer for Foods. I: Instrument Evaluation and Initial Tests', *Food Technol.* **9**, 471.

Ross, L. R. and Porter, W. L.: 1966, 'Preliminary Studies in Application of Objective Tests to Texture of French Fried Potatoes', *Am. Potato J.* **43**, 177.

Ross, L. R. and Porter, W. L.: 1968, 'Interpretation of Multiple-Peak Shear Force Curves Obtained on French Fried Potatoes', *Am. Potato J.* **45**, 461.

Shannon, S. and Bourne, M. C.: 1971, 'Firmness Measurement of Processed Table Beets', *J. Texture Studies* **2**, 230.

Szczesniak, A. S.: 1963, 'Objective Measurements of Food Texture', *J. Food Sci.* **28**, 410.

Szczesniak, A. S. and Bourne, M. C.: 1969, 'Sensory Evaluation of Food Firmness', *J. Texture Studies* **1**, 52.

Szczesniak, A. S., Humbaugh, P. R., and Block, H. W.: 1970, 'Behaviour of Different Foods in the Standard Shear-Compression Cell of the Shear Press and the Effect of Sample Weight on Peak Area and Maximum Force', *J. Texture Studies* **1**, 356.

Voisey, P. W.: 1971, 'The Ottawa Texture Measuring System', *Can. Inst. Food Technol. J.* **4**, 91.

Voisey, P. W. and Nonnecke, I. L.: 1971, 'Measurement of Pea Tenderness. I: An Appraisal of the FMC Pea Tenderometer', *J. Texture Studies* **2**, 348.

CHAPTER VIII

MECHANICAL DAMAGE TO THE PROCESSED
FRUITS AND VEGETABLES

STEVENSON W. FLETCHER

Food and Agricultural Engineering Department, University of Massachusetts, Amherst, Mass., U.S.A.

The reduction in the grade classification of processed fruits and vegetables is of grave concern to the food industry. Far too often this is caused by mechanical damage during handling procedures associated with processing and distribution.

Current research by engineers, food scientists and plant scientists is concerned with non-mechanical injury aspects of grading (i.e., foreign material, inferior raw products, poor trimming, color and size). Associated with the current increasing mechanization in the area of production, harvesting, processing and distribution is increased exposure of the product to the hazards of mechanical damage. Mechanical damage is one of the major reasons for products scoring low in the Factors of Quality (i.e., defects and character) with the resulting reduction in grade classification according to current United States Standards (1963).

It is obvious that mechanical damage does exist, but the published information is almost non existent. Occasionally you read of things that indicate the industry is aware of the problem. For example, a recent item from NCA says that:

The revised standard for grades of canned green beans and canned wax beans for No. 10 cans has recently been published that has reduced the drained weight from 59.0 ounces to 57.5 ounces. This was in part due to the difficulty in consistently obtaining this weight and "the studies revealed slight impairment of quality because of leakage of the whole bean pod when the cans were filled to 59 ounces."

These 'studies' rarely appear in professional journals and mainly consist of visual observation tests. This chapter discusses how significant the problem of damage during handling and transportation is, what causes damage, how resistant the products are to damage, and what can be done to protect them.

One of the first steps undertaken in this study was to do some preliminary tests with processed pears in 303 cans to observe what physical damage may occur when subjected to several conditions of vibration and impact similar to what may occur during actual physical handling and shipping. Pears were selected simply because they damage easily, are a fairly regular shape, and somewhat homogeneous.

First looking at the effect of vibration. The bruising of processed pears due to vibration showed up as soft or mushy spots within the pears. This is different from the visual damage due to impact discussed later. The visual damage of simulated vibration tests showed, in this experiment, that splits and partitioning did occasionally occur

ChoKyun Rha (ed.), Theory, Determination and Control of Physical Properties of Food Materials, 163–179.

but were less frequent than bruising. The damaging force due to vibration is of low intensity, but of long duration. And it is very likely that the small force is not enough to cause splits and partitioning; therefore, the frequency of ocurrence should be small. The extent of bruising depended largely on the relative motion of the pear halves, probably on the relative orientation, too.

As shown in Table I, the average visual damage value per can was 2.5 for five minutes of vibration and increased to 4.6 per can for one hour of vibration.

TABLE I

Visual condition ratings and drained weight changes due to physical damage after vibrational and impact stressing conditions

Test description	Average damage value/can	Average drained weight loss (grams)
Horizontal vibration 1 h	4.6	5.02
Horizontal vibration 5 min	2.5	2.68
24″ Drop 3 replications	3.7	10.18
36″ Drop 1 replication	6.4	12.03

Visual damage value in this case was the result of a standard grading system used by Mr Hwang and was based on the size of the splits, bruises and degree of partition.

The drained weight loss changed from an average 2.68 g after a short duration (5 min) to 5.02 g after the long duration test (one hour). Drained weight loss indicates that part of the product has broken off and is in the syrup. The higher the weight loss, the more physical damage is observed when all other factors is held constant. It is significant to note that the longer the product was subjected to a vibratory condition, the larger the amount that is broken off and later found in the syrup. Likewise, the degree of visual damage increases with duration of vibration.

Relative to impact loading, it is significant that physical damage due to drop testing was higher than that due to vibration testing. During impact tests (i.e., dropping the container), the impact force required in a short period loading must be of higher intensity than in a long period of loading as in the vibration tests.

During the short period of impact, it is likely that the cell destruction and resulting plastic or permanent deformation could not extend too far. On the other hand, if for a visible bruising, sufficient layers of fruit cell were to be broken down during the short impact period, impact stresses of higher intensity had to be applied (Mohsenin, 1968). Therefore, as would be expected, most of the observable damage was in the form of splits and partitioning. Bruising did occur but only occasionally in this phase of the experiment. From this, it could be concluded that it was breaking situation, rather than that of crushing or scraping, which was most probably caused by impact.

Again, from Table I, it is seen that average damage value per can was 3.7 for the 24″ (3 replications) drop and 6.4 for the 36″ drop, while the average drained weight loss was 10.2 g for the 24″ drop and 12.0 g for the 36″ drop. Thus both the degree of visual damage and the drained weight loss are more serious for the 36″ height free fall than it is for the 24″ height free fall, replicated three times. It is concluded that the higher the drop the more the damage. On the other hand, it revealed that the height of drop or level of shock was a more important factor in causing the damage than was the frequency of drop, or shock, experienced by the product.

TABLE II

Various mechanical properties of canned
pears before and after having been subjected
to vibrational and impact stressing

Test description	Rupture force (lb)	Elastic modulus (lb in⁻¹)
Initial product	0.66	9.09
Horizontal vibration 1 h	0.36	2.32
Horizontal vibration 5 min	0.32	6.25
24″ Drop 3 replications	0.38	8.43
36″ Drop 1 replication	0.43	8.75

Rupture force or strength of the pear and elastic modulus or firmness of the pear were also determined for the various simulation tests. Table II shows that the elastic modulus of the unstressed product was found to be 9.1 lb in.$^{-1}$ and the decrease in value to a level of 6.25 lb in.$^{-1}$ after five minutes vibrational stressing; however, the vibrational stressing for one hour was found to be only 2.32 lb in.$^{-1}$. Based on the above information, the firmness of canned process pears due to one hour vibration stressing is less than that after five minutes of vibrational stressing. It indicates that the longer the vibration lasts, the higher the damage level is.

In general, the rupture force level likewise was at a lower level after vibration, although the duration of the vibration did not seem to make a significant difference.

The strength of product is dependent on the structural elements, and the rupture force is an indicator of product strength. It is apparent that impact due to vibration causes structural damage in the product and will cause texture degradation or quality reduction.

In the drop tests, there was no significant difference between the two drops relative to firmness and strength which were expressed in terms of rupture force and elastic modulus. And also, it is evident and important to see that vibration (5 min or one hour vibration) had more effect on elastic modulus change than two single impacts did, while for the change of rupture force they had almost the same effect.

Table III shows another interesting result of the study, that of the percent of canned pears damaged relative to the location in the can. During vibration the most damage seems to occur in the top portion of the can which agrees with damage to fresh produce in pallet boxes due to vibration (O'Brien *et al.*, 1964 and 1968). Impact results are less conclusive but it appears that there is more damage in the upper layers of the product.

TABLE III

Percent of canned pears damaged in relation
to location in the can

Test description	Position of pear half in cans					
	Bottom					Top
	1	2	3	4	5	6
Horizontal vibration 1 h	0	8	17	17	50	92
Horizontal vibration 5 min	0	0	10	30	40	50
24″ Drop 3 replications	33	17	8	42	42	58
36″ Drop 1 replication	10	30	10	50	40	60

This study was of course only for processed Bartlett pears and cannot be broadly carried over into every product, but it does point out a few of the possible causes of physical damage to processed fruits and vegetables.

Recently, significant work has been done in the development of mechanized systems. Associated with this has been research, not only to determine the mechanical forces which raw fruits and vegetables can withstand, but to design machines and processes to protect them from these forces.

Information is now available on the static and dynamic properties of selected raw fruits and vegetables. In contrast, once a product has been processed and many of its physical, as well as chemical and biological characteristics have been changed, little is known about its ability to resist damage (Mohsenin, 1963).

Determining mechanical properties of biological materials during static and dynamic loading is an essential initial step in any work relating to the reduction of mechanical damage to processed foods during physical handling. While the specific causes of mechanical damage in biological materials have not yet been thoroughly investigated, there is evidence that it is a function of the mechanical behavior of the material. If a product's mechanical behavior is known over the entire range of loading rates it experiences during handling, then the processor can design his handling system to reduce the external stresses that would cause the product to be damaged. Information on the rate sensitivity of the mechanical properties is also necessary in any work dealing with damage-prediction procedures.

Some preliminary rate sensitivity studies were performed with apples (Malus domestica) cultivar Baldwin. Apples were selected for this investigation due to their availability, their relative structural homogeneity in comparison to many other biological materials, the large amount of the product that is processed each year, and their susceptibility to physical damage. Fresh apples have been widely investigated and a great deal of information is available on the mechanical behavior of the raw product.

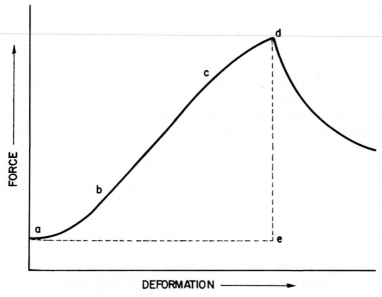

Force-deformation curve: *a d e*
Rupture point: *d*
Rupture force: *e d*
Rupture deformation: *a e*
Rupture energy: area under curve *a d e*
Elastic modulus: slope of straight line through *b c*

Fig. 1.   Mechanical properties derived from typical force-deformation curves of processed Baldwin apples under slow rates of loading.

The variety Baldwin was selected due to its availability in New England, its popularity as a cooking apple, and its high recommendation as a canning apple (Smock and Neubert, 1950).

Figure 1 shows a typical stress-strain curve obtained for processed apples which have been subjected to loading with the $\frac{1}{4}$ in. diameter plunger. The force-deformation relationship was not strictly linear, for it had a slight curvature in it. As the rate of loading increased, the curvature became less pronounced, and, above a loading rate of 10 in. min^{-1}, the relationship became directly linear. In general, the curves exhibited nonlinear behavior up to the point of material surface failure. Cooper (1962) and others suggested that apples exhibit a phenomenon of initial tissue breakdown prior to surface failure. This initial tissue breakdown, or so-called bio-yield point, is normally shown by a sudden discontinuity in the force-deformation curve, either in the form of a force drop or slope change. This situation was not observed in any portion of the study dealing with the processed product. The bio-yield point on apples is an indication of the initial fracture in the individual cell structure or bruising, and can be visually observed by the resulting browning effect due to the oxidation of cellular contents. On fresh fruit the bio-yield point is generally assumed to be the end of the elastic portion of the stress-strain curve. Any loading of the product past this point will result in a permanent set of physical damage to the material. A bio-yield point has previously been observed in materials such as fresh apples and pears that exhibit a strict linear stress-strain relationship during the initial portions of the curve (Cooper, 1962).

As stated previously, the force-deformation portion of the curve was not completely linear so a bio-yield condition was not expected. Heat treatment of apples changes the texture of the product and inactivates the plant enzymes (Griswold, 1962). The enzyme inactivation will eliminate the possibility of using enzyme browning as an indicator of browning.

Likewise, by treating a fruit with heat, certain chemical and physical changes took place that caused a sensory change of product firmness to softness. During heat treatment much of the substance in the cell wall changes from a water-insoluble protopectin to a water-soluble pectin capable of forming a gel (Mohsenin, 1968). This feature along with the reduction in hemicellulose, causes the cell wall to lose its rigidity, thus making it less elastic (Griswold, 1962).

During heating, the cell content's ability to exert an internal pressure called 'turgor pressure' is affected due to the coagulation of the protoplasm. While the cell contents are a less significant factor than the cell wall, they both affect product rigidity (Mohsenin, 1968). After heating, the cells become less rigid and will accept more deformation prior to failure. Processed apple cells as compared to fresh apple cells, tend to flow away from the stress location and disassociate from each other before they fail (Griswold, 1962), thus the absence of any observable bio-yield condition.

The force-deformation curve increases up to a maximum force at which material failure occurs. In many fields this situation is called the ultimate strength, or failure strength, but in the area of mechanical properties of biological materials this has generally been called the rupture point (Cooper, 1962). This point is defined as the point of maximum force on the force-deformation curve. After this point the force will decrease in an erratic pattern. This point can be observed visually in the material as a failure in the structure. In processed apple plugs this observed as a shear, or a localized, failure. Associated with the rupture point are rupture stress, rupture strain and rupture energy.

Rupture stress is defined as the maximum stress the product can resist without failure, and the corresponding strain is the rupture strain (Mohsenin, 1968).

The final parameter determined from the force-deformation curve was the modulus of elasticity ('elastic modulus', 'Young's modulus' or 'modulus'). For materials that exhibit linear force-deformation behavior, it is the slope of the initial portion of the curve or the ratio of force to deformation in the elastic portion of the curve. In non-linear behavior, modulus can be defined as an apparent modulus or an initial tangent modulus, secant modulus or tangent modulus. The modulus is commonly referred to in the food industry as the stiffness, firmness or rigidity of the material.

For this study the modulus measured and reported was taken as the tangent modulus which was the slope of the line drawn parallel to the linear portion of the curve which normally appeared at a point located at $\frac{2}{3}$ maximum stress.

As expected, the rupture parameters were sensitive to the time of heat treatment the apples received. Preliminary tests were run at the start of the program to determine an optimal firmness that would be applicable for future studies. Recommendations for optimal process times vary widely depending on the recommending source, type of apple and condition of the fruit. In general, the recommendations call for a precook in a 30% sucrose solution at 212°F for 1–5 min and a process in quart jars placed in boiling water for 5–25 min. The duration of heat treatment seems to depend on the textural level of firmness desired, the original degree of ripeness and the size container. Most processed fruits and vegetables seem to have a similar firmness, so the final duration of heat treatment selected was one that gave a processed apple at a stress level equivalent to commercially available processed apples.

Figure 2 shows the relationship of the rupture parameters to the total time that Baldwin apples were subjected to a heat treatment of 212°F. The product test temperature was held constant at 72°F and the loading rate remained at 0.385 in. min^{-1}. The total heat treatment included both the precook and process time. The graph was plotted from data obtained from 24 replications at each of 5 different levels of heat treatment.

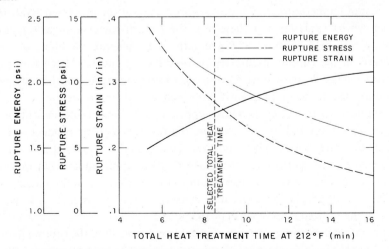

Fig. 2.   Rupture strain, stress, and energy for processed Baldwin apples subjected to varying total heat treatment times at 212°F. (Rate of loading 0.286 in. min^{-1}.)

As the total heat treatment increased, the rupture force and rupture energy decreased. This indicated that the longer the apple is subjected to a heat treatment, the more susceptible it is to physical damage. Likewise, as the total time that the apple was exposed to high temperatures increased, the rupture strain increased at the linear rate.

Density changes were also experienced by the processed apple during heat treatment, but the changes within the time range were not significant. In general, the longer the sample was exposed to heat the higher was its specific gravity or density within the limits of 0.85 gm cm^{-3} and 0.93 gm cm^{-3}.

The phenomena of rupture parameter change due to total heat exposure would warrant further study under more controlled conditions.

Processed Baldwin apples exhibited a slightly viscous relationship when subjected to temperatures within the range that they might experience during normal handling and distribution situations. The limits of specimen temperature for this experiment were from 35°F to 95°F. The samples had been stored 180 days at 36°F prior to testing. Any situation where this range is exceeded would be abnormal and needs not to be considered.

TABLE IV

The effect of temperature on the
rupture parameters of processed Baldwin apples

Temperature °F	Rupture stress (psi)	Rupture strain (in. in^{-1})	Rupture energy (psi)
35 ± 1	12.42	0.228	2.02
50 ± 1	11.78	0.247	1.97
72 ± 1	11.68	0.247	1.92
95 ± 1	10.62	0.272	1.84

The effect of temperature on the rupture properties of the processed Baldwin apple plug is shown in Table IV. For these tests the ¼ in. diameter plunger was used and the loading rate held constant at 0.4 in, min^{-1}. In general, the effect of specimen temperature on the rupture parameters is of some significance. Both rupture stress and rupture energy decrease linearly with increased temperature of the test specimen, while rupture strain increased linearly with increased temperature.

Since damage to processed products appears to be heavily dependent on shock loading or force level and time of loading, the resistance of the product to forces at various loading rates seems to be important. To obtain this fundamental information, a series of tests were made on the processed Baldwin apple under what we call compression plunger and impact plunger loading. Compression plunger loading using a ¼ in. plunger, utilized plunger movement rates from 0 to 60 in. min^{-1}, while impact plunger loading utilized rates from 24 to 135 in. s^{-1}.

Under compression plunger testing of processed pears, all of the rupture parameters generally increased as the strain rate increased within the strain rate range of 0.158 in. min^{-1} to 60 in. min^{-1}. Both rupture stress and rupture energy increased linearly with increased strain rate (Figure 3). The rupture strain behavior seems to be inconsistent, in that it does not consistently increase with the increased strain rate. The curve can best be plotted as a quadratic polynomial, but this type of curve is highly irregular for a viscoelastic material (Mohsenin, 1968). It is proposed that, because of the small change in strain due to strain rate, the inconsistencies result from textural variations or replication limitations rather than an actual behavioral trend.

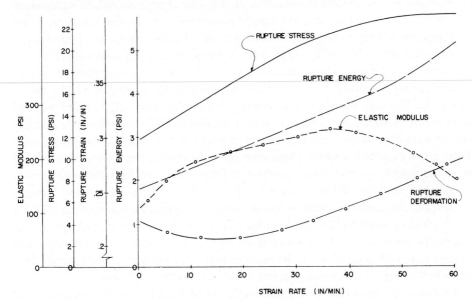

Fig. 3. Influence of strain rate upon the stress-strain parameters for processed Baldwin apples (slow rates of loading).

The textural variation between samples, as previously discussed, becomes more pronounced in the apple after processing. Throughout the data, the processed product generally has a higher coefficient of variation than does the fresh product.

These significant variations indicate, as studies involving all biological materials indicate, that numerous samples should be utilized in each treatment and that all information reported on biological materials should be in terms of ranges or trends. The high coefficients of variation are a common occurrence with biological materials, especially when dealing with the physical behavior of biological materials (Finney, 1963).

At the present time, scientists have not agreed as to whether force or energy should be used as the criterion for either a biological or non-biological material's resistance to damage. The processed material in this study had the properties of being able to absorb larger amounts of energy without physical breakdown, as well as being able to resist higher levels of force or stress without breakdown, as the rate of loading increased. Whichever criterion of damage one selects, the sample will withstand higher levels of stress with increased rate of stressing.

The elastic modulus increased as the strain rate increased within the loading rate range of 0.158 in. min^{-1} to approximately 36 in. min^{-1}. Within this range the increase was by a linear relationship. Unlike the rupture parameters, the elastic modulus exhibited a limiting point within the strain rate range used in compression loading. At a rate of loading of 33.5 in. min^{-1}, a change in slope or a reversal in the direction of the behavior of the property occurred with increased rate of loading. A more detailed investigation of this important occurence of behavior change should be undertaken.

Since the elastic modulus is a mechanical parameter that can, and has been, correlated with the subjective measurements of firmness (Finney, 1968), it is important to recognize the significance of loading on this variable. The elastic modulus is the most rate sensitive of the parameters measured. The level varied approximately 300% from a static loading rate of 0.158 in. min^{-1} to a relatively moderate rate of 33.5 in. min^{-1} per minute. This high percent change is due to the mathematical calculations required to get elastic modulus and does not seem to be due to any accumulated or additive error or textural variation since neither type of error is significantly different from some of the other parameters which experience less of a change with increasing rate. If the elastic modulus is highly rate sensitive, it follows that any firmness measurement is highly rate sensitive. When tests are being made to determine the relative firmness of a soft product, such as most processed fruits and vegetables, it is important that a special effort be made to hold constant the rate at which the force is applied and that all tests that are to be correlated with one another be subjected to stresses at a common stressing rate. Unless this is done, the experimental variation will be much greater than the biological variation, causing the test results to be of limited value.

The elastic modulus had the highest coefficient of vatiation within the treatments of the mechanical parameters. This can be taken either as a phenomenon that enhances the importance of elastic modulus determination or a phenomenon that hinders the valid data that can be readily applied to practical problems. It may be important in determining significant biological variation within a given set of samples and thus may be the most meaningsful measurement that one can take. Conversely, it may be taken as a property that is too variable and difficult to use, without making an exorbitant

TABLE V

Mean values of the results of the rate of loading tests on processed Baldwin apples during compression plunger testing. Total heat treatment time 14 min. (10 samples per test)

Rate of loading (in. min^{-1})	Plug length (in.)	Rupture stress (psi)	Rupture strain (in. in^{-1})	Rupture energy (psi)
0.286	0.60	5.60	0.310	1.36
2.00	0.60	6.10	0.316	1.38
10.00	0.58	7.34	0.324	1.61
24.00	0.60	9.62	0.328	1.96
48.00	0.60	10.31	0.337	2.40
60.00	0.60	9.86	0.352	2.10

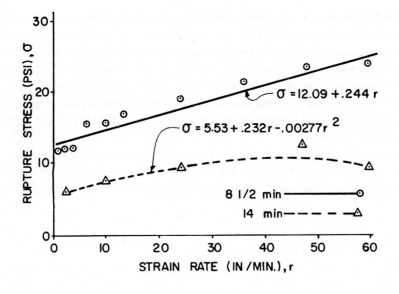

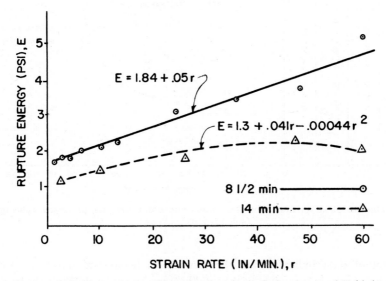

Fig. 4. Influence of strain rate upon rupture stress and energy for processed Baldwin apples at different total heat treatments.

number of replications in determining significant trends or sample correlations. Since this is a study of resistance to physical damage and not one of material firmness measurement, a detailed study of this contradiction was left for future investigations.

The strain rate effect on the rupture parameters for processed Baldwin apples for samples processed a total of 14 min are shown in Table V. A graphical representation of these data for rupture stress and energy compared to the apple samples processed $8\frac{1}{2}$ min is shown in Figure 4. Although a limited number of different strain rates were

evaluated for the samples processed 14 min, it is apparent that the same general trends
exist for both heat treatments, although at different stress and energy levels. While
the apple that was processed the longest exhibited the lowest rupture stress and rupture energy levels at every strain rate, the trends were generally one of increasing
levels with increased rate of loading. Up to 40 in. min^{-1}, a linear relationship occured.
It must be noted that somewhere around a 40 in. min^{-1} loading rate, a critical point
existed. At this point there appears to be a change in the slope of the curve with
respect to rate of loading.

The rupture strain increased with increased strain rate throughout the test range
for each heat treatment. While the behavioral trends were similar, the samples that
were subjected to longer heat processing exhibited higher levels of rupture strain
throughout.

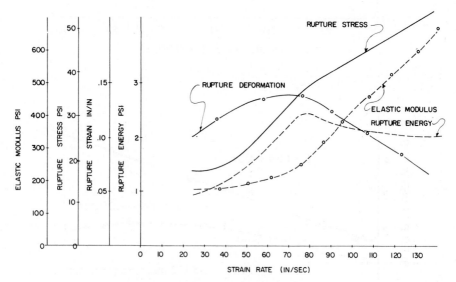

Fig. 5. Influence of strain rate upon the stress-strain parameters for processed Baldwin apples
(high rates of loading).

This set of data indicates that the mechanical behavioral trends of processed
Baldwin apples are independent of the total heat treatment level. While the level of the
parameter at any given strain rate varies, it follows the general pattern of increasing
with increased rate of loading. The only contradiction to this general statement may
be in the presence or position of a critical point where trend inconsistencies were
indicated.

At impact plunger loading the rupture stress increased as the rate of loading increased
within the strain rates reported (Figure 5). This relationship for rupture stress was
found to be linear. The behavior characteristics did not increase at the same rate as
during the compression loading test, and the initial stress at 25 in. s^{-1} was less than
the final stress at 60 in. min^{-1}. This indicated that, within the range of 60 in. min^{-1}

and 25 in. s^{-1}, a trend reversal occurred, and that the rupture stress behavior experienced a characteristic critical point.

The rupture strain and the rupture energy, due to its dependency on the strain, are inconclusive, but they seem to be experiencing another trend reversal. Apparently, the rupture strain experiences at least two critical points where behavior reversals occur, although the possibility exists that this is, again, a severe textural variation.

The textural variations between samples become more pronounced in the apple after processing although not as significant at the high rates as at the low rates. The variations for the fresh product are of similar magnitude for high rates of loading while the processed product has, in general, lower variations at high rates than it does at the low rates of loading. This seems to indicate that the processed product is less texturally rate sensitive than the fresh product under high rates of loading since the errors measured were not significantly different between the compression and impact plunger tests.

Apparently, beyond a free fall drop height of six inches, the resistance of the processed Baldwin apple to rupture continues to increase with increased strain rate, thus making it more resistant to damage by impact than by quasi-static or low rate compression loading. This is a significant factor that must be considered when designing equipment and systems that handle processed biological materials. Impact loading, which is, perhaps, the most common type of stress placed on the body during physical handling, causes such observable damage as cracks, splits, permanent deformation and structural collapse.

The trend reversal characteristic of the rupture strain and energy curves interjects a new consideration. Unlike the trend at lower rates of loading, a general increase in the material's ability to absorb energy or to resist deformation as the rate of loading increases cannot be expected. This indicates that there is a logical rate of loading level which should not be exceeded due to additive physical damage considerations. The specific value of this rate would have to be determined from numerous tests concentrated in the critical region of any given product.

The elastic modulus increased as the strain rate increased. Unlike the rupture strain or energy, the elastic modulus exhibited no limiting factor within this range of strain rates. Like the elastic modulus at slow loading rates, this firmness index (Finney, 1968) exhibited a high degree of rate sensitivity as well as a high level of variation among individual samples. Again, this shows that rate of loading must be considered when using any measurement involving firmness and its correlation to other parameters or indices of quality. The elastic modulus behavior under increasing rates of loading is consistent with normal viscoelastic theory (Mohsenin, 1968).

Other processed products were evaluated to a limited extent. Mean results of the effect of rate of loading on Northern Spy apples processed in the laboratory and commercially processed Bartlett pears, yellow cling peaches, carrots, and white potatoes are presented in Tables VI through X. Mechanical properties at various rates of loading from 0.4 in. min^{-1} to 60 in. min^{-1} were obtained.

TABLE VI

Mean values of the results of the rate of loading test on processed Northern Spy apples during compression plunger loading. (12 samples per test)

Rate of loading (in. min⁻¹)	Plug length (in.)	Rupture stress (psi)	Rupture strain (in. in⁻¹)	Rupture energy (psi)	Elastic modulus (psi)
0.4	1.11	14.38	0.25	2.59	73.9
2.0	1.09	15.12	0.24	2.52	81.0
10.0	1.01	15.96	0.24	2.58	89.6
30.0	1.12	19.67	0.26	2.66	108.2
60.0	1.07	22.04	0.31	3.21	140.4

TABLE VII

Mean values of the results of the rate of loading test on processed cling peaches during compression plunger loading. (Processed by Sweet Life Brands, Inc., 12 samples per test)

Rate of loading (in. min⁻¹)	Plug length (in.)	Rupture stress (psi)	Rupture strain (in. in⁻¹)	Rupture energy (psi)	Elastic modulus (psi)
0.4	0.76	10.90	0.51	4.11	41.4
2.0	0.88	12.06	0.54	4.53	50.1
10.0	0.77	13.01	0.57	4.72	63.6
30.0	0.81	14.26	0.60	5.06	93.5
60.0	0.85	15.89	0.61	5.18	96.2

TABLE VIII

Mean values to the results of the rate of loading test on processed Bartlett pears during compression plunger loading. (Processed by Sweet Life Brands, Inc., 12 samples per test)

Rate of loading (in. min⁻¹)	Plug length (in.)	Rupture stress (psi)	Rupture strain (in. in.⁻¹)	Rupture energy (psi)	Elastic modulus (psi)
0.4	0.66	4.95	0.32	1.28	61.9
2.0	0.68	5.06	0.32	1.22	63.6
10.0	0.74	5.43	0.30	1.46	70.0
30.0	0.71	6.16	0.34	1.63	76.2
60.0	0.67	7.21	0.37	1.84	91.1

Like the processed Baldwin apple, all showed a strain rate dependency although the degree of change and location of the critical velocity (or lack of it), within the range used, was dependent on the material. No consistent factor other than the loading rate dependency could be observed.

In general, the trends of the processed Northern Spy apple were consistent with the trends of the processed Baldwin apple, indicating a possible consistency in the beha-

TABLE IX

Mean values of the results of the rate of loading test on
processed carrots during compression plunger loading.
(Distributed by Draper-King Cole, Inc., 12 samples per test)

Rate of loading (in. min^{-1})	Plug length (in.)	Rupture stress (psi)	Rupture strain (in. in.$^{-1}$)	Rupture energy (psi)	Elastic modulus (psi)
0.4	1.11	8.97	0.17	3.06	56.6
2.0	1.07	9.23	0.18	3.24	58.4
10.0	1.08	11.44	0.20	3.64	67.6
30.0	1.02	16.08	0.20	3.71	94.0
60.0	1.14	13.86	0.15	3.56	155.3

TABLE X

Mean values of the results of the rate of loading test on processed
white potatoes during compression plunger loading. (Distributed
by Sweet Life Brands, Inc., 12 samples per test)

Rate of loading (in. min^{-1})	Plug length (in.)	Rupture stress (psi)	Rupture strain (in. in.$^{-1}$)	Rupture energy (psi)	Elastic modulus (psi)
0.4	1.06	7.69	0.14	1.84	73.5
2.0	1.01	7.80	0.14	1.87	74.1
10.0	0.98	7.92	0.14	1.86	73.6
30.0	1.05	7.98	0.13	1.78	79.4
60.0	1.08	8.12	0.12	1.81	79.6

vior of processed apples. The level at any particular loading rate of the rupture stress
and elastic modulus for the Northern Spy was higher than for the Baldwin, but this
was primarily due to the difference in variety and ripeness.

Processed white potatoes exhibited a somewhat unique behavior under varied rates
of loading. Between the loading rates of 0.375 in. per minute and 60 in. min^{-1} the
potato seemed to exhibit little if any change in mechanical behavior with increased rate
of loading. This effect, which is somewhat of a contradiction to normal viscoelastic
theory, has been observed by Finney (1963) with raw white potatoes and Wright and
Splinter (1968) with raw sweet potatoes. This behavior for a high starch material
warrants fruther investigation.

On the basis of the results obtained from this study on the characterization of the
processed Baldwin apple through its mechanical behavior, the following conclusions
are justified:

(1) Processed Baldwin apples exhibit a non-linear stress-strain relationship. Their
behavior can be characterized as viscoelastic. A bio-yield point does not exist for
processed apples, the only distinct point observable on the stress-strain curve
being the rupture point, which is displayed when the product fails at some localized
point.

(2) The mechanical properties of processed apples associated with resistance to physical damage (rupture stress, strain and energy) and the property associated with firmness (elastic modulus) increase with increased rate of loading.

(3) Processed Baldwin apples have a critical rate of loading region at which the rupture parameters and the elastic modulus experience a rapid change in slope and direction as the rate of loading is further increased.

(4) The duration of heat treatment during thermal processing and the temperature of the product at time of testing will affect the rate dependency of the mechanical parameters evaluated in this study.

(5) Processed Baldwin apples exhibit a rate dependency similar to other processed fruits and vegetables.

The information reported here is, of course, very basic and raises more questions than are answered. The real purpose of this work is to determine how resistant a processed fruit or vegetable is to small impacts due to vibration, or large impacts due to dropping, similar to what they will experience during physical handling and distribution. It is hoped that the next few years will see significant advances in understanding the mechanical behavior of processed products, and how we can use this information to protect them from damage. It appears that the efforts for the future will or should fall into the following areas.

(1) Investigations into the basic mechanical behavior and the causes of physical damage for other processed fruits and vegetables, especially those that damage easily. Special attention should be given to those products that have a high problem of reduction in grade classification due to handling and distribution methods.

(2) The measurement of the actual impacts that a product is exposed to during physical distribution and not just the determination of the impacts to which the cans or jars are exposed.

(3) Investigations into what protection the syrup offers a product and what benefit syrup viscosity modification would have on the protection of the processed product to damage such as splits, partition or bruising.

(4) Development of a way that the resistance of a product to physical damage at impact loading levels can be predicted from simple, easy to obtain mechanical behavior at relatively slow rates of loading.

(5) Develop ways to package either the processed fruit and vegetables or the containers in which they are packed, from the vibrations and impacts that appear to cause the damage to the product. This, of course, could take many froms such as new containers, cushioning or strapping.

(6) Investigate the effect that the processing time or duration of cooking has on the ability of the product to resist damage produced by impacts and vibrations. The interesting question will have to be answered whether processing method changes that may make the product more resistant to damage will cause texture or consumer protection changes that would be unacceptable.

(7) Studies to determine if it would be beneficial to grade the product according to a property such as firmness, prior to processing, in order to minimize biological

variation after processing and thus increase the product's ability to resist damage, even within a single can.

The past published work in this area is very limited, but the problems are real, interesting and with a conscientious effort and scientific knowledge and information, these problems can be solved.

References

Anonymous: 1953, 'United States Standards for Grades of Canned Apples', USDA, Agricultural Marketing Service, Washington, D.C.

Cooper, H. E.: 1962, 'Influence of Maturation on the Physical and Mechanical Properties of the Apple Fruit', Unpublished M.S. Thesis. The Pennsylvania State University, University Park, Pennsylvania.

Finney, E. E.: 1963, 'The Viscoelastic Behavior of the Potato, Solanum Tuberosum, Under Quasi-Static Loading', Unpublished Ph.D. Thesis, Michigan State University.

Finney, E. E.: 1968, 'Mechanical Resonance Within Red Delicious Apples and Its Relationship to Fruit Texture', Presented at 1968 Annual Meeting of ASAE, Chicago, Illinois, December 10–13.

Finney, E. E.: 1969, 'To Define Texture in Fruits and Vegetables', *Agricultural Engineering* **50**, 462–465.

Fletcher, S. W.: 1969, 'The Mechanical Behaviors of Processed Apples', Paper presented at Winter Meeting of ASAE, Chicago, Illinois, December 9–12.

Fletcher, S. W.: 1970, 'Characterization of the Processed Apple Through Mechanical Behavior Studies Under Quasi-Static and Dynamic Loading Conditions', Unpublished Ph.D. Thesis, University of Massachusetts.

Griswold, R. M.: 1962, *The Experimental Study of Foods*, Houghton Mifflin Company, Boston, Massachusetts.

Hamann, D. D.: 1969, 'Dynamic Mechanical Properties of Apple Fruit Flesh', *Transactions ASAE* **12**, 170–174.

Hammerle, J. R. and Mohsenin, N. N.: 1966, 'Some Dynamic Aspects of Fruits Impacting Hard and Soft Materials', *Trans. ASAE* **9**, 484–488.

Hwang, C. T.: 1971, 'Damage to Canned Pears During Simulated Handling Procedures', Unpublished Special Problem. University of Massachusetts.

Jones, J. W.: 1961, *Tensile Testing of Elastomers at Ultra-High Strain Rates High Speed Testing*, Vol. II. Interscience Publishers, Inc., New York.

Mohsenin, N. N.: 1968, 'Physical Properties of Plant and Animal Materials. Parts I and II', The Pennsylvania State University, University Park, Pennsylvania.

Mohsenin, N. N., Cooper, H. E., and Tukey, L. D.: 1963, 'An Engineering Approach to Evaluating Textural Factors in Fruits and Vegetables', *Trans. ASAE*, **6**, 85–88, 92.

O'Brien, M.: 1964, 'Vibrational Characteristics of Fruits as Related to the Transit Injury', Paper presented at Annual Meeting of ASAE, Colorado State University, Fort Collins, June 21–24.

O'Brien, M. and Guillou, R.: 1968, 'An In-Transit Vibration Simulator for Fruit Handling Studies', Paper presented at Annual Meeting of ASAE, Logan, Utah, June.

Smock, R. M. and Neubert, A. M.: 1950, *Apples and Apple Products*, Interscience Publishers, Inc., New York.

Whittenberger, R. T.: 1951, 'Measuring the Firmness of Cooked Apple Tissues', *Food Technology*, January 17–20.

Wiley, R. C. and Lee, Y. S.: 1970, 'Modifying Texture of Processed Apple Slices', *Food Technology* **24**, 1168–1170.

Wright, R. S. and Splinter, W. E.: 1967, 'The Mechanical Behavior of the Sweet Potato Under Slow Loading Rates and Impact Loading', *Trans. ASAE* **12**, 765–770.

CHAPTER IX

ALTERATION OF APPARENT PHYSICAL PROPERTIES OF FRUIT FOR HARVEST

LESTER F. WHITNEY

Food and Agricultural Engineering Department, University of Massachusetts, Amherst, Mass., U.S.A.

1. Introduction

Fruit for the processing industry, for the most part, are harvested by mass removal techniques and handled in bulk as a matter of common practice. For the fresh market, fruit continues to be hand picked. Yet there are many obvious limitations to both methods and purposes linked to the physical properties of the fruit, which suggest the need for new approaches.

Trim losses from bulk handling and the in-plant labor required to perform the trimming operation result in greater costs to the consumer and less return to the farmer (Mohsenin, 1963). Mounting labor costs for hand picking is compounded by the rapidly decreasing seasonal labor supply with continued trends toward bulk handling. The overall problem is so serious that the continued availability of many fresh fruits at reasonable prices is threatened.

Several approaches are in the offing by many researchers: (1) Gentler mechanical mass harvesting and handling techniques. (2) Genetic alteration, such as thicker skinned tomatoes (Mohsenin *et al.*, 1965). (3) Electronic scanning of fruit surfaces with

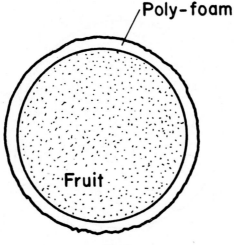

Fig. 1.

ChoKyun Rha (ed.), Theory, Determination and Control of Physical Properties of Food Materials, 181–196.

rejection of off-grade fruit for processing (Whitney and Lord, 1970). (4) Partial processing, such as peeled, sliced oranges in tubes of gelatin; *ad infinitum*.

In an effort to 'alter' the physical properties of apples, encapsulation with a polyfoam plastic of the growing matured fruit just prior to the shake and catch harvesting
operation has been conceived as a procedure to absorb the energies of impact and
reduce bruising.

2. Theoretical Analysis

The ideal treatment would be the complete encapsulation in a shell of structural design
so as to provide complete protection, with all energies of impact absorbed by the shell.

The problem has been analyzed by Ling (see appendix) in which a spherical shell
(inside radius r_1, outside radius r_2) is acted upon by two concentric forces, F, using
energy methods. The success of the analysis depends on correctly identifying the zero
stress components of a point which results in great simplification of the problem and
makes the energy method easier to apply.

The results of analysis presented in the appendix are the moment expressed as

$$M_0 = \frac{(r_1^2 + r_1 r_2 + r_2^2) F}{3\pi (r_2 - r_1) (r_1 + r_2)^2}.$$

Thus it can be shown that the maximum moment, M_0 is a function of the thickness
of the shell and the force applied. Using expressions for the elastic energy, i.e.,
Castigliano's theorem, and knowing the elastic modulus of the plastic foam material,
one can solve for the thickness of foam covering. For example, styrofoam requires
a thickness of approximately $\frac{1}{8}''$ to completely absorb the energies of impact for a 3 in.
diameter apple dropping six inches.

From a practical standpoint, complete coverage of the growing fruit with a uniform
thickness is out of the realm of possibility. Field studies on a limited scale have been

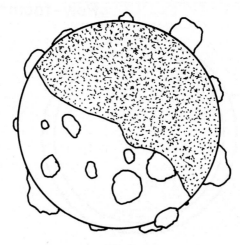

Fig. 2.

conducted with McIntosh apples on trees to determine the feasibility of such a practice. Using a hand-held spray gun, the material was found to form globules of varying sizes, shapes and percentage of surface coverage over the apple. Statistical determination of sizes, placement and shape are necessary to more nearly characterize the random conditions.

But dynamic analysis of the stack-up of a series of hemispheres of attached random sized plastic foam globules and spherical apples can be shown to yield appreciable absorption of the energies of impact, due largely to the cellular breakdown of the plastic foam, and not the elastic energy of the material.

A mathematical model of the system could have been developed with dashpots in series with the appropriate rheological model, such as the Burger model. However, the breakdown of the cellular structure of the plastic foam material was envisioned such that the apparent constants of the model elements would change appreciably during impact. This was not considered further in these studies.

3. Methods and Materials

Even with the relatively crude application method used, and the non-uniformity of the material, a significant reduction in bruises is indicated as determined by bruise analysis. A statistical analysis yields results shown in Figures 3 and 4, as reported previously (Wright and Splinter, 1967). Here, the criteria of interest is the number of U.S. Fancy grade fruit for apples established by the USDA and accepted by law in most states. Thus the feasibility is established and questions of quantity and cost become of interest. This relates to the optimum size range of globules which will effect the greatest energy absorption with the most economical procedure. The research reported herein is an initial experimental effort toward this end.

A spray-on plastic foam material has been tested – 'Insta-Foam (T.M.) as a Froth Pak'* The Poly-foam results from a chemical reaction of polymeric isocyanate with Polygol, which contains amines and fluorocarbons. These two liquid chemicals combine within a mixing chamber of the spray gun, and as it is applied to a surface, an instantaneous expansion of approximately 30:1 occurs with a corresponding instantaneous set. Thus the plastic foam achieves its optimum energy absorption properties immediately, and the tree can be shaken with the fruit harvested on catching frames.

For laboratory tests, individual globules were difficult to obtain and classify, so that wafers of various diameters were formed using a cork punch from selected thicknesses. Dimensional combinations were arbitrary based on observation of field application. Wafer sizes were selected to provide relationships of volume to energy absorption and protection afforded to fruit. Four sizes were investigated: $\frac{3}{8}''$ diameter $\times \frac{1}{2}''$ thick, $\frac{1}{2} \times \frac{1}{4}$, $\frac{3}{8} \times \frac{3}{16}$ and $\frac{1}{4} \times \frac{1}{8}$.

A single wafer was affixed to the topmost surface of a 3 in. diameter silicone rubber ball, onto which a flat plate (0.585 lb) was dropped (6''). The elastic modulus of the

* Manufactured by Insta-Foam, Inc., 880 S. Fiene Drive, Addison, Illinois 60101.

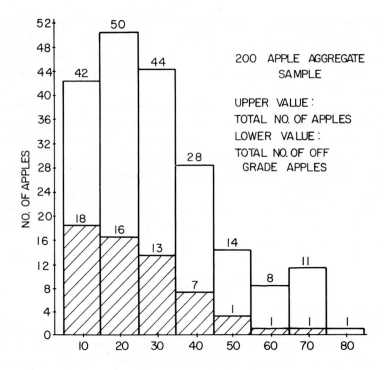

Fig. 3.   Percent coverage of spray on apple surface.

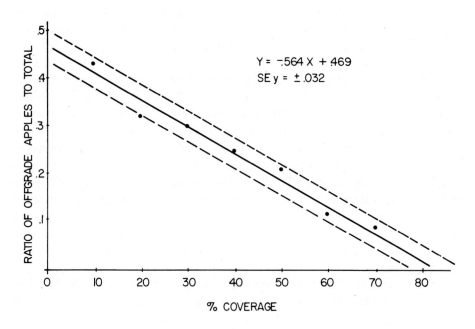

Fig. 4.   Ratio of off grade apples related to percent coverage of spray on poly-foam plastic.

silicone rubber was adjusted to that of a McIntosh apple at harvest – approximately 700 psi (Mohsenin *et al.*, 1965). The physical properties of this material permitted a reasonable modeling material so that plant related variables can be eliminated for repetitive drop tests. The size of the orb was commensurate with large apples which suffer the greatest damage on impact.

The test apparatus used was similar to that developed by Wright and Splinter (1967), Figure 8, in which a wire-guided drop weight was released by a magnet onto a test specimen which rested directly upon a force transducer (Quartz Load Cell, Kistler Model 912) with direct readout through a storage oscilloscope with auto erase (Tektronic Type R564B). Typical readouts are shown in Figure 6.

Moduli of the polyfoam material were determined at various rates of loading using a compression tester developed at PSU (5). Two conditions of the test material were selected: (1) as sprayed and set with a detectable skin, and (2) with the skin carefully removed with a sharp blade. A ¼ in., round plunger, cut off and turned square, was used as the penetrating member. Twenty compression tests were made for each of seven plunger speeds on two conditions of material, randomly selected for a total of 280 tests.

4. Results and Discussions

Upon dissection, this polyfoam as formed under field conditions by the crude method described earlier, exhibited a wide variety in cell size. This was borne out by the erratic behavior of the elastic moduli determination shown in Figure 5. No discernible relationship to strain rate was possible, yet this type of material can logically be expected

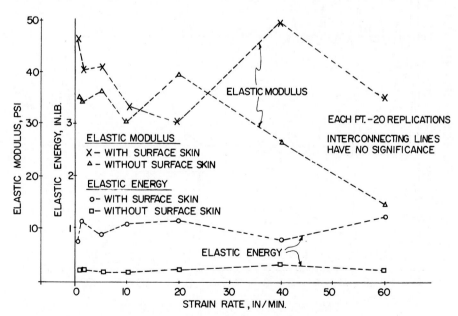

Fig. 5.   Influence of strain rate on stress strain parameters for poly-foam with and without surface skin effects.

to behave predictably. What is not shown is the large variances as revealed by statistical analysis, unacceptable for meaningful generalization.

Also, the effects of the skin as the material is formed can be expected to significantly affect the apparent modulus over that of similar material with the skin carefully cut away. With the exception of the 4th rate (20 in. min^{-1}) there appears to be a rough family of curves indicated, but again not well enough defined for conclusion.

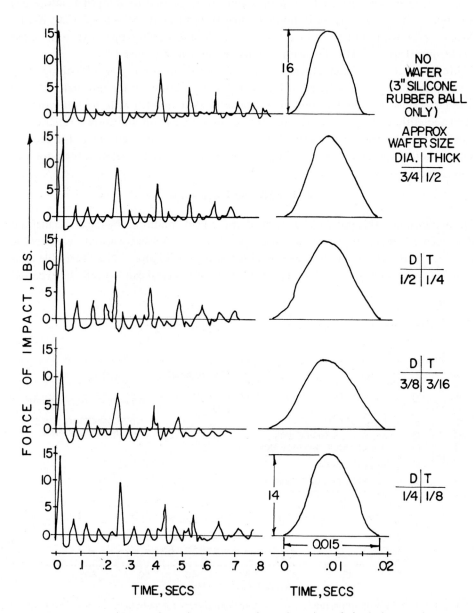

Fig. 6.   Typical impact force-time responses for various sizes of plastic foam wafers.

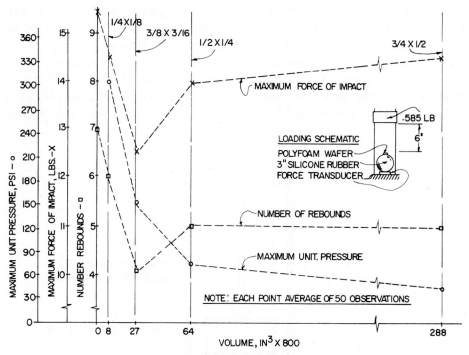

Fig. 7.   Effects of wafer volume on impact factors.

The elastic energies for all rates of strain were nearly constant for both material treatments, but with the exhibited inconsistencies, only workable ranges are suggested rather than a statistically derived regression curve. Here again, the polyfoam without skin appears as the better energy absorbent with one third of the elastic energy exhibited by the unaltered material. In other terms, cellular breakdown is much more likely for the skinless polyfoam, a desirable property as demonstrated further on.

The impact tests are more indicative, shown graphically in Figure 7 and as typical readouts of raw data in Figure 6. Two time scales were employed: (1) to determine the number of rebounds as an indication of secondary damage after impact, and (2) to indicate acceleration forces and characteristics. The silicone ball did not behave exactly as the apple, and the resulting forcetime responses are related to the natural frequency and resonance peculiar to the test specimen developed. However, the differences as indicated by the energy absorbed by the polyfoam wafer are of interest. A definite damping can be noted, seen clearly by the number of spikes of the left hand column of signals in Figure 6. The right hand column of signals at the faster time scale also indicates a variable time base which can be translated into forces of acceleration using the basic equations of motion. The time for deceleration was taken as one half of the time span, with the largest span calculable as the least force of impact. This can readily be seen by examining the force data for the $\frac{3}{8} \times \frac{3}{16}$ in. wafer in comparison with any other wafer or the silicone ball alone.

While single wafer test readouts are illustrated in Figure 6, the consolidated data are shown in Figure 7, which illustrates a definite break in the relationship for the wafer size mentioned above. In fact, both the number of rebounds and the force of impact can be shown to be significantly reduced, presumably as a result of the discernible, but not complete, crushing of the cell walls. Thus, for this combination of impact and material, an optimum size of wafer is indicated. Further tests for other specific materials and associated fruit drops are needed to make similar disclosures.

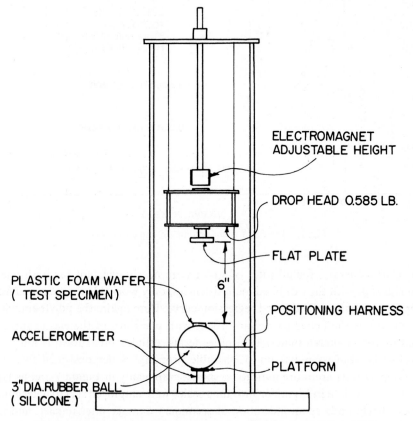

Fig. 8.   Schematic of impact drop apparatus (Wright and Splinter, 1967).

Complete crushing of the cells of the smallest wafer indicate that there is a lower limit which should be adhered to because of the lack of effectiveness in absorbing energy and the intolerably high unit pressures attained. Mathematical modeling would not necessarily reveal this phenomena.

5. Conclusions

These initial tests with polyfoam point the way to procedures which can lead to a feasible solution to reduce fruit damage significantly during the shake and catch harvesting operation. Further, any damages incurred in transport, storage, and handling can be correspondingly reduced. The material as used can easily be wiped off as globules by a soft brush during the pack-out as part of a polishing operation.

However, polyfoam as used would not be practical from a cost standpoint. It is not recyclable, further adding to the cost, and may accumulate on and under the fruit tree over a period of years. This of itself may not be undesirable, as any buildup on limbs would tend to reduce damage to fruit on impact.

The problem of material adhesion onto the fruit and instant set led to the original selection of this material. However, it may be possible to use techniques of 'spraying' a mixture of contact adhesive and appropriately sized wafers directly onto the fruit with modified conventional spray equipment. These techniques are already developed by commercial applicators of cellulose as an insulation material for buildings. The bulk use and reuse of appropriately sized plastic foam particles of suitable material could make the harvesting procedure economic.

References

Mohsenin, N. N.: 1963, 'A Testing Machine for Determination of Mechanical and Rheological Properties of Agricultural Products', Penn. State Univ., Exp. Sta. Bul. 701.

Mohsenin, N. N., Cooper, H. E., Hammerle, J. R., Fletcher, S. W., and Tukey, L. D.: 1965, 'Readiness for Harvest of Apples as Affected by Physical and Mechanical Properties of the Fruit', Penn. State Univ., Ag. Exp. Sta. Bul. 721.

Whitney, L. F. and Lord, W. T.: 1970, 'Energy Absorption of Plastic Foam on Apples', ASAE Paper No. 70-653. Presented at Chicago, Ill., Dec. 8-11, 1970.

Wright, R. S. and Splinter, W. E.: 1967, 'The Mechanical Behavior of the Sweet Potato under Slow Loading Rates and Impact Loading', *Trans. ASAE* **12**, 765–770.

APPENDIX. ANALYSIS OF SPHERICAL SHELL ACTED ON BY TWO CONCENTRIC FORCES IN OPPOSITE DIRECTIONS

by

CHI-YUAN LING and LESTER F. WHITNEY

1. Introduction

This is an attempt to analyze the problem of a spherical shell acted on by two concentric forces by energy method. The success of the analysis depends on correctly identifying some of the zero stress (moment) components of a point which results in great simplification of the problem and makes the energy method easy to apply.

2. Procedure

As shown in Figure 1, X, Y, Z axes intersect at the center of the sphere with radius r_1, r_2, the two forces are on the line of Z axis. It is obvious that the configuration,

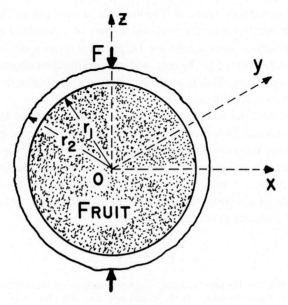

Fig. 1.

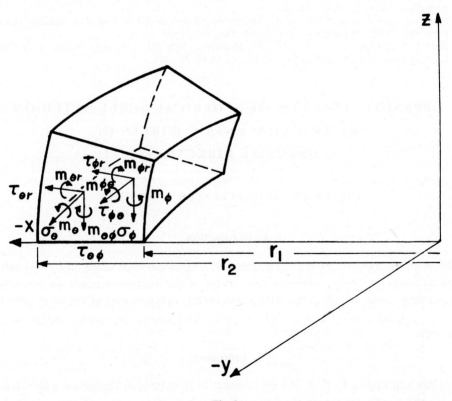

Fig. 2.

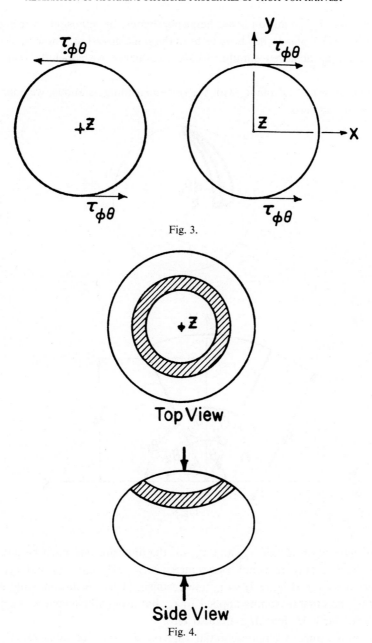

Fig. 3.

Top View

Side View
Fig. 4.

and hence the stress (moment), are symmetrical with respect to the XY plane, the Z axis, and any plane passing perpendicular to the Z axis.

A volume element with various force and moment components is shown on Figure 2. Suppose the element lies on XY plane*. By action and reaction, $\tau_{\phi r}$ acting on upper

* This condition is not necessary for below.

hemisphere and $\tau_{\phi r}$ acting on lower hemisphere must be oppositely directed. But by symmetry w.r.t. XY plane, they have to be in the same direction. These two conditions certainly can't be achieved simultaneously and hence we conclude that $\tau_{\phi r}$ must vanish.

By symmetry w.r.t. XZ plane, with the same reasoning as above, we conclude that $\tau_{\theta r}$, $\tau_{\theta \phi}$, m_{θ} are all zero.

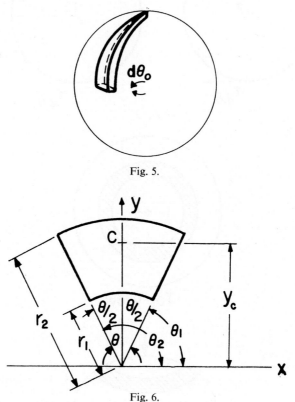

Fig. 5.

Fig. 6.

By symmetry w.r.t. Z axis $\tau_{\phi \theta}$ and $\tau_{\phi \theta}$ on the opposite side must be equal and in opposite direction (Figure 3) but by symmetry w.r.t. YZ plane $\tau_{\phi \theta}$ and $\tau_{\phi \theta}'$ must be in the same direction (Figure 3) so $\tau_{\phi \theta}$ must vanish. (This conclusion could have been obtained by the same reasoning applied to $\tau_{\phi r}$ but using a hemisphere is sufficient to prove it.) Similarly M_{ϕ} and $M_{\phi r}$ are also zero.

Furthermore, since a circumferential ring on the shell remains exactly circular (Figure 4), we conclude that $M_{\theta r} = M_{\theta \phi} = 0$.

Now we have only 3 components left, namely σ_{ϕ}, σ_{θ}, $M_{\phi \theta}$ which do exist for the present case.

To start with our analysis, we cut a slice from the sphere by two planes passing through the Z axis (subtracting $d\theta$), and take half of it as a free body, as in Figure 5.

We need to calculate the centroid of area of a sector: (Figure 6).

Area of the sector subtending an angle

$$\theta = \frac{\theta}{2\pi} (\pi r_2^2 - \pi r_1^2) = \frac{\theta}{2} (r_2^2 - r_1^2).$$

Let the distance from the centroid to X axis by y_c then

$$A y_c = \int_{\theta_1}^{\theta_2} y \, dA$$

$$\frac{\theta}{2} (r_2^2 - r_1^2) y_c = \int_{\theta_1}^{\theta_2} \int_{r_1}^{r_2} (r \sin \theta) r \, dr \, d\theta$$

$$= \int_{\theta_1}^{\theta_2} \int_{r_1}^{r_2} r^2 \sin \theta \, dr \, d\theta$$

$$= \int_{\theta_1}^{\theta_2} \left[\frac{r^3}{3} \Big|_{r_1}^{r_2} \right] \sin \theta \, d\theta$$

$$= \int_{\theta_1}^{\theta_2} \left[\frac{r_2^3 - r_1^3}{3} \right] \sin \theta \, d\theta$$

$$= \left[\frac{r_2^3 - r_1^3}{3} \right] (- \cos \theta) \Big|_{\theta_1}^{\theta_2}$$

$$= \tfrac{1}{3} (r_2^3 - r_1^3) (- \cos \theta_2 + \cos \theta_1)$$

$$= \tfrac{1}{3} (r_2^3 - r_1^3) \left[- \cos \left(\frac{\pi}{2} + \frac{\theta}{2} \right) + \cos \left(\frac{\pi}{2} - \frac{\theta}{2} \right) \right]$$

$$= \tfrac{1}{3} (r_2^3 - r_1^3) \left(\sin \frac{\theta}{2} + \sin \frac{\theta}{2} \right)$$

$$= \tfrac{2}{3} (r_2^3 - r_1^3) \sin \frac{\theta}{2}$$

$$\frac{\theta}{2} (r_2 + r_1) (r_2 - r_1) y_c = \tfrac{2}{3} (r_2 - r_1) (r_2^2 + r_2 r_1 + r_1^2) \sin \frac{\theta}{2}$$

$$y_c = \frac{4 \sin \theta/2}{3} \frac{r_1^2 + r_1 r_2 + r_2^2}{\theta} \frac{}{r_1 + r_2}$$

when $\theta = d\theta$. (This is the actual condition. Figure 5)

$$\sin \frac{d\theta}{2} \approx \frac{d\theta}{2} \left(\lim_{\theta \to 0} \frac{\sin \theta}{\theta} = 1 \right)$$

$$y_c = \frac{4}{3} \frac{\tfrac{1}{2}\mathrm{d}\theta_0}{\mathrm{d}\theta_0} \frac{r_1^2 + r_1 r_2 + r_2^2}{r_1 + r_2} = \frac{2}{3} \frac{r_2^2 + r_1 r_2 + r_1^2}{r_1 + r_2}$$

$$I_x = \int y^2 \, \mathrm{d}A$$

$$= \int_{\theta_1}^{\theta_2} \int_{r_1}^{r_2} (r \sin \theta)^2 \, r \, \mathrm{d}r \, \mathrm{d}\theta$$

$$= \int_{\theta_1}^{\theta_2} \left. \frac{r^4}{4} \right|_{r_1}^{r_2} \sin^2 \theta \, \mathrm{d}\theta$$

$$= \int_{\theta_1}^{\theta_2} \tfrac{1}{4} (r_2^4 - r_1^4) \sin^2 \theta \, \mathrm{d}\theta$$

$$= \tfrac{1}{4} (r_2^4 - r_1^4) \int_{\theta_1}^{\theta_2} \frac{1 - \cos 2\theta}{2} \, \mathrm{d}\theta$$

$$= \tfrac{1}{8} (r_2^4 - r_1^4) \left(\theta - \frac{\sin 2\theta}{2} \right) \bigg|_{\theta_1}^{\theta_2}$$

$$= \tfrac{1}{8} (r_2^4 - r_1^4) \left[(\theta_2 - \theta_1) - \tfrac{1}{2} (\sin 2\theta_2 - \sin 2\theta_1) \right]$$

$$= \tfrac{1}{8} (r_2^4 - r_1^4) \left\{ \theta - \tfrac{1}{2} \left[\sin 2 \left(\frac{\pi}{2} + \frac{\theta}{2} \right) - \sin 2 \left(\frac{\pi}{2} - \frac{\theta}{2} \right) \right] \right\}$$

$$= \tfrac{1}{8} (r_2^4 - r_1^4) \left\{ \theta - \tfrac{1}{2} \left[\sin (\pi + \theta) - \sin (\pi - \theta) \right] \right\}$$

$$= \tfrac{1}{8} (r_2^4 - r_1^4) \left[\theta - \tfrac{1}{2} (- \sin \theta - \sin \theta) \right]$$

$$= \tfrac{1}{8} (r_2^4 - r_1^4) (\theta + \sin \theta)$$

when $\theta = \mathrm{d}\theta_0$, $\sin \mathrm{d}\theta_0 \approx \mathrm{d}\theta$

$$I_x = \tfrac{1}{8} (r_2^4 - r_1^4) (\mathrm{d}\theta_0 + \mathrm{d}\theta_0) = \tfrac{1}{4} (r_2^4 - r_1^4) \, \mathrm{d}\theta_0 \, .$$

By the parallel-axis theorem

$$I_x = I_c + A y_c$$

$$\therefore I_c = I_x - A y_c$$

$$= \tfrac{1}{4} (r_2^4 - r_1^4) \, \mathrm{d}\theta_0 - \frac{\mathrm{d}\theta_2}{2} (r_2^2 - r_1^2) \left(\frac{2}{3} \frac{r_2^2 + r_1 r_2 + r_1^2}{r_1 r_2} \right)^2$$

$$= (r_2^2 - r_1^2) \left[\frac{r_2^2 + r_1^2}{4} + \frac{2}{9} \left(r_2^2 + \frac{r_1 r_2 + r_1^2}{r_1 + r_2} \right)^2 \right] \mathrm{d}\theta_0$$

$$= K \mathrm{d}\theta_0 \ \text{(where } K \text{ is a constant).}$$

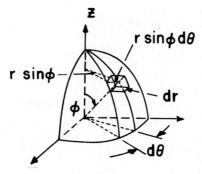

Fig. 7.

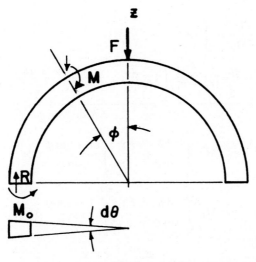

Fig. 8.

Since $d\theta$ varies with ϕ along the slice (Figure 7) and

$$d\theta = \frac{r_1 \sin\phi \, d\theta_0}{r_1} = \frac{r_2 \sin\phi \, d\theta_0}{r_2}$$

$$= \sin\phi \, d\theta_0$$

$$I_c = K \sin\phi \, d\theta_0 .$$

Now we are in a position to calculate $M_{\phi\theta}$. For simplicity we neglect the energy caused by σ_θ and σ_ϕ and consider bending energy caused by $M_{\phi\theta}$ only.* We take the slice as a beam, assume an M_0 of intensity M_0 acting as shown (Figure 8).

---

* Without this assumption, we can also analyze this problem with essentially the same procedure as above by curved beam method and using Poisson's ratio. Although it will be accurate, it is lengthy and very complicated. In fact, we can't solve it in closed form.

$$R = \frac{F}{\pi(r_2^2 - r_1^2)} \frac{d\theta_0}{2} (r_2^2 - r_1^2)$$

$$= \frac{F\, d\theta_0}{2\pi}$$

$$M = M_0 \frac{d\theta_0}{2} (r_2^2 - r_1^2) - R(y_c - y_c \sin\phi)$$

$$= M_0 \frac{d\theta_0}{2} (r_2^2 - r_1^2) - \frac{F\, d\theta_0}{2\pi} y_c (1 - \sin\phi)$$

$$\frac{\partial M}{\partial M_0} = \frac{d\theta_0}{2} (r_2^2 - r_1^2)$$

$$O = \frac{\partial U}{\partial M_0} = \int_0^{\pi/2} \times$$

$$\times \frac{\dfrac{d\theta_0}{2} (r_2^2 - r_1^2) \left[M_0 \dfrac{d\theta_0}{2} (r_2^2 - r_1^2) - \dfrac{F\, d\theta_0}{2\pi} y_c (1 - \sin\phi) \right] y_c\, d\phi}{E K\, d\theta_0 \sin\phi}$$

$$O = (r_2^2 - r_1^2) M_0 \int_0^{\pi/2} \frac{d\phi}{\sin\phi} - \frac{Fy_c}{2\pi} \int_0^{\pi/2} \frac{1 - \sin\phi}{\sin\phi}\, d\phi$$

$$= (r_1^2 - r_1^2) M_0 \int_0^{\pi/2} \frac{d\phi}{\sin\phi} - \frac{Fy_c}{2\pi} \int_0^{\pi/2} \frac{1 - \sin\phi}{\sin\phi}\, d\phi$$

$$= (r_2^2 - r_1^2) M_0 \log\tan\frac{\phi}{2}\Big|_0^{\pi/2} - \frac{Fy_c}{2\pi}\left[\log\tan\frac{\phi}{2}\Big|_0^{\pi/2} - \frac{\pi}{2} \right]$$

$$M_0 (r_2^2 - r_1^2) \log\tan\frac{\phi}{2}\Big|_0^{\pi/2} = \frac{Fy_c}{2\pi}\left[\log\tan\frac{\phi}{2}\Big|_0^{\pi/2} - \frac{\pi}{2} \right]$$

$$M_0 = \frac{Fy_c}{2\pi(r_2^2 - r_1^2)}\left[1 - \frac{\pi/2}{\log\tan\phi/2\,|_0^{\pi/2}} \right]$$

$$= \frac{Fy_c}{2\pi(r_2^2 - r_1^2)}\left[1 - \frac{\pi/2}{\infty} \right]$$

$$= \frac{F\dfrac{2}{3}\left[\dfrac{r_2^2 + r_1 r_2 + r_1^2}{r_1 + r_2} \right]}{2\pi(r_2 - r_1)(r_2 + r_1)}$$

$$\boxed{M_0 = \frac{(r_1^2 + r_1 r_2 + r_2^2) F}{3\pi(r_2 - r_1)(r_1 + r_2)^2}}$$

CHAPTER X

SORPTION PHENOMENA IN FOODS:
THEORETICAL AND PRACTICAL ASPECTS

THEODORE P. LABUZA

Dept. of Food Science and Nutrition, University of Minnesota, St. Paul, Minn., U.S.A.

1. Definition of Water Activity – Water Content

The control of moisture content in the processing of foods by various techniques is an ancient method of preservation, but only in recent years has the physical-chemical basis of the methods been studied and understood. When food is preserved the first and foremost principle is the destruction of control of deleterious or pathogenic microorganisms. With respect to water, the methods of preservation that operate on this principle are based on the fact that the water is made unavailable for microbial growth. These methods include: salting and sugaring of foods whereby the water present is tied up in some way so as to be made unavailable; freezing by which process the liquid water is converted into a solid state and thus is removed from being available to the organisms; and drying by which process the water is removed directly by vaporization from the food material. In most of the foods after the processing, micro-biological growth is thus prevented during storage, however, chemical deterioration does occur to affect quality. As has been shown by Salwin (1959), Scott (1957) and reviewed by Labuza (1971) the relationship between the loss of quality of the food and the moisture content of the food is best represented by the term *water activity* or A_w.

The A_w defines the degree to which the water present in the food is tied up or bound and thus unavailable for certain reactions. From a physical chemistry point of view water activity is defined as: A_w is the equilibrium vapor pressure that the water in a food exerts (P_{H_2O}) divided by the vapor pressure of pure water (P_0) at the temperature of the food. Water activity also can be defined as the relative humidity of air $(\% \text{ RH})$ at which a food if held would neither gain nor lose moisture. In equation form this becomes:

$$A_w = \frac{P_{H_2O}}{P_0} = \frac{\% RH}{100}. \tag{1}$$

With respect to a food at different water activities a different amount of water is held or bound. This relationship is called the water sorption isotherm since it defines the moisture content (m) in equilibrium with the different values of A_w at constant temperature. Figure 1 shows a typical plot of these values in which moisture is calculated on a grams H_2O per gram dry solids basis. Thus, at 3 g H_2O per gram solids the moisture content on a wet basis is 75%. As can be seen in Figure 1, in most foods such as fruits, vegetables, meats, fish and liquids the water activity is very high and is not too different from that of pure water. It is not until the moisture content is reduced to less than 50% on a wet basis does the A_w get reduced. This lowering is due to many factors which will be discussed subsequently. Firstly, the methods of determination of the water sorption isotherm should be examined.

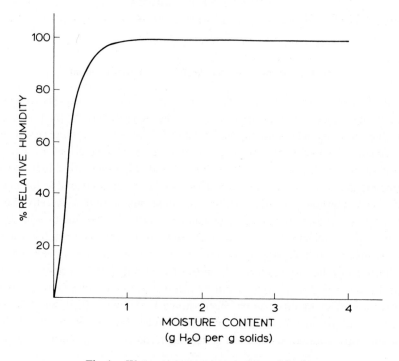

Fig. 1.   Water content – water activity of foods.

2. Measurement

Figure 2 shows the general shape of the water sorption isotherm for foods. It is just the expansion of the region in which A_w is lowered from Figure 1. As seen it represents the moisture content in equilibrium at each A_w. The curve also shows hysteresis, i.e., the equilibrium moisture at a given A_w depends on the direction the isotherm was made from. A higher m is usually observed as one does the experiments from the desorption direction by drying out the food as opposed to humidification.

In order to establish the isotherm both A_w and moisture content must be determined. Many methods are available; the important ones are discussed below. It should be noted, as the term isotherm implies, it is assumed that the values are established for the product at some constant temperature.

2.1. MOISTURE

Moisture content means the determination of the water held in the food. Since the water is bound by some energy either one must apply this energy to remove the water from the food and measure the weight change or measure the energy level by some electronic technique. Each method used gives a different basis making comparisons between various researchers sometimes difficult.

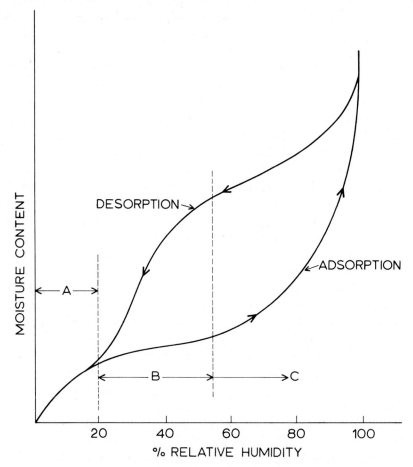

Fig. 2.   Typical sorption isotherm of a food showing hysteresis.

In terms of weight change techniques, most of the common methods involve drying the food at some temperature for a specified time period. The air oven technique is probably the most crude of these methods and consists of placing the product in a circulating (or noncirculating) air chamber at a relatively high temperature (about 100–200 °C) for 18 to 48 h. The A.O.A.C. has set specifications for various foods by this technique. The basic problem is that the high temperature can cause reactions which could change the weight of the food, volatiles may be lost, and the method is destructive to the sample. The method is also relatively slow; even after 24 h the weight may still be changing. The precision of the method is based on the precision of the weighing device used and one must weight both before and after. The accuracy will vary due to heterogeneity within the sample, the sample size, and the time used.

The vacuum oven technique overcomes some of the problems by using a lower temperature (60–70 °C) and drying the sample in a vacuum so that the rate of diffusion of water is faster. However, volatile loss would still occur reducing the accuracy of the technique. In industry infrared oven devices are used quite extensively for moisture determination. Many companies manufacture these devices which are basically enclosed balances on which the sample is tared and then dried in 1 to 20 min by a heat lamp. Usually the temperature is over 400 °F. This gives a tremendous error in accuracy, but as a daily quality control device for the same type of food such as meat emulsions it is useful. It should not be used for isotherm determinations.

A very accurate technique is to use freeze-drying at room temperature. It has the advantage of the lower vacuum to speed the rate of drying, but since a temperature of 20–25 °C is used to prevent volatile loss it takes at least 24–48 h. Again, the sample must be weighed before and after. The use of P_2O_5 desiccant is one of the most accurate techniques but takes a long time (5–7 days). It is based on the fact that P_2O_5 has a high absorptive capacity for water with a very low vapor pressure. The technique uses the low vapor pressure of P_2O_5 instead of heating the air to a low vapor pressure or using a condensor as in freeze-drying. Since there is no hot air circulating over the sample, or a vacuum continuously being pulled, little or no volatiles are lost making it the most accurate technique.

Other methods are based on extracting or volatilizing the water in the food by means of an organic solvent. One technique, toluene distillation, is based on mixing the food with toluene and then boiling the mixture. The water is removed at a lowered temperature by steam distillation with the toluene. The method is rapid but is not very accurate. It is used basically for quality control techniques. Much more accurate is the Karl Fisher Technique in which the food is extracted with methanol. The extraction is done in an oxygen-free atmosphere and the mixture is titrated with a reagent (sulfite) to measure the water. Many rapid technique devices have been built for this test which require about 20 min per sample. The basic theory is that the methanol replaces the water bound to the food, however, not all of it is replaced so the method is not as accurate as the P_2O_5 technique. However, it is fast even when done at room temperature. The other problems that occur however are interfering reactions of the reagents with food components and the extremely obnoxious odor of the sulfite reagent.

To eliminate these problems the methanol extract (after $\frac{1}{2}$ to 1 h) can be injected directly into a gas chromatograph using Pora-Pak Q as a column packing. The water is separated and detected by a thermal conductivity detector. This method is accurate and rapid and is being used by many researchers and companies.

Other methods are based on the physical-chemical properties of bound and free water. Sudhakar et al. (1970) has documented the use of NMR techniques for measuring water. Many other instruments devised for moisture content are measurement of dielectric strength, absorption spectroscopy, and infrared spectroscopy. These techniques besides giving some information as to the structure of water in foods can also be adapted to on-line process control of moisture content without destruction of the product. It should be remembered, however, that all these methods require that the instrument be calibrated by some other moisture measurement technique. Therefore any inherent error in the moisture determination is carried over into the instrument reading. A review of the various moisture determination techniques has been made also by Heiss (1968), Hofer and Mohler (1962), Gal (1967) and Stitt (1958).

2.2. WATER ACTIVITY

As with moisture content, many techniques both direct and indirect are available for the measurement of water activity. The oldest technique, that of the hair hygrometer was in fact reported by Leonardo da Vinci (*Hygrodynamics Technical Bulletin* #4). A hair being basically a protein will absorb water and change in both weight and length as the humidity increases. The hair can be tied to a pointer to indicate the degree of saturation of the vapor space in terms of % relative humidity. The device is very inaccuarte below 30% and above 80% RH but is used (usually with synthetic fibers) in home air conditioning units for humidity control. It is not accurate for research purposes.

A very accurate technique which cannot be applied to A_w measurement of small samples is wet bulb psychrometry. The method is described in detail by Van Arsdel (1963) and involves spinning two thermometers, one of which is immersed in a wet wick, in the vapor space. Based on the properties of the air the relative humidity can be determined. This technique was used by the U.S. National Weather Service. It is still used in the measurement of the % RH in large warehouses such as in potato sheds where there is a large air volume. The method is fast but cannot be applied to small samples. However, a dew point device employing the same principle can be used. In this case a surface is cooled and the temperature at which water condenses is measured optically. This then can be related to the A_w. Instruments available have an accuracy of $\pm 3\%$ RH.

One of the best direct measurements of A_w is the direct measurement of the vapor pressure of water in the vapor space surrounding the food by manometric techniques. Devices based on this have been described by Taylor (1961), Karel and Nickerson (1964) and Labuza *et al.* (1972). The device as shown in Figure 3 consists of a sample flask connected to a manometer. The system is thoroughly evacuated (to less than 200 μ Hg) with the vacuum unconnected to the sample. The vapor space around the food is then evacuated for less than two minutes to remove the gases present and then the stopcock across the manometer is closed. The whole system is kept at constant temperature and the food sample will lose water to reequilibrate with the vapor space. This will be indicated by the difference in height on the manometer. A low density, low vapor pressure oil should be used for maximum precision. The sample size to vapor space should be large so that a loss in water from the food will not change its A_w or moisture centent significantly. After equilibration, usually 40–60 minutes for a

5–10 g sample, the stopcock to the sample is closed, the one to the desiccant flask is opened and after 30 min the difference in oil manometer height indicates any leaks into the system as well as volatiles or gases lost from the sample. This system is extremely accurate ($\pm 0.002\ A_w$ units) as long as temperature is accurately controlled. Liquid samples usually should be frozen first and, therefore, require more equilibration time as they warm up.

Another technique which directly measures water in the vapor space is the use of gas-liquid chromatography (GLC). A sample is equilibrated in a jar, and an air sample

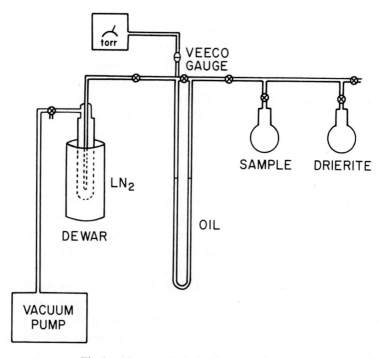

Fig. 3.   Manometric device for measuring A_w.

is treated similarly to the liquid methanol-water determination method. Unfortunately, at high A_w the suction and compression of the gas tight syringe can cause condensation of water vapor leading to inaccuracies.

Electrical devices have been designed to do the same thing indirectly. Schmidhofer and Egli (1972) have described the use of the Sina-Scope for A_w measurement of foods. This measures the absorption of water vapor on a filament and can be directly related to water activity in the vapor space. This device is expensive and is not used by many researchers. The most used devices are the relative humidity sensors based on electrical resistance. A probe which contains a filament coated with a salt is placed in a chamber containing the food. After equilibration, usually 1 h to 24 h, a current is passed through the filament and a readout is made on a dial. Depending on the probe used, the filament salts absorb different amounts of water and thus the resistance to current flow changes. A chart then is used to convert the readout to % RH or A_w. Although good to within $\pm 0.5\%$ RH when new, the probes age and become less accurate so they must be recalibrated. They also are subject to errors due to absorption of volatiles from the food such as glycerol. These types of instruments should not be used for accurate research work, but since they are relatively inexpensive, they are useful for quality control and on-line measurements.

The last direct method of A_w measurement is by measurement of freezing point depression as described by Strong et al. (1970). This can only be used for high A_w systems (low solute concentration) and not for solid foods. It is based on the Clasius Clayperon Equation for dilute systems and assumes the activity coefficient to be equal to one (Equation (2)). This method is used basically by microbiologists for studies of microbial growth as a function of A_w when mixed solute systems are used as media.

$$n_2 = \frac{\text{grams solvent}}{1000} \cdot \frac{\Delta T_f}{1.86} \tag{2}$$

$$A_w = \gamma \frac{n_1}{n_1 + n_2}, \tag{3}$$

where

$n_1 = $ moles of water in system,
$n_2 = $ moles of theoretical solute in system
$\Delta T_f = $ measured freezing point depression °C
$\gamma = $ activity coefficient $= 1$.

Because of either the expense of equipment or difficulty in operation most researchers use the indirect method of controlling water activity. This is done by using a chamber system in which the A_w can be directly controlled by some means. The moisture content after equilibrium is then measured rather than the water activity as it is a much simpler technique. The basic technique is the use of a saturated salt solution slurry. In a desiccator various salts at saturation give a definite A_w which varies little with temperature. Tables of salt solutions vs. A_w have been compiled by Wink and Sears (1950), Richardson et al. (1955), Rockland (1960) and in the *Hygrodynamics Technical Bulletin* #5. Table I lists some of the more common salt systems used. The method has some problems however:

(a) The salt must be saturated at the temperature of the experiment.

(b) There must be a large surface area and small air space volume available to the adsorbing or desorbing sample.

(c) The vapor space should be the same temperature as the liquid (a 2°F difference can, for example, lead to a 5% RH error).

(d) The salts and water used must be pure.

(e) The solution should be a slurry since if the sample loses water it will dissolve fresh salt. If the sample gains water the surface should not dry out.

(f) Salts are not available to give a point at every A_w desired.

To get around the latter problem both glycerol-H_2O and sulfuric acid-H_2O mixtures have been used. These allow for a continuous number of water activities but since at low A_w if the samples lose or gain water the constant humidity solution changes composition and therefore the A_w changes. This can be compensated for by using a large solution volume to sample volume and also measuring the specific gravity of the solution afterwards to determine the equilibrium A_w. Lastly, a mechanical chamber, of which there are many, can be used to provide a constant humidity. The better the

TABLE I

Typical salts used for
constant humidities 20 °C

Salt used	A_w
Desiccant	< 0.001
$LiCl_2$	0.12
$MgCl_2$	0.34
K_2CO_3	0.49
$Mg(NO_3)_2$	0.55
$NaNO_2$	0.65
NaCl	0.76
$CdCl_2$	0.82
K_2CrO_4	0.88
KNO_3	0.94
Na_2HPO_4	0.99

accuracy the more expensive the device with prices going over $ 5000. The latter devices are rarely used because a chamber would be needed for each A_w. With the salt solutions inexpensive desiccators or even jars can be used. Most researchers use vacuum desiccators which allow equilibrium in under 24 h. Thus, with a number of desiccators a complete isotherm can be made in one day. Some anomalous results can occur, however, if nitrite or halogen salts are used (Chou and Labuza, 1972).

3. Factors Responsible for Lowering of Water Activity

The physical-chemical factors responsible for the lowering of A_w have been reviewed extensively by Heiss (1967), Van Arsdel (1963), Labuza (1968) and Labuza (1971). The important principles should be reemphasized as it explains the fundamental interactions of water within a food in terms of storage stability.

3.1. SOLUTE-WATER INTERACTIONS

The basic factor lowering A_w when a solute such as sodium chloride is dissolved in water is the fact that these solutes associate with the water to form a hydration shell. Depending on the amount present, the availability or vapor pressure of water is decreased according to Equation (3). Most solutes, especially as concentration is increased, behave non-ideally and bind or structure water more than predicted. This is true for solutes of importance to the food industry such as salt and sugar. Bone (1969) has listed the actual A_w for various solute solutions and has pointed out the extreme non-ideal behavior of some large molecular weight molecules. Table II summarizes some of the typical A_w values for salt solutions. Sucrose and sodium chloride are ideal substances for use in making intermediate moisture foods (those with an A_w of 0.80 to 0.88), however, the amount that can be used is limited by their taste in the product. A mixture is usually used therefore.

TABLE II

Water activities of solutions of common
salt and sugar for use in
intermediate moisture foods

Salt	Solution concentration by weight	A_w
Sodium chloride	3%	0.98
	5%	0.97
	10%	0.93
Sucrose	5%	0.999
	10%	0.994
	20%	0.993
	40%	0.96
	60%	0.89

With any solid system a second important factor in lowering A_w is the capillary effect in a food. According to the Kelvin Equation (Equation (4)), the vapor pressure or

$$A_w = \exp\left[-\frac{2\gamma \cos\theta\, v}{rRT}\right],$$ (4)

where
  γ = surface tension of liquid in capillary,
  θ = contant angle,
  v = molar volume of liquid,
  r = capillary radius,
  R = gas constant,
  T = absolute temperature,
activity of a liquid present in a capillary is reduced as the radius of the capillary decreases. For pure water the values in Table III have been calculated. As seen, the effect becomes important only in capillaries of a radius less than 1000 Å. Bluestein and Labuza (1972) have shown that most of the capillaries in a food are of greater than 10 μ size but as water is removed the water present in small capillaries (<100 Å) comprises a significant amount of the total water. Thus, these capillaries do control partially the lowering of A_w. The actual contribution is difficult to assess, however,

since the Kelvin Equation is inconsistent at low capillary size (Shereshefsky *et al.*, 1950) and the values of γ and θ for pure water cannot be used in foods (Salas and Labuza, 1968; Labuza and Rutman, 1968; Labuza and Simon, 1969).

A third factor responsible for lowering of water activity can account partially for the fact that hysteresis occurs. A different path is followed depending on whether the isotherm is approached by adsorption or desorption. Labuza (1968) and Gregg and Sing (1967) have reviewed some of the basic reasons for hysteresis, however, another possibility not presented previously is the supersaturation of solutes as moisture is

TABLE III

Effect of capillary size
on lowering A_w

Radius		A_w
μ	Å	
0.1	1000	0.99
0.01	100	0.91
0.001	10	0.89

decreased, i.e., the solutes, as they lose water, form a glass that holds more water than expected and so not crystallize out at the true A_w of crystallization but must be dried to a lower A_w. This also suggests that the desorption branch is not the true equilibrium. This principle is illustrated in Figure 4.

The last and most important factor in controlling the vapor pressure of water is the interaction of water with solid surfaces as well as with high molecular weight colloidal systems. Water molecules usually interact with the polar groups on surfaces and are held very tightly. The energy to remove these water molecules is greater than the energy to vaporize a water molecule from the surface of pure water. If one starts from a dry state, a moisture content-A_w is approached where there is one water bound per polar group to form a monolayer (McLaren and Rowen, 1952). This occurs close to the inflection on the isotherm which gives a monolayer value of about 0.2 A_w to 0.3 A_w for most foods. Water above this monolayer is usually thought of as being the same as pure water, but long-range effects do occur which structure the water in such a fashion that it is also 'bound' to a certain degree to have a lower A_w. Ling (1965) has studied this effect in living tissues and Nemethy (1968) calculated the effect of large molecules such as proteins on structuring water in solutions. Duckworth (1972) recently demonstrated that large molecules by themselves can prevent water from freezing at temperatures down to $-20\,^\circ$C. This was shown to be directly related to the A_w lowering effect on the water.

In summary, there are many aspects of a food system which causes the lowering of the availability or activity of water. At normal tissue moisture these effects are very small, but as moisture is removed the ratio of solid to liquid increases and the water activity is greatly reduced. Many people have tried to quantitate the degree to which water is bound or free in a food, however, as shown by Labuza *et al.* (1970) and Labuza (1971) water even below an $A_w = 1$ still has the properties of that of bulk water. This water can still dissolve solutes, act as a medium for their mobilization, allow reactions to occur within the structure itself and be available as a reactant for reactions such as hydrolysis.

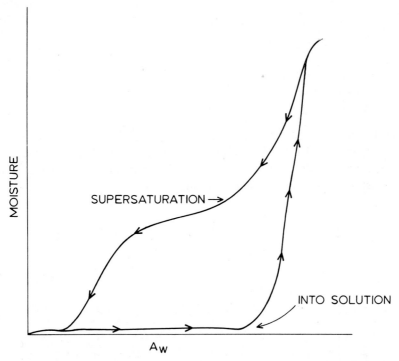

Fig. 4.   Water binding effect due to supersaturation of solutes and hydrate formation.

4. Isotherm Equations

Many theories have been developed to describe the shape of the sorption isotherm both from a theoretical and a mathematical standpoint. The need for an equation is obvious within food processing operations such as in drying, shelf life predictions, storage moisture content for maximum stability, etc. Adamson (1960) and Gregg and Sing (1967) have reviewed the theoretical basis of the major isotherm equations. Labuza (1968) has discussed the use of these equations within the food field. The basis of these equations will not be discussed here; the reader is referred to the above references. It should be noted that most equations are usually based on fitting a theory to a sigmoid curve and as Gregg and Sing (1967) have pointed out, most of the thories fall apart if one tries to apply them to the sorption of water. The major isotherm equations will be presented below.

4.1. BET ISOTHERM (Brunauer *et al.*, 1938)

The BET isotherm is the most popular isotherm of use in the food field as it gives very simply the value of the monolayer of water sorption. This is important since as shown by Salwin (1959), Rockland (1957) and Labuza (1971) the monolayer value of moisture is the moisture content at which the food is most stable. The basic equation was derived from the kinetic gas theory and has the form of:

$$\frac{A_w}{(1 - A_w)\,m} = \frac{1}{m_0 c} + \frac{(c - 1)\,A_w}{m_0 c},$$

(5)

where

m = moisture content at given A_w,

m_0 = monolayer value,

c = constant.

Plotting the left-hand side of Equation (5) vs A_w as in Figure 5 gives a straight line for values up to A_w 0.3–0.5. Above that the theory no longer holds, but since three or four isotherm points can be determined below the maximum using the saturated salt solution method this is quite useful for foods.

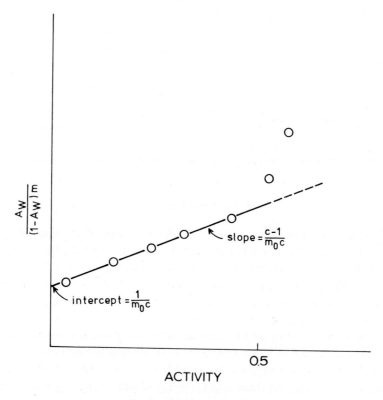

Fig. 5.   BET isotherm plot.

4.2. HARKINS JURA ISOTHERM (Harkins and Jura, 1944)

Many people have advocated the use of this isotherm since it is simpler than the BET equation. The basic form of the equation is as in Equation (6) which is based on the two dimensional gas theory.

$$\ln A_w = B - Am^{-2}, \tag{6}$$

where A and B are constants.

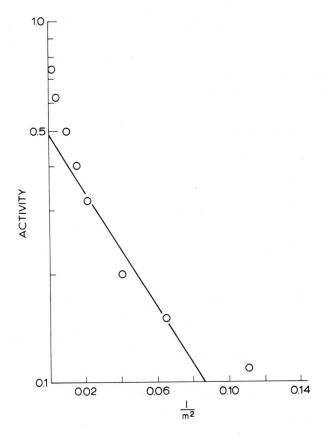

Fig. 6.   Harkins-Jura isotherm plot.

Plotting $\ln a$ vs m^{-2} should give a straight line (Figure 6) but in fact since the theory only applies up to a monolayer a curved line results. From the value of A (the slope of the line) the monolayer moisture can be determined by Equation (7). The basic problem is that most of those who use the

$$m_0 = 10^{-2} A^{1/2} \tag{7}$$

isotherm apply it over the whole range of the isotherm as in Figure 6 which is theoretically impossible (Kapsalis et al., 1968).

4.3. Capillary condensation isotherm (Zsigmondy, 1911)

Use of the Kelvin Equation (Equation (4)) has been advocated for the isotherm equation, however, all the problems discussed previously with respect to the properties of water and capillary size tend to make it unacceptable. Even of more difficulty is the fact that there has been no food for which a true pore size distribution has been determined.

4.4. Other equations

Several other equations have been used for isotherm description both based on theoretical as well as empirical derivations. Rockland (1957) has favored the Henderson Isotherm (Equation (8)) based on the kinetic gas law and Kuhn (1964) has derived a general isotherm (Equation (9)). Labuza *et al.* (1972b) have used a general linear equation (Equation (10)), a Pearson's Series Expansion as derived by Oswin (1946) (Equation (11)) and a purely statistical equation (Equation (12)).

$$\ln\left(1 - A_{\mathrm{w}}\right) = Am^B \tag{8}$$

$$m = \frac{A}{\ln\left(1/A_{\mathrm{w}}\right)} + B, \tag{9}$$

$$m = BA_{\mathrm{w}}, \tag{10}$$

$$m = A\left[\frac{A_{\mathrm{w}}}{1 - A_{\mathrm{w}}}\right]^B, \tag{11}$$

$$A_{\mathrm{w}} = \frac{A + m}{B + m}, \tag{12}$$

where A and B are constants derived by curve fitting.

The latter three equations are most amenable to use in mathematical solutions for drying and packaging because they can be easily integrated in unsteady state transport equations.

5. Practical Applications of Food Isotherms

5.1. Storage stability

As stated previously, the moisture content-water activity of a food can be used to predict the storage stability of a food. The basis for this has been reviewed by Labuza *et al.* (1970) and Labuza (1971) from the standpoint of the solvent properties of water and the degree to which it is bound in food.

The control of water content of a food is a basic food processing technique. It is based on the fact that the water content is decreased to a level to which microbial growth is prevented. This can be done by drying which completely removes the water or freezing which converts the liquid water into a solid state both of which make the liquid water unavailable for microbial growth. Water does not have to be completely removed, however, and as in the process of salting or sugaring which are age old processes, and in the new intermediate moisture food technology, chemical agents are added which bind the water to make it unavailable for microbial growth. This binding in fact is measured by the degree to which the A_w is lowered. Table IV lists the water activities below which common micro-organisms are inhibited in growth. Much work is needed in the area of the interaction of, for example, heat processing and A_w (Murrel and Scott, 1966), the interaction with pH, and the interaction with antimetabolites.

TABLE IV[a]

A_w growth minima for micro-organisms

A_w	Bacteria	Yeast	Molds
0.96	Pseudomonas		
0.95	Salmonella		
	Eschericia		
	Bacillus		
	Clostridium		
0.94	Lactobacillus		
	Pedioccus		
	Microbacterium		
0.93			Rhizopus, Mucor
0.92		Rhodotorula	
		Pichia	
0.90	Micrococcus	Saccharomyces	
		Hansenula	
0.88		Candida, Torulopsis	Cladosporium
0.87		Debaryomyces	
0.86	Staphylococcus		
0.85			Penicillium
0.75	Halophilic bacteria		
0.65			Aspergillus
0.62		Zygosaccharomyces	
0.60			Xeromyces

[a] adapted from Leistner (1970).

If microbiological problems are eliminated by control of A_w the storage life of a food becomes limited due to chemical reactions. The rates of these reactions can be predicted as a function of water acitivity of the food. For example, below the monolayer moisture content value, very few reactions can proceed which require the solubilization of reactants and an aqueous phase for reaction. This is because water is tightly bound at the monolayer and does not behave as bulk water (Duckworth and Smith, 1961).

Above the monolayer hydrolytic reactions increase with increasing A_w. Reactions of importance that have been studied are enzyme hydrolysis (Acker, 1969) chlorophyll degradation (LaJolla et al., 1972), sucrose hydrolysis (Schobell, 1969), anthocyanin destruction (Erlandson and Wrolstad, 1972) and ascorbic acid destruction (Karel and Nickerson, 1964). In all cases it can be shown that above the monolayer, the water, even though still having a low A_w, is available as a solvent, solubilizer and reactant.

Thus, for prevention of these reactions it is best to keep dehydrated foods as close to the monolayer A_w as possible.

A more complicated reaction that occurs in storage of dehydrated foods is that of non-enzymatic browning through the Maillard Reaction. This is a reaction of reducing sugars under the influence of either free amino acids or protein side chains leading to darkening, off flavor and loss of solubility of proteins. For long-term storage this also

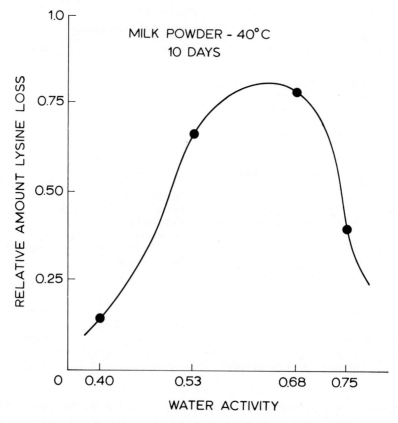

Fig. 7.   Lysine loss due to non-enzymatic browning of milk powder.

means a reduction in the biological value of the food since lysine, an essential amino acid, becomes tied up to the pigment produced in the reaction. Water plays a very important role in the non-enzymatic browning reaction in that it dissolves the substrates and mobilizes them for reaction. The reaction usually does not occur below the monolayer and proceeds at a linear rate at constant A_w. A maximum in rate occurs, however, at an intermediate moisture as is illustrated in Figure 7 from the data of Loncin et al. (1968) for lysine loss. This maximum is attributed to the fact that at some A_w a greater increase in moisture content causes a decrease in the dissolved solute concentration leading to a lowering of the browning rate. Eichner and Karel (1972)

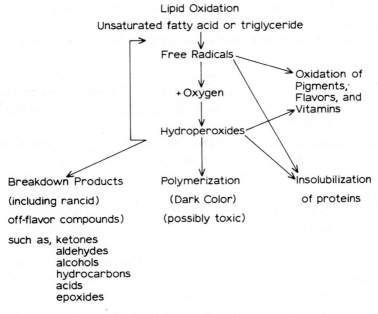

Fig. 8.   Lipid oxidation pathway.

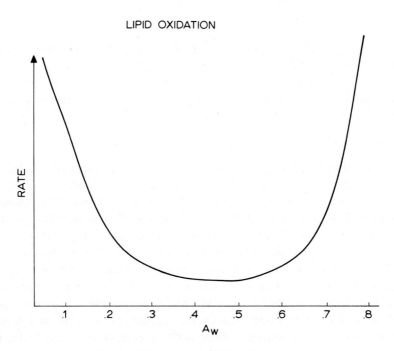

Fig. 9.   Effect of A_w on lipid oxidation rate.

has also shown this to be due to product inhibition by water since water is a product of the browning reaction.

Another major reaction limiting storage stability of dehydrated foods is lipid oxidation. The major pathways of lipid oxidation are shown in Figure 8 and the general influence of water activity is illustrated in Figure 9. As seen, moisture content once again plays a major role in controlling the oxidation of lipids with a minimum in rate occurring somewhat above the monolayer. Below the monolayer A_w, the rate increases rapidly as it also does in the intermediate moisture range. Since other reactions, as discussed above, also increase in rate above the monolayer, this A_w is indicated at the best condition for maximum storage stability of dehydrated foods.

With respect to lipid oxidation, the reason for the minimum curve (Figure 9) can be explained on the basis of the interaction of water in the system. Water has both antioxidant and prooxidant properties. At low A_w as moisture increases it hydrates metal catalysts making them less active and hydrogen bonds with the peroxides produced taking them out of the reaction. Thus, the rate decreases as A_w increases (Labuza et al., 1969). At a certain A_w above the monolayer, however, these catalysts become more mobile and even though they are less active their effective concentration is greater and thus the rate increases (Heidelbaugh and Karel, 1970). New catalysts may also be dissolved to enhance the reaction. In addition, swelling of the polymeric matrix of the food should open up new capillaries making more catalyst sites available for reaction (Chou et al., 1972). In some cases a maximum should occur in the intermediate moisture range if swelling is not important since increased moisture would dilute the metal catalyst concentration (NASA Contract Report 9–12560, November 1972).

Overall, as illustrated in Figure 10, a stability map can be drawn which directly relates the stability of a food to its A_w. Once this is known for a particular food and the isotherm is determined then as shown by Salwin and Slawson (1959) one can predict 'a priori' what would be the ideal combination of ingredients in making a dried food mixture such as for soup.

5.2. PACKAGING PREDICTIONS

A unique use of the isotherm equations and the storage stability map has been made by the research group from MIT in a series of publications (Mizrahi et al., 1970a, 1970b; Simon et al., 1970; Labuza et al., 1972; Quast and Karel, 1972; Quast et al., 1972) on the prediction of the shelf life of foods stored in semipermeable films. In these studies it was shown how to use reaction kinetics of deterioration as determined under steady state conditions and apply them to the condition where moisture is being slowly transported into a package. The food is thus continually increasing in moisture or A_w, if the instantaneous reaction rate for deterioration is known then the amount of deterioration can be summed up. The only other factor that has to be determined is the cutoff value. This is the amount of reaction that is allowed up to the point of unacceptability. Thus, the total time to reach that value can be predicted.

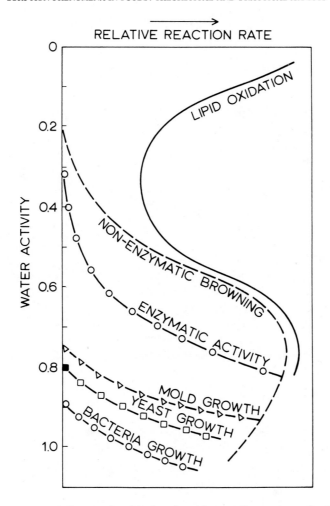

Fig. 10.   Stability map for dehydrated and intermediate moisture foods.

As a simple example one can take the case of moisture adsorption into a package of grain. One limiting acceptability factor would be the critical water activity at which mold would grow on the grain. This A_w could be related to the food through the isotherm to give a critical moisture content. The rate at which the package of food picks up water is:

$$\frac{dw}{dt} = \frac{k}{x} A (P_{out} - P_{in}),$$
(13)

where

    w = weight of water gained by package,
    k = package permeability to water,
    x = film thickness,

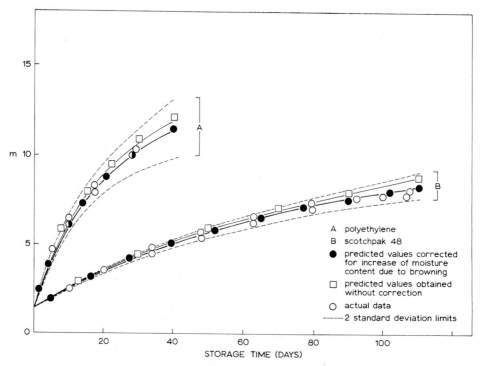

−Comparison of predicted increases in moisture content of samples stored in 2 types of packages, and experimental results.

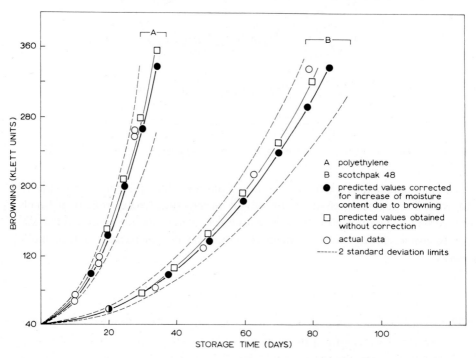

−Comparison of predicted increases in browning of samples stored in 2 types of packages, and experimental results.

Fig. 11.   Moisture and browning predictions for dehydrated cabbage.

A = film area,

P_{out} = outside water vapor pressure,

P_{in} = vapor pressure of water in equilibrium with the food.

If the assumption is made that the major resistance to water vapor flow is the film, then the water in the package should equilibrate rapidly. Thus, P_{in} can be determined from the isotherm. Simply, if one assumes a linear isotherm them:

$$m_{in} = BP_{in}/P_0, \qquad (14)$$

where m_{in} is the instantaneous moisture content of the food. Also, the value of P_{out} is related to the equilibrium moisture content of the food if left in the outside atmosphere with no package. Dividing both sides of Equation (13) by w_s, the weight of dry solids, and substituting for the linear isotherm we get:

$$\frac{dm}{dt} = \frac{k}{x} \frac{A}{w_s} \frac{P_0}{B} (m_e - m) \qquad (15)$$

which when integrated gives:

$$\phi = \frac{k\,A}{x\,w_s} = \frac{B}{P_0\theta_c} \ln \frac{m_e - m_i}{m_e - m_c},$$

where

m_c = outside equilibrium moisture content,

m_i = initial moisture content,

θ_c = critical shelf life desired.

In this form one can, by substituting in the moisture values, the isotherm slope (B) and the shelf life desired, get the value of ϕ necessary to reach those conditions. For a given weight of food (w_s) and package size (A) thus one could get the film permeability and thickness necessary for this. Then one could choose a film that does this at the least expensive cost. This methodology works and has been verified by a number of food companies.

For more complicated reactions such as browning the methodology referenced above should be studied. Figure 11 shows a study made by Mizrahi *et al.* (1970a), for example, in which both the predicted moisture increase and predicted browning come very close to the actual values.

6. Summary

The importance of the water activity-moisture content concept of foods cannot be over emphasized. This paper has tried to summarize some of the approaches to understand and measure these values. More importantly, the value of water activity has been shown to control the stability of dehydrated and semi-moist foods.

References

Acker, L. W.: 1969, 'Water Activity and Enzyme Activity', *Food Tech.* **23**, 1241.

Adamson, A.: 1960, *Physical Chemistry of Surfaces*, J. Wiley & Sons, New York.

Bone, D.: 1969, 'Water Activity: Its Chemistry and Applications', *Food Prod. Devel.* (August) 81.

Brunauer, S., Emmett, P. H., and Teller, E.: 1938, 'Adsorption of Gases in Multimolecular Layers', *J. Am. Chem. Soc.* **60**, 309.

Chou, H. E. and Labuza, T. P.: 1972, 'Anamolous Effects of Humidity Control', *J. Ag. Food Chem.* **20**, 1288.

Chou, H. E., Acott, K., and Labuza, T. P.: 1972, 'Sorption Hysteresis and Chemical Reactivity: Lipid Oxidation', *J. Food Sci.* **38**, 316.

Duckworth, R. B.: 1972, 'Properties of Water around Surfaces of Food Colloids', Dept. of Food Science. University of Strathclyde. Scotland.

Duckworth, R. B. and Smith, G.: 1963, 'Environment for Chemical Change in Dried and Frozen Foods', *Proc. Nutr. Soc.* **22**, 182.

Eichner, K. and Karel, M.: 1972, 'Influence of Water Content and Water Activity on the Sugar-Amino Browning Reaction in Model Systems', *J. Ag. Food Chem.* **20**, 218.

Erlandson, J. A. and Wrolstad, R. E.: 1972, 'Degradation of Anthocyanins at Limited Water Concentration', *J. Food Sci.* **37**, 592.

Gal, S.: 1967, *Die Methodik der Wasserdampf-Sorptions-Messungen*, Springer-Verlag, Berlin.

Gregg, S. J. and Sing, K. S.: 1967, *Adsorption Surface Area and Porosity*, Academic Press, New York.

Harkins, W. D. and Jura, G.: 1944, 'Vapor Adsorption Method for Determining the Area of a Solid without Assumption of a Molecular Area', *J. Am. Chem. Soc.* **66**, 1366.

Heidelbaugh, N. D. and Karel, M.: 1970, 'Effect of Water Binding Agents on Catalyzed Oxidation of Methyl Linoleate', *JAOCS* **47**, 539.

Heiss, R.: 1967, *Haltarbeit und Sorptionsverhalten wasserarmen Lebensmittel*, Springer-Verlag, Berlin.

Hofer, A. A. and Mohler, H.: 1962, 'Zur Aufnahmetechnik von Sorptionsisothermen und ihre A_w wendung in der Lebensmitteln Industrie', *Trav. Chim. aliment. et d'Hygiène*, **53**, 274.

Hygrodynamics Tech. Bull. #4, 'Hair Hygrometry', Hygrodynamics Inc., Silver Spring, Maryland.

Hygrodynamics Tech. Bull. #5, 'Creating and Maintaining Humidities by Salt Solutions', Hygrodynamics Inc. Silver Spring, Maryland.

Kapsalis, J. G., Wolf, M., Driver, M., and Walker, J.: 1968, 'The Moisture Isotherm as a Basis for Study of Sorption and Stability Characteristics in Dehydrated Foods', *Proc. 16th Res. Conf. AMIF*, Chicago.

Karel, M. and Nickerson, J. T. R.: 1964, 'Effects of Relative Humidity, Air and Vacuum on Browning of Dehydrated Orange Juice', *Food Tech.* **18**, 104.

Kuhn, I.: 1964, 'A New Theoretical Analysis of Adsorption Phenomena', *J. Colloid Sci.* **19**, 685.

Labuza, T. P.: 1968, 'Sorption Phenomena in Foods', *Food Tech.* **22**, 15.

Labuza, T. P.: 1971, 'Properties of Water as Related to the Keeping Quality of Foods', in *Proc. 3rd Int. Cong. Food Sci. and Tech.* p. 618.

Labuza, T. P. and Rutman, M.: 1968, 'Effect of Surface Active Agents on Sorption Isotherms of a Model System', *Can. J. Chem. Eng.* **46**, 364.

Labuza, T. P. and Simon, I.: 1969, 'Surface Tension of Food Juices', *Food Tech.* **23**, 96.

Labuza, T. P., Tsuyuki, H. and Karel, M.: 1969, 'Kinetics of Lipid Oxidation in Model Systems', *JAOCS* **46**, 409.

Labuza T. P., Tannenbaum, S. R., and Karel, M.: 1970, 'Water Content and Stability of Low Moisture and Intermediate Moisture Foods', *Food Tech.* **24**, 543.

Labuza, T. P., McNally, L., Gallagher, D., Hawkes, J., and Hurtado, F.: 1972a, 'Stability of Intermediate Moisture Foods', *J. Food Sci.* **37**, 154.

Labuza, T. P., Mizrahi, S., and Karel, M.: 1972b, 'Mathematical Models for Optimization of Flexible Film Packaging of Foods for Storage', *Trans. A.S.A.E.* **15**, 150.

Leistner, L.: 1970, *Arch. Lebensmittelhyg.* **21**, 121 and 264.

Lajolla, F., Labuza, T. P., and Tannenbaum, S. R.: 1971, 'Reactions at Limited Water Concentration. 2. Chlorophyll Degradation', *J. Food Sci.* **36**, 850.

Ling, G.: 1965, 'The Physical State of Water in the Living Cell and Model Systems', *Ann. New York Acad. Sci.* **125**, 401.

Loncin, M., Bimbenet, J., and Lenges, J.: 1968, 'Influence of Water Activity on the Spoilage of Foods', *J. Food Tech.* **3**, 131.

McLaren, A. D. and Rowen, J. W.: 1952, 'Sorption of Water Vapor by Proteins and Polymers: a Review', *J. Polymer Sci.* **7**, 289.

Mizrahi, S., Labuza, T. P., and Karel, M.: 1970a, 'Computer Aided Prediction of Food Storage Stability: Extent of Browning in Dehydrated Cabbage', *J. Food Sci.* **35**, 799.

Mizrahi, S., Labuza, T. P., and Karel, M.: 1970b, 'Feasibility of Accelerated Storage Tests for Browning of Cabbage', *J. Food Sci.* **35**, 804.

Murrel, W. G. and Scott, W. J.: 1966, 'The Heat Resistence of Bacteria Spores at Various Water Activities', *J. Gen. Micro.* **43**, 411.

Nemethy, G.: 1968, 'The Structures of Water and Aqueous Solutions in *Low Temperature Biology of Foodstuffs*', *Rec. Adv. Food Sci.* **4**, 1.

Quast, D. G. and Karel, M.: 1972, 'Computer Simulation of Storage Life of Foods Undergoing Spoilage by two Interacting Mechanisms', *J. Food Sci.* **37**, 679.

Quast, D. G., Karel, M., and Rand, W. M.: 1972, 'Development of Mathematical Model for Oxidation of Chips as a Function of Oxygen Pressure, Extent of Oxidation and Equilibrium Relative Humidity', *J. Food Sci.* **37**, 673.

Oswin, C. R.: 1946, 'The Kinetics of Package Life. III: Isotherm', *J. Chem. Ind. London* **64**, 419.

Richardson, G. M. and Malthus, R. S.: 1955, 'Salts for Static Control of Humidity at Relatively Low Levels', *J. Appl. Chem.* **5**, 557.

Rockland, L. B.: 1957, 'A New Treatment of Hygroscopic Equilibric', *Food Res.* **2**, 604.

Rockland, L. B.: 1960, 'Saturated Salt Solutions for Static Control of Relative Humidity between 5 and 40 °C. *Anal. Chem.* **32**, 1375.

Salas, F. and Labuza, T. P.: 1968, 'Surface Active Agents. Effects on Drying Characteristics of Model Food Systems', *Food Tech.* **22**, 1576.

Salwin, H.: 1959, 'Defining Minimum Moisture Contents for Dehydrated Foods', *Food Tech.* **13**, 594.

Salwin, H. and Slawson, V.: 1959, 'Moisture Transfer in Combinations of Dehydrated Foods', *Food Tech.* **13**, 715.

Schmidhofer, T. and Eoli, H.-R.: 1972, 'Zur Wasseraktivität von Fleischwaren', *Alimenta* **5**, 169.

Schobell, T., Tannenbaum, S. R., and Labuza, T. P.: 1969, 'Reaction at Limited Water Concentration. I: Sucrose Hydrolysis', *J. Food Sci.* **34**, 324.

Scott, W. J.: 1957, 'Water Relations of Food Spoilage Microorganisms', in *Adv. Food Res.*, vol. III, Academic Press, New York.

Shereshefsky, J. L. and Carter, C. D.: 1950. *J. Am. Chem. Soc.* **72**, 3682.

Simon, I. B., Labuza, T. P., and Karel, M.: 'Computer Aided Prediction of Food Storage Stability: Oxidation', *J. Food Sci.* **36**, 280.

Stitt, F.: 1958, 'Moisture Equilibrium and the Determination of Water Content of Dehydrated Foods', in *Fundamental Aspects of the Dehydration of Foodstuffs*, Soc. Chem. Ind., London.

Strong, D. H., Foster, E., and Duncan, C.: 1970, 'Influence of A_w on the Growth of *Clostridium perfringens*, *Appl. Micro.* **19**, 980.

Sudhakar, S., Steinberg, M. P., and Nelson, A. I.: 1970, 'Bound Water Defined and Determined at Constant Temperature by Wide Line NMR', *J. Food Sci.* **35**, 612.

Taylor, A. A.: 1961, 'Determination of Moisture Equilibrium in Dehydrated Foods', *Food Tech.* **15**, 536.

Wink, W. A. and Sears, G. R.: 1950, 'Equilibrium Relative Humidities above Saturated Salt Solutions at Various Temperatures', *Tappi.* **33**, 96A.

VanArsdel, W. B.: 1963, *Food Dehydration*, Avi Pub. Co. Westport, Conn.

Zsigmondy, R.: 1911, 'Über die Struktur der Gelsder Kieselsaver Theorie der Entwässerüng', *Z. Anorg. Chem.* **71**, 356.

CHAPTER XI

PROPERTIES CONTROLLING MASS TRANSFER IN FOODS AND RELATED MODEL SYSTEMS

MARCUS KAREL

Dept. of Nutrition and Food Science, Massachusetts Institute of Technology, Cambridge, Mass., U.S.A.

1. Introduction

A number of food operations involve mass transfer. Extraction, distillation, dehydration, infusion and leaching are among such operations, and water, salt, oils, proteins and flavors among food components most commonly involved in such transfer operation. In addition mass transfer is important in storage of foods since food stability often is determined by biological, chemical and physical processes dependent on transfer of various components notably water, oxygen and carbon dioxide.

Description of foods in terms of simple physical properties is often difficult and mass transfer is particularly difficult to treat in this manner, because:

(a) Many food components show complicated, non-ideal dependence of mass transfer properties on concentration, temperature and other parameters.

(b) Foods are often heterogeneous, and mass transport properties often show dependence on geometrical orientation of the food materials with respect to the directions of transport.

(c) In many processes of importance in food engineering, mass transport is coupled with heat transfer.

(d) In many instances the properties of foods change with time, and often this change is promoted by the mass transport of some components into the food.

2. Mass Transfer in Homogeneous Media

It is instructive, however, to begin our discussion with the consideration of mass transfer in homogeneous materials. If we consider unidirectional flow in such homogeneous materials we find that J, the flux of a component, is directly proportional to the local concentration and to the chemical potential gradient (Equation (1)).

$$J = -D(c/RT)\left(\frac{\mathrm{d}\mu}{\mathrm{d}Z}\right),\tag{1}$$

where

μ = chemical potential,
J = flux (moles) (cm^{-2}) (s^{-1}),
D = diffusivity $\mathrm{cm}^2\ \mathrm{s}^{-1}$,
c = concentration moles cm^{-3},

ChoKyun Rha (ed.), Theory, Determination and Control of Physical Properties of Food Materials, 221–250.

R = gas constant,
T = absolute temperature,
Z = distance in direction of flow, cm.

Under idealized conditions we assume that activity of component is equal to its concentration, and we can then simplify Equation (1) to Equation (2) which is Fick's First Law.

$$J = -D \frac{dc}{dZ}. \tag{2}$$

Of course analogous equations can be written for flow in two or three dimensions, but we shall limit ourselves to one-directional flow.

The unsteady state situation, in which the local concentrations change with time is given by Fick's Second Law (Equation (3)).

$$\frac{dc}{dt} = D \frac{d^2c}{dZ^2}, \tag{3}$$

where t = time.

Under ideal conditions therefore and in homogeneous solids, all that is required for analysis of mass transport is knowledge of concentration gradients and of diffusivities. Typical diffusivities of some components in selected liquids are shown in Table I. In most liquids and for relatively small molecules the diffusion coefficients are of the order of 10^{-6} to 10^{-5} cm^2 s^{-1}. Diffusivities in liquids depend on viscosity of liquid strongly, and ideally are inversely proportional to it. (Sherwood and Pigford, 1952). However, as we shall see later, deviations from ideal behavior are common in food systems.

TABLE I

Diffusivity of selected solutes at room temperature

Solute	Solvent	Diffusivity (cm^2 s^{-1})
Hydrogen	water	4.6×10^{-5}
Sulfuric acid	water	1.7×10^{-5}
Sodium chloride	water	1.1×10^{-5}
Acetic acid	water	9.5×10^{-6}
Ethanol	water	9.5×10^{-6}
Sodium acetate	Methanol	9.5×10^{-6}
Water	Peanut oil	2.5×10^{-6}

In gases the diffusivities depend on both temperature and pressure, and ideally follow the correlation:

$$D = (K_1) \left(\frac{T^{3/2}}{P_\theta} \right), \tag{4}$$

where

K_1 = constant depending on gas composition,
P_θ = total pressure.

Typical diffusivities in air at 25°C are:

carbon dioxide	0.16
ethanol	0.12
hydrogen	0.41
water	0.26

In gaseous diffusion, it is often advantageous to express driving forces in terms of pressure differences. For ideal gases we have:

$$p = RTc \tag{5}$$

and Fick's First Law becomes:

$$J = \frac{-D}{RT} \frac{dp}{dZ}. \tag{6}$$

When a component can be assumed to diffuse through a stagnant layer of another gas, the following equation applies:

$$J = \frac{DP_\theta}{RT} \frac{\Delta p}{\Delta Z} (P_\theta - \bar{p})^{-1}, \tag{7}$$

where

$\bar{p}$ = mean partial pressure of diffusing gas in layer ΔZ,

Δp = partial pressure difference of diffusing gas across distance ΔZ.

When partial pressure differences across ΔZ are relatively small, arithmetic average may be used for $\bar{p}$. When these become larger, logarithmic mean should be used.

In homogeneous *solids* the diffusion is very much more complicated. Diffusion in amorphous polymers for instance depends strongly on temperature and on polymer – penetrant interactions. In semicrystalline solids the diffusion depends on the volumetric ratio of crystalline to amorphous regions. The usual assumption made is that the crystallites are impermeable to diffusion and that only the amorphous fraction of the polymer matrix participates in diffusion (Michaels and Bixler, 1968). The diffusion coefficient for materials interacting with the polymer is usually strongly concentration dependent. In the case of foods, which contain hydrophilic polymers, water shows this concentration-dependence.

Food materials are more complex than the homogeneous systems discussed above, and there is less data available for them. Diffusivities, and permeabilities for water and water vapor have been studied in connection with dehydration processes and these will be discussed later. Less is known about diffusivities of other compounds. Sodium chloride diffusion into muscle of fish has been studied by Del Valle (1965) who found that the diffusivity was concentration-dependent. Typical data are shown in Table II.

TABLE II

Summary of measured and calculated diffusion coefficients for NaCl
in fish muscle (from Del Valle, 1965)

Temperature °C	Salt concentration in diffusion coefficient	
	Muscle, Mole l^{-1}	$\times 10^5$, cm^2 s^{-1}
5	0.0	0.65
	0.7	0.64
	1.4	0.67
	2.2	0.54
	2.7	0.72
25	0.0	1.25
	0.08	1.18
	0.7	1.05
	1.4	0.95
	2.3	1.37
	3.1	1.45

Rate of diffusion of salts, acid, and of sugars are also of critical importance in various processes of pickling, salting and curing. Unfortunately the processes are complex and there are no published correlations between effective diffusivities and process conditions. In most studies only penetration at selected conditions is reported, and no attempt made to analyze the mass transfer in terms of engineering properties. Data for cure diffusion in pork are given by Arganosa and Henrickson (1969), and some information on the extent of simultaneous extraction of water and infusion of sugar in *osmotic dehydration* is given in a study by Farkas and Lazar (1969). Osmotic dehydration by immersion of fruit pieces in concentrated sugar solution is an interesting mass transfer situation. Diffusivity of sugars is much lower than that of water, and as consequence it is possible to design processes which result in substantial water removal, with only marginal sugar pick-up. Slow processes, on the other hand, approach equilibrium for both sugar and water and result in production of 'candied' sugar-rich fruits.

3. Mass Transfer in Porous Solids

In *porous solids* we have the situation in which mass transfer can occur within the solid phase, or within the pores and channels, which are normally gas-filled.

Transport in solids may occur by diffusion in the solid itself, in which case Fick's Law Equation (2) may be used.

In porous solids, however, various mechanisms may be involved including:

(1) diffusion in the solid itself;

(2) diffusion in gas-filled pores;

(3) surface diffusion along the walls of pores and capillaries;

(4) capillary flow due to gradients in surface pressure; and

(5) convective flow in capillaries due to total pressure differences.

It is often convenient to establish an overall coefficient for flow in porous solids without attempting to establish the contribution of each of the above factors. One equation often used for this purpose is the following:

$$J = b \frac{dp}{dZ},$$
(8)

where

b = an overall permeability coefficient,

p = partial pressure of transported component

The permeability factor b is dependent on various conditions and may be related to diffusivity in the gaseous phase. Attempts have been made to relate the permeability factor b to various properties of the gas and of the solid. One such correlation is through Darcy's equation and the Kozeny Carman equation (Carman, 1956) applicable to laminar flow of gases:

$$J = - A \frac{\varepsilon^3}{(1 - \varepsilon)^2} \cdot \frac{\varrho}{K_k \cdot \sigma^2 \cdot \eta} \cdot \frac{dp}{dZ},$$
(9)

where

A = crossectional area,

ε = porosity of the solid,

σ = internal surface area of the solid,

K_k = a constant (in consistent units),

ϱ = gas density (in consistent units),

η = gas viscosity (in consistent units).

Unfortunately, studies in foods often show lack of correlation with the above equation, and mass transfer has to be characterized in terms of permeabilities or 'effective diffusivities' which are dependent on transport conditions.

4. Mass Transfer between Phases

Phases in contact are usually assumed to be in equilibrium at the plane of contact, that is, the interface. This assumption means that the controlling factor in transfer of components from one phase to another is the rate of transport to and from the interface. The analysis of transfer between phases requires, therefore, consideration of transport in each phase, and of equilibrium *between* phases *at the interface*.

4.1. GAS-LIQUID EQUILIBRIA

In equilibrium between liquids and gases, in the case of pure liquids the vapor pressure data are all that is required, and these are readily available in literature, but the relations are more complex in solutions and in actual foods.

The partial pressure of a solvent, for example: water, in equilibrium with an aqueous solution is lower than vapor pressure of pure water. An estimate of the pressure of solvent over a solution is given by Raoult's Law:

$$p = P^\circ x \tag{10}$$

or

$$y = \frac{p}{P_\theta} = \frac{P^\circ}{P_\theta} x, \tag{11}$$

where

   y = mole fraction of the solvent in vapor,

   x = mole fraction of solvent in liquid,

   P° = vapor pressure.

Raoult's Law is accurate only in predicting vapor-liquid equilibria for ideal solutions in equilibrium with ideal gases. It is most applicable to dilute solutions. At moderate and high concentrations substantial deviations may occur.

In certain systems where Raoult's Law does not apply, the relation between partial pressure and liquid phase composition may be estimated by Henry's Law:

$$p = Sx, \tag{12}$$

where S = a constant. (When $S = P^\circ$, Henry's Law is identical with Raoult's Law).

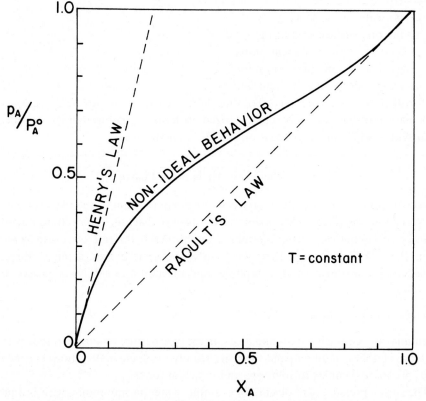

Fig. 1.   Dependence of partial pressure of a component on its mole fraction in liquid.

The relationship between partial pressure, composition and temperature may be represented graphically in different ways, each of which is useful for particular purposes.

At constant temperature, the partial pressure of one component (usually the more volatile component) may be plotted against its mole fraction in the solution. A directly related plot is that of the activity of the component against its mole fraction in solution (Figure 1).

$$a = p/p^\circ. \tag{13}$$

This representation is useful for isothermal processes and for evaluation of relative humidity in equilibrium with solutions.

For use in evaluation of evaporation and distillation, it is often more convenient to evaluate the relationships between gas phase and liquid phase composition at the boiling point, that is, at a temperature at which the sum of partial pressures of the two volatile components equals atmospheric pressure, usually arbitrarily taken at 760 mm Hg. A diagram relating boiling point to composition of gas and liquid phases of a

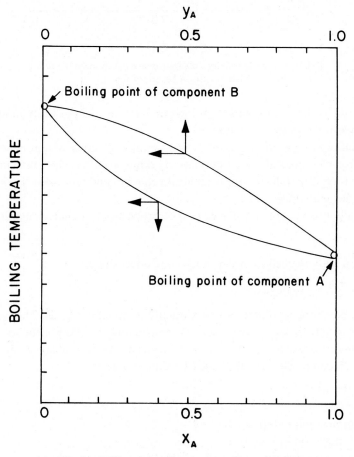

Fig. 2.   Boiling point diagram for a binary system.

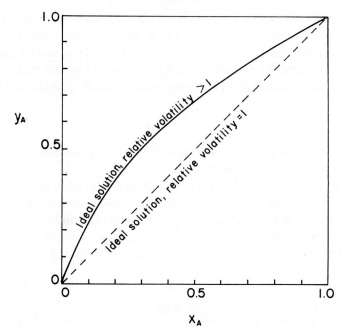

Fig. 3.   Relation of mole fraction in vapor phase (y) to mole fraction in liquid
phase (x) for a binary system.

binary mixture is shown in Figure 2. A diagram relating x to y at the boiling point for
a binary mixture is shown in Figure 3.

Most solutions of importance in food show nonideal behavior at high concentration,
and some behave nonideally even in relatively dilute systems. The solid curve shown
in Figure 1 is typical of deviations from ideal behavior, and represents 'positive deviation' from Raoult's Law.

Equation (14) is suitable for correlation of properties of nonideal solutions:

$$a = x\gamma, \tag{14}$$

where $\gamma =$ an activity coefficient dependent on concentration.

4.2. Liquid-Liquid Equilibria

Consider a component soluble in two mutually immiscible liquids. The two liquids,
when brought into contact, constitute two phases and the component will be transported from one phase to the other until the concentrations in phase 'L' and in phase
'$\varLambda$' will reach equilibrium levels defined by Equation (15).

$$C_L = K_p C_A, \tag{15}$$

where
  $C_L =$ concentration in liquid 'L',
  $C_A =$ concentration in liquid '$\varLambda$',
  $K_p =$ partition coefficient.

4.3. Gas-solid equilibria

Food dehydration and in general the preservation of foods by reduction of water activity depend critically on the equilibrium relationships between water vapor pressure in the atmosphere and the amount of water in the food.

Gas-solid equilibria are usually presented in the form of relations between partial pressure of a given component in the gas phase and its concentration in the solid phase. Three representations are often made, of which the isotherm is the most common.

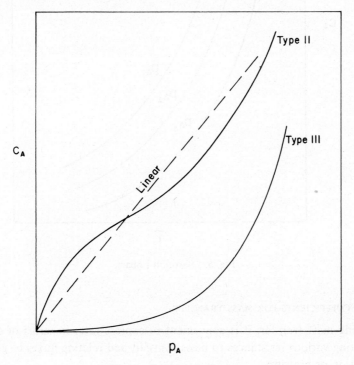

Fig. 4.   Sorption isotherms.

These representations are as follows.

Isotherms are analytical or graphical representations of partial pressure of a component in equilibrium with its concentration in a solid, at constant temperature:

$$p = f(C), \quad T = \text{constant}, \tag{16}$$

where C = concentration of the component in the solid phase.

The simplest case is one in which Henry's Law applies. In most cases, however, isotherms are nonlinear (Figure 4).

Isobars represent relations between concentration and temperature while partial pressure of a component is held constant as shown in Figure 5.

Isosteres are curves or functions relating equilibrium pressure of an adsorbed component to temperature, when the amount adsorbed is held constant as shown in Figure 6.

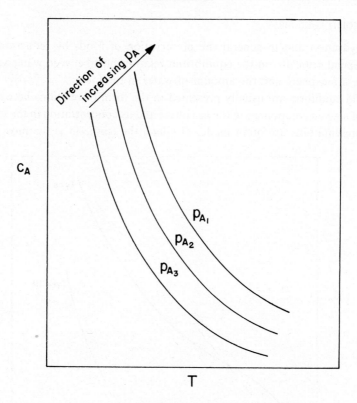

Fig. 5.   Sorption isobars.

4.4. FILM COEFFICIENTS FOR MASS TRANSFER

In analyzing mass transfer, it is convenient to express rates in terms of coefficients incorporating various resistances to mass transfer, and relating fluxes to gradients of concentration or pressure.

For instance, consider a liquid phase in contact with a gaseous phase as shown in Figure 7. Let us assume that a component is transferred from liquid to gas. In the bulk of liquid, the concentration of the component is uniform and equal to C. The gradient of concentration is assumed to occur in a film adjacent to the interface with the gas phase. At the interface, the concentration (C_i) is in equilibrium with the partial pressure (p_i) of the component at the interface. The partial pressure in the bulk of the gas phase is also considered uniform and equal to p, with all the gradient in partial pressure occurring in a gaseous film adjacent to the interface. Given steady state conditions, the following equation gives the rates of transfer in the liquid:

$$J = k_L (C - C_i), \tag{17}$$

where k_L = film transfer coefficient in the liquid in units consistent with those of J and C.

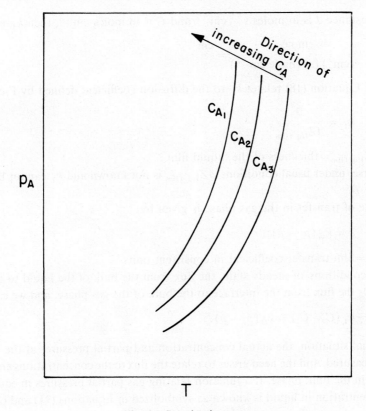

Fig. 6.   Sorption isosteres.

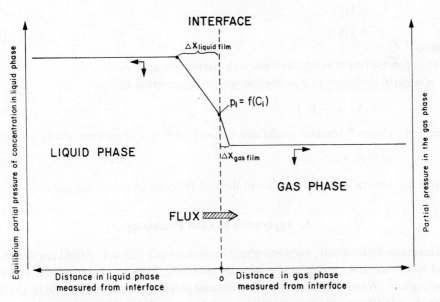

Fig. 7.   Schematic representation of mass transfer from a liquid to a gas phase.

If for instance J is in moles s^{-1} cm^{-2} and C is in moles cm^{-3}, then k_L is in

$$\frac{\text{moles}}{(\text{cm}^2)\,(\text{s})\,(\text{mole cm}^{-3})} = \frac{\text{cm}}{\text{s}}.$$

Note that Equation (18) relates k_L to the diffusion coefficient defined by Fick's Law.

$$k_L = \frac{D}{\Delta Z_{\text{liq. film}}}, \tag{18}$$

where $\Delta Z_{\text{liq. film}}$ = thickness of the 'liquid film'.

Of course, under usual conditions $\Delta Z_{\text{liq. film}}$ is not known and k_L cannot be simply related to D.

The rate of transfer in the gas phase is given by:

$$J = k_G(p_i - p), \tag{19}$$

where k_G = film transfer coefficient in consistent units.

Under conditions of steady state, the flux from the bulk of the liquid to the interface equals the flux from the interface to the bulk of the gas phase, and we can write:

$$k_L(C - C_i) = k_G(p_i - p). \tag{20}$$

In the usual situation, the actual concentration and partial pressures at the interface are not measured, and the need arises to relate the flux to the concentrations and partial pressures in the bulk phase. If a function relating gas partial pressures in equilibrium with concentration in liquid is known as symbolized in Equations (21) and (22), then overall transfer coefficients can be defined as shown in Equations (23) and (24).

$$C' = f(p), \tag{21}$$
$$p' = f(C), \tag{22}$$

where

C' = concentration in equilibrium with partial pressure p,
p' = partial pressure in equilibrium with concentration C.

$$J = K_L(C - C'), \tag{23}$$

where K_L = overall transfer coefficient defined in terms of concentrations in liquid

$$J = K_G(p' - p), \tag{24}$$

where K_G = overall transfer coefficient defined in terms of partial pressures.

5. Application in Food Processing

As mentioned previously, mass transfer is prominent in a number of food applications, and both rate and equilibrium properties of foods, have to be considered in these applications. We already mentioned curing and pickling. In extraction, and in absorption the equilibrium properties are very important and the choice of solvents is governed primarily by distribution properties. The choice of solvents on the basis of polarity is well understood, and is practiced in many applications. In some situations food technologists have pioneered unusual solvent use. For instance in the case of flavor recovery by extraction the use of liquid carbon dioxide offers unusual advantages because of its distribution coefficient which is shown for some flavor compounds in Table III (Schultz and Randall, 1970).

TABLE III

Distribution coefficient between liquid CO_2 and
aqueous solutions for selected compounds
(after Schultz and Randall, 1970)

Compound	Distribution coefficient
Isoamyl acetate	850
Ethyl acetate	42
Pentanol	5
Methanol	0.4

In dehydration, gas-solid equilibria, and diffusivity of water and water-vapor in porous solids and in air are the important mass transfer properties. Freeze-drying is the operation in which mass transfer can be analyzed with greater ease than in other drying operations, because all of the water transport occurs as vapor flow and capillary flow is considered inconsequential.

Ideally, the distribution of moisture during sublimation is that shown in Figure 8a. There is an interface at which the moisture content drops from the initial level (m_0) in the frozen layer, to the final moisture content of the dry layer (m_f) which in turn is determined by equilibrium with partial pressure of water (p_s) in the space surrounding the dry layer. Actually, this ideal representation is not true and the process is represented more accurately by Figure 8b, which shows the existence of some gradients in the 'dry layer'. The figure shows that there exists a 'transition region' in which there is no longer any ice, but in which the moisture content is still substantially higher than the final content of the dry layer (m_f). Recent detailed studies on the nature of this transition layer show it to be relatively narrow, and the assumption of uniformly receding interface of zero thickness does not lead to large errors in the engineering analysis of sublimation (King, 1970).

Freeze-drying is an operation involving both mass transfer and heat transfer, and the rate of drying depends on the magnitude of resistances to these transfers. Heat and mass transfer rates must continuously be balanced and when all of the supplied heat goes into energy of sublimation we have:

$$(\Delta H_s)(G) = q, \tag{25}$$

where
$\Delta H_s =$ latent heat of sublimation,
$q =$ rate of heat transfer.

Three cases represent the three basic types of possibilities in vacuum freeze-drying (Figure 9).

(1) Heat transfer and mass transfer pass through same path (dry layer), but in opposite directions.

(2) Heat transfer through the frozen layer, mass transfer through the dry layer.

(3) Heat generation within the ice (by microwaves), mass transfer through the dry layer.

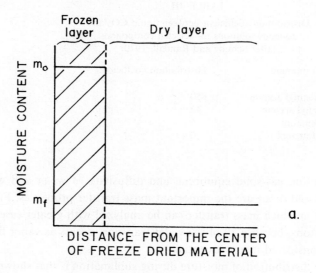

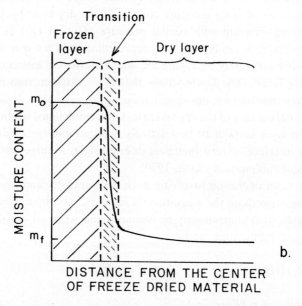

Fig. 8. Schematic representation of the moisture gradient in freeze-drying material: (a) idealized gradient; (b) probable gradients occurring in freeze-drying.

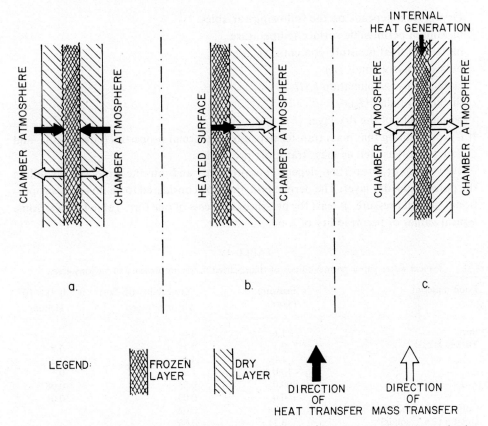

Fig. 9.   Types of heat transfer defining different types of freeze-drying: (a) heat transfer by conduction trough dry layer; (b) heat transfer through frozen layer; (c) internal heat gencration by micro waves.

Consider the first case which is simplified, but nevertheless typical of most vacuum freeze-drying operations. The material to be dried is heated by radiation to the dry surface, and its internal frozen layer temperature is determined by the balance between heat and mass transfer. For simplicity's sake, we shall consider a slab geometry with negligible end effects, and slab thickness L. We shall assume that the maximum allowable surface temperature T_s is reached instantaneously and that the heat output of the external heat supply is adjusted in such a manner as to maintain T_s constant throughout the drying cycle. We also assume that the partial pressure of water in the drying chamber, p_s, is constant, and that all of the heat is used for sublimation of water vapor.

It can be shown that under these idealized conditions the drying time is given by Equation (26):

$$t_d = \frac{L^2 \varrho (m_0 - m_f)}{8b(p_i - p_s)}, \tag{26}$$

where

ϱ = bulk density of solids,

t_d = drying time.

Drying time depends on the following variables:
maximum permissible surface temperature,
initial and final moisture contents (m_0, m_f),
bulk density of solids (ϱ),
latent heat of sublimation (ΔH_s),
thickness of slab (L),
permeability of the dry layer (b).

It is worth noting that mass transfer depends on thermal properties (latent heats, and conductivities) as well as mass transport properties.

The transport properties depend on pressure, and on the kind of gas filling the pores of the dry layer. The permeability may be considered to be inversely proportional to total pressure, at least for pressures in excess of one torr. Table IV gives some typical values of permeability of foods.

TABLE IV

Typical water vapor permeabilities of freeze-dried foods, in presence of various gases

Food material	Pressure (Torr)	Permeability (lb Torr^{-1} ft^{-1} h^{-1}) $\times 10^3$	
		Air or nitrogen	Helium
Beef	1.6	4.6	–
Turkey breast	0.1	5.1	4.9
	1.0	4.6	3.2
	10.0	1.7	0.7
	100.0	0.25	0.08
	760.0	0.03	0.01
Coffee (10% solids)[a]	0.15	0.02	–
Gelatin (2.6% solids)[a]	0.21	0.002	–
Coffee (20% solids)[a]	0.93	0.01	–

[a] Solids content before freeze-drying

It should be noted that the thermal conductivities of dry layers of foods are extremely low, and compare with conductivities of insulators such as cork and strofoam. As a consequence, the temperature drops across the dry layer are large, and with surface temperatures often limited to values below 150°F because of danger of discoloration, and in some cases to values below 100°F because of danger of denaturation, the resultant ice temperatures are in most cases well below 0°F. Except for foods with very low eutectic melting points, it is the surface temperature that limits drying rate.

The mass transport in the dry layer of foods varies not only with pressure and temperature, but it also depends on the structure and orientation of the drying materials. King (1970), for instance, reports that the effective diffusivity for turkey meat drying with the grain perpendicular to the direction of vapor flow was about 2.5 times lower than that for the same material oriented with the grain parallel to flow. The importance of structure and orientation on mass transfer was also pointed out by Spiess et al. who attempted to correlate mass transfer resistance with structural parameters (Spiess et al., 1969).

The temperature and moisture gradients in the freeze-drying product depend strongly on the drying layer's properties which are partially 'set' during freezing. For instance, if solute migration is possible during freezing, an impermeable film which impedes drying by substantially lowering its rate can form at the surface (Quast and Karel, 1968). Slush freezing can prevent the formation of such a film or it may be removed mechanically. Slow freezing produces bigger crystals, hence usually bigger pores and better mass flow during drying and reconstitution, if the solute migration

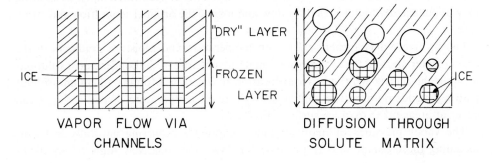

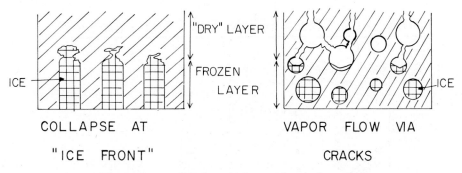

Fig. 10.   Representation of types of vapor flow in freeze-drying.

and film formation are prevented. The beneficial effect of slow freezing on mass transfer rates is often offset by the structure's collapse if the frozen-layer temperature is too high.

Ideally, the vapor flows through the pores and channels left by ice crystals. However, if freezing produces isolated ice crystals surrounded by a solid matrix, then the vapor must diffuse through the solids (Figure 10). A similar situation results if the matrix collapses at the ice front and seals the channels. However, the matrix can be cracked, particularly in rapidly frozen systems.

Collapse often occurs at a fixed temperature similar to recrystallization temperature (MacKenzie, 1966; Ito, 1971) when the matrix is sufficiently mobile to allow flow under the influence of various forces. Freezing causes a separation of the aqueous solutions present in foods into a two-phase mixture of ice crystals and concentrated aqueous solution. The properties of the concentrated aqueous solution depends on temperature, concentration and composition. If drying is conducted at a very low temperature, then mobility in the extremely viscous concentrated phase is so low that no structural changes occur during drying, and the resultant structure consists of pores in the locations which contained ice crystals, surrounded by dry matrix of insoluble components and precipitate compounds originally in solution. If, on the other hand, the temperature is above a critical level, mobility in the concentrated aqueous solution may be sufficiently high to result in flow and loss of the original separation and struc-ture.

During freeze-drying there exist both temperature and moisture gradients in the drying materials, and the mobility and therefore collapse of the concentrated solutions forming the matrix may vary from location to location. Mobility of an amorphous matrix depends on moisture content as well as on temperature, hence collapse can occur at areas other than ice surface (Figure 11).

The vapor pressures of many organic compounds occurring naturally in foods, including those compounds responsible for flavor, are relatively high. Many flavor compounds show vapor pressures, which over the range of temperatures involved in freeze-drying are above the vapor pressure of water. A natural consequence is that many of them are subject to evaporation and loss during drying. The laws governing

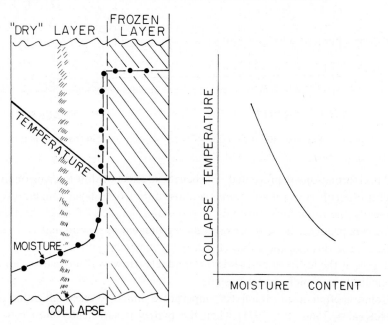

Fig. 11.   Potential collapse of structure in the partially dry layer of freeze-drying materials.

their removal, however, are complex. A number of differing models can be made including simple batch distillation, assuming constant volatility, and various assumptions about deviations for some of these models are shown in Table V. It is interesting to note that all of these relationships underestimate the retention of volatile in actual freeze-drying operations.

TABLE V

Retention of pyridine in coffee powder

A. *Predicted from theoretical relations*	Retention (%)
Raoult's Law	7.5
Volatility in infinitely dilute solutions, data of Thijssen[a]	0.01
Relation suggested by Sivetz[b]	5.0
Calculated using volatility determined experimentally for 0.01 % aqueous solution at $-20\,°C$[c]	$< 10^{-5}$
B. *Observed retention*	
Coffee powder[b]	50
Freeze-dried model (1 % glucose)[c]	85
Freeze-dried model (1 % starch)[c]	50

[a] Thijssen (1970)
[b] Sivetz and Foote (1963)
[c] Fritsch *et al.* (1971)

The retention of organic volatiles has been considered to result from surface adsorption of the volatile on the dry layer of the freeze-drying sample (Rey and Bastien, 1962) or from an entrapment mechanism which immobilizes the volatile compounds within the amorphous solute matrix (Flink and Karel, 1970; Thijssen and Rulkens, 1968). A simple experiment shows that the retention phenomena depend on local entrapment rather than adsorption (Flink and Karel, 1970). Maltose solutions were frozen in layers, some of which were volatile-containing, others, volatile-free. After freeze-drying, the layers were separated and analyzed separately for the volatile. The results (Table VI) show that volatile retention is localized in those areas at which the volatile is initially present. This experiment demonstrates that adsorption is not a retention

TABLE VI

Retention of 2-propanol in specified layers of freeze-dried maltose solutions[a]

	Sample A		Sample B	
	Before freeze-drying	After freeze-drying	Before freeze-drying	After freeze-drying
Top layer	0	0	4	2.52
Middle layer	0	0.05	0	0.05
Bottom layer	4	2.73	0	0.02

[a] n-propanol content is g/100 g solids

mechanism since volatile escaping from the lowest layer of Sample A was not retained on the upper dry layers. Further, volatile retention is not affected by the passage of water vapor through the volatile-free maltose layers as shown for Sample B. In a companion experiment, sections of the freeze-dried cake were cut from a sample perpendicular to the mass transfer axis. Essentially uniform retention was observed for the whole sample which supports the observations that volatile retention is determined locally in the food material and not by surface adsorption. The gross structure of the freeze-dried material is freely permeable to the flow of volatile from the lower freeze-drying levels, and the retained volatile is not located on the surface of the dry layer but within the amorphous solute matrix.

The physical aspects by which the volatile is entrapped within the amorphous solute matrix are only partially understood. Two mechanisms, selective diffusion (Thijssen and Rulkens, 1968) and microregions (Flink and Karel, 1970) represent perhaps macro- and micro-views of the same basic phenomenon. The size of the entrapments is small since grinding and evacuation of the dry material does not release any volatile. Recently the size of the microregions has been shown to vary with, among other things, the solubility of the organic volatile in the aqueous solution. Retained hexanal (about 1 g hexanal per 100 g maltodextrin) freeze-dried from an aqueous 20% maltodextrin solution appeared in the optical microscope within the amorphous solute matrix as 2- to 6-μ droplets (Flink and Gejl-Hansen, 1972). Concurring evaluations have been made with a scanning electron microscope. The numbers and sizes of droplets observed in the optical microscope for a series of n-alcohols qualitatively indicated that droplet size and number increased with alcohol molecular weight.

The addition of sufficient water to the dry material will cause volatile loss, the extent of which depends on the amount of water added and the particular solute matrix, since water has the capability of disrupting and/or plasticizing various hydrophilic substances composing food solids.

Most recently Lambert working with Flink and Karel has studied n-butanol loss from frozen aqueous solutions. When equilibrated at $-10°C$ over activated charcoal in the presence of ice (to prevent water loss from the samples), only 25% of the butanol was lost. The remaining 75% was strongly retained and was not removed by transferring the samples to a desiccator which contained fresh activated charcoal. While the mechanism for this retention in the absence of a solute phase is not fully understood, apparently, ice crystals are fully capable of entrapping organic volatile, though this entrapment cannot persist through any freeze-drying process.

In dehydration processes other than freeze-drying structural changes in the food materials occur freely, since there is possibility for liquid flow, solute redistribution and shrinkage. As a result the mass transport properties change drastically during the dehydration process.

Fish (1958) reports that in scalded potatoes the diffusion coefficient for water is 10^{-8} to 10^{-7} cm^2 s^{-1} when the moisture is still at a level of 15 to 20% (dry basis), but that it drops to less than 10^{-11}, when moisture is less than 1%. Similar data were reported for other materials, but in some foods diffusion coefficients do not fall off as sharply with decreasing water content as in the case of potatoes noted by Fish (1958). Thus Jason (1958) observed that fish muscle dehydration behavior could be described by the use of two diffusion coefficients, the first (D') valid down to about 5% free water and the second (D'') down to almost complete dehydration. Jason (1958) found that D' was about 2 to 4×10^{-6} cm^2 s^{-1} for various species of fish, and D'' 0.1 to 1.0×10^{-6} cm^2 s^{-1}.

However, mass transport in air drying depends on drying conditions and cannot be described adequately by constant diffusion coefficients.

6. Mass Transport in Emulsion, Foams and Dispersions

In heterogeneous systems including foams and emulsions the transport is complicated by the formation of interfacial films, and by surface energy effects which become significant when the size of dispersed particles or droplets is in colloidal range. Emulsions

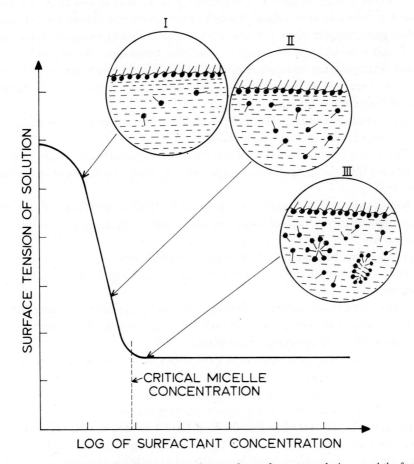

Fig. 12. Adsorption of amphipolar compounds at surfaces of aqueous solutions, and the formation of micelles at critical micelle concentration.

are an example of important heterogeneous systems present in foods, and food emulsions are stabilized by the adsorption at oil-water interface of surface active molecules. As a consequence of differences in attractive forces between molecules of different liquids, or differences in molecular density between phases, surface molecules possess extra energy, the free surface energy. Water surface energy (or as it is also called, surface tension) or 72 erg cm^{-2}. The presence of amphipolar molecules including polar lipids and proteins lowers surface tension as shown in Figure 12, due to adsorption and orientation of these molecules in the surface; molecules of this type are termed surface active, or surfactants. Typically, this adsorption shows three phases with increasing concentration:

(a) saturation and orientation in the surface;

(b) region of constant surface excess (surface excess is approximately equal to surface concentration) but increasing bulk concentration. In this stage the decrease in free surface energy is proportional to the logarithm of the surfactant concentration;

(c) region of micelle formation. The formation of micelles as well as orientation of amphipolar molecules at interfaces between water and air or between water and oil are of key importance in many colloidal phenomena, including emulsion stabilization, foaming and solubilization of otherwise insoluble compounds within micelles. For each type of amphipolar molecule, a critical micelle concentration range exists, and a very comprehensive tabulation of these was published recently by the National Bureau of Standards (Mukerjee and Mysels, 1971).

Micelle formation and the adsorption of amphipolar molecules at oil-water interfaces have important effects on mass transfer:

(1) The micelles themselves can entrap various components and either hinder or facilitate their transport; depending on whether the micelles themselves can be readily transported.

(2) Similar orientation of surface active agents at oil-water interfaces stabilize emulsion, and may or may not retard mass transport significantly depending on interfacial film properties.

Figure 13 shows how interfacial films stabilize emulsions:

(a) Film formation at the oil-water interface with strong steric hindrance to oil coalescence.

(b) Electrostatic repulsion between charged groups located in the oil-water interface.

(c) Formation of hydration layers outside the oil droplet, because water-orienting hydrophilic groups are present at the surface.

(d) Low interfacial tension can stabilize emulsions by allowing large drop deformations.

The resistance of interfacial films varies with conditions. For transport of SO_2, Plevan and Quinn (1966) found that only the densely packing hexadecanol monolayer showed some resistance (Table VII) and Bernett et al. (1970) found that nitromethane evaporation was only marginally retarded by thick layers of silicone (Table VIII). On the other hand thick polymeric films, especially gelatin films, were found to offer appreciable interfacial resistance to drug transport by Ghanem et al. (1969).

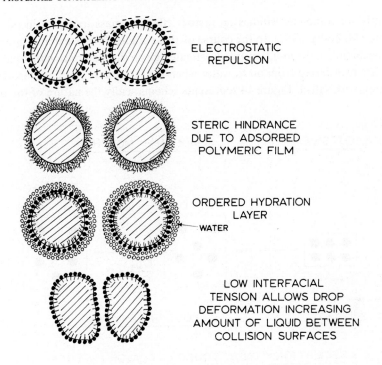

ELECTROSTATIC
REPULSION

STERIC HINDRANCE
DUE TO ADSORBED
POLYMERIC FILM

ORDERED HYDRATION
LAYER
WATER

LOW INTERFACIAL
TENSION ALLOWS DROP
DEFORMATION INCREASING
AMOUNT OF LIQUID BETWEEN
COLLISION SURFACES

MECHANISMS OF ANTI-COALESCENCE

Fig. 13.   Emulsion stabilization due to formation of interfacial films:
mechanisms of anti-coalescence.

TABLE VII

Surface resistance of monolayers
after Plevan and Quinn (1966)
transport of SO_2

Monolayer	R_s s cm^{-1}
Hexadecanol	~ 200
Stearic acid	35
Oleic acid	0
Hexadecylamine	0
Cholesterol	0

TABLE VIII

Reduction of evaporation of Nitromethane by silicone layers at
nitromethane-air interface (after Bernett et al., 1970)

Thickness of silicone layer mm	Number of molecular layers	Reduction %
7.5×10^{-2}	10^4	3
31×10^{-2}	4×10^4	28

Recently we completed studies on factors affecting mass transport in oil-in-water emulsions (McNulty, 1972). In the course of these studies the extent and rate of flavor release occurring in the mouth were investigated given the assumption that flavor compounds are transferred from oil to water when the interphase equilibria are disturbed by dilution with saliva. Figure 14 represents schematically the nature of the problem.

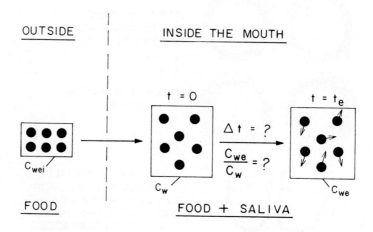

OUTSIDE | INSIDE THE MOUTH

ASSUMPTION : ONLY AQUEOUS FLAVOR IS PERCEIVED

QUESTIONS

1. HOW MUCH FLAVOR CAN BE TRANSFERRED ?

2. HOW FAST IS THE RATE OF TRANSFER ?

Fig. 14. Schematic representation of a simple model of flavor transfer in an oil-in-water emulsion in the mouth.

The concentration of the transported compound in the aqueous phase immediately upon dilution (C_w) and upon restablishing equilibria (C_{we}) depend on various factors as follows:

$$C_{we} = \frac{C_{emo}}{\phi_0 (K_p - 1) + DF_{em}} \tag{27}$$

$$C_w = \frac{C_{weo} (1 - \phi)}{DF_{em} - \phi_0}, \tag{28}$$

where:

C_{emo} = initial concentration in emulsion;

K_p = partition coefficient C_{oil}/C_{we};

C_{oil} = volumetric fraction of oil phase, initially;

DF_{em} = dilution factor = ϕ_0/ϕ_f;

C_{weo} = initial aqueous concentration.

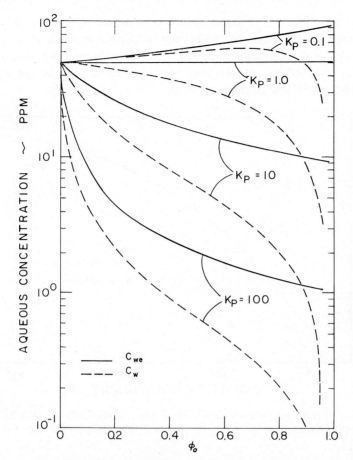

Fig. 15.   Effects of K_p and of ϕ_0 on the aqueous concentration immediately upon dilution (C_w) and at equilibrium (C_{we}), for $DF_{em} = 2$ and $c_{emo} = 100$ ppm.

Figure 15 shows how C_w and C_{we} depend on ϕ_0 and on K_p, when the dilution factor is 2, and the initial flavor concentration is 100 ppm.

The rates of transport were studied in a stirred diffusion cell using a variety of conditions, as well as actual release and uptake in model emulsions. The following major conclusions were made:

(1) The rate of transfer of normal alcohols and other relatively small molecules from oil to water is strongly dependent on K_p, but surfactants adsorbed in the oil-water interface do not normally form significant mass transfer resistance. This effect is shown in Figure 16 which presents results for transport of alcohols from Wesson oil to water in presence and in absence of surfactants (Span 60 and Tween 60).

(2) Rate of flavor release and uptake from and by the dispersed phase of liquid o/w emulsions is normally rapid, with 50% of the equilibrium attainment in less than 15 seconds. This conclusion was qualitatively supported by the rapid in vivo perception of such primarily oil-soluble compounds as anethole and benzaldehyde (2–5 s), and

caffeine and menthol (5–15 s) found when o/w emulsions were tasted in the mouth.

(3) The transfer rate of model compounds in liquid o/w emulsions is not significantly retarded by (a) increasing dispersed phase viscosity by using silicone oils with viscosities up to 93 000 cp, (b) increasing the vegetable oil solid fat index from zero to 100, and (c) using mixed surfactant systems.

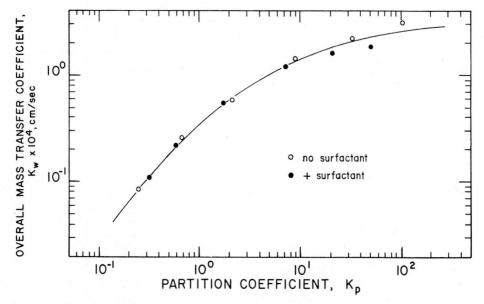

Fig. 16. Effect of partition coefficient on the overall mass transfer coefficient (K_w) during transfer of n-alcohols from Wesson oil to water in a stirred diffusion cell at 24 °C.

(4) Flavor-surfactant association appears to produce a major reduction in transfer rate only, when organic compounds are present in the range close to their limit of aqueous solubility. We have demonstrated such association phenomena in a cholesterol-Tween 60 system.

(5) Flavor release in the mouth can be greatly retarded by increasing dispersed phase volume until the emulsion is 'plastic' because mixing with saliva will probably be slow. In these cases, flavor release rate will depend not only on the physical-chemical emulsion properties but also on the bite size, the extent of mastication and mixing, and the residence time in the mouth.

In some food systems very complicated interfacial phenomena can occur. In milk, cream and butter, for instance, the milk-fat-globule-membrane can be affected by processing conditions and result in changes in mass transport properties.

7. Mass Transfer as a Factor in Food Stability

Food stability is affected by reactions which involve environmental components such as water, oxygen and carbon dioxide the mass transport of which can affect rates of reaction. In other cases mass transport of some reactant or product may control the outcome of reactions in the food in as far as food acceptability is concerned. Thus nonenzymatic browning may be affected by ability of sugars to diffuse in the food (Duckworth and Smith, 1963; Eichner and Karel, 1972). In other cases contamination of foods by toxins or carcinogens produced by micro-organisms may be a factor (Armbrecht, 1972).

One example is that of oxygen diffusion in oxidation-susceptible foods.

Headspace oxygen has a deleterious effect to the quality of many products. To minimize this effect, the food industry uses processes such as deaeration, inert-gas packaging, inert-gas blanketing, vacuum-breaking with an inert gas, and packaging in materials with low permeability of oxygen.

Oxygen uptake by a food product may be due to respiration, enzymatic browning, lipid oxidation, or oxidation of proteins, vitamins, and many other food components. Lipid oxidation is probably the most common mechanisms of oxygen uptake in dried foods and is responsible for the very high oxidation rates observed in such items as fish meal and potato chips. It may also result in significant flavor deterioration even when the oxygen uptake is relatively low, as in potato flakes.

TABLE IX

Effect of diffusion on oxidation of some food products in air
(after Quast and Karel, 1971)

Product	Temperature	Effective oxygen diffusivity (cm^2) (s^{-1})	Product thickness (cm)	Reduction of oxidation due to diffusion resistance (%)
Potato chips (past induction)	37°C	2×10^{-1}	100	0
Fish meal	37°C	8×10^{-2}	10	0
Fish meal	37°C	8×10^{-2}	100	50
Butter	~5°C	1.6×10^{-8}	10	60

Marcuse (1967) and others have suggested that oxygen diffusion through the food product towards the reaction site may be a rate-limiting factor in some instances.

Quast and Karel (1971) studied the diffusion of oxygen in foods for conditions that may occur during processing, handling, packaging, and storage. It was found that oxygen diffusion can affect rates of oxidation in two types of situations: (Table IX)

(1) extremely rapid rates of oxidation (such as fish meal during curing in bulk bins);
(2) Very dense products (such as butter).

Oxygen diffusion is also important in various aspects of meat color due to myoglobin derivatives. Among studies on this subject there is a recent study by Morley (1971).

List of Symbols

(Units are given for some quantities. These are intended as examples. Other consistent units in either English or metric systems may be used.)

A	area
C	concentration (mole l^{-1}) or (lb ft^{-3})
C'	equilibrium concentration
D	diffusion coefficient (cm^2 s^{-1}) or (ft^2 h^{-1})
DF_{em}	dilution factor (defined in Equation (28))
G	drying rate (lb h^{-1})
J	flux (mole cm^{-2} s^{-1}) or (lb ft^{-2} h^{-1}) or (g cm^{-2} s^{-1})
K_p	partition coefficient
K_G	overall mass transfer coefficient based on partial pressure difference (lb ft^{-2} h^{-1} atm) or (mole cm^{-2} s^{-1} atm^{-1})
K_L	overall mass transfer coefficients based on concentration difference [lb ft^{-2} h^{-1} (lb ft^{-3})$^{-1}$] or [mole cm^{-2} s^{-1} (moles cm^{-3})$^{-1}$]
K_k	a constant
K_1	constant defined in Equation (4)
L	thickness of slab
P	pressure
P^0	vapor pressure
P_θ	total pressure
R	gas constant
S	constant in Henry's Law as defined in Equation (13)
T	temperature
V	volume
Z	distance
a	activity
b	permeability of dry layer (lb ft^{-1} h^{-1} Torr^{-1})
c	concentration
k_L	mass transfer coefficient in liquid phase
k_G	mass transfer coefficient in gas phase
p	partial pressure
$\bar{p}$	mean partical pressure
q	heat flux (BTU h^{-1})
t	time
x	mole fraction in liquid phase
y	mole fraction in gas phase
γ	activity coefficient
ΔC	concentration difference
Δp	pressure difference
ΔH_s	latent heat of sublimation (BTU/lb)
ΔZ	distance interval
$\Delta Z_{liq.\,film}$	thickness of liquid film
$\Delta Z_{gas\,film}$	thickness of gas film
ε	porosity
μ	chemical potential
η	viscosity
ϱ	density
σ	internal surface area
ϕ	volumetric fraction of dispersed phase

Subscripts

L, Λ	specific liquid phases	i	interface between two phases
d	drying	0	initial
e	equilibrium	oil	oil phase
em	emulsion	s	space in freeze-drier
	final	w	aqueous phase

Acknowledgment

The research was supported in part by U.S. Public Health Service Grant No. FD-00140 from the Food and Drug Administration.

References

Arganosa, F. C. and Henrickson, R. L.: 1969, 'Cure Diffusion through Pre- and Post-Chilled Procine Muscles', *Food Technol.* **23**, 1061.

Armbrecht, B. H.: 1972, 'Aflatoxin Residues in Food and Feed Derived from Plant and Animal Sources', *Residue Rev.* **41**, 13.

Bernett, M. K., Halper, L. A., Jarvis, N. L., and Thomas, T. M.: 1970, 'Effect of Adsorbed Monomolecular Films on Evaporation of Volatile Organic Liquids', *I&EC Fund.* **9**, 150.

Carman, P. C.: 1956, *Flow of Gases through Porous Media*, Academic Press Inc., New York.

Crank, J. and Park, G. S.: 1968, *Diffusion in Polymers*, Academic Press, London.

Del Valle, F. R.: 1965, *Factors Affecting the Salting and Drying of Animal Muscle Tissues*, Ph.D. Thesis, Massachusetts Institute of Technology June, 1965, Cambridge, Mass.

Duckworth, R. B.: 1962, 'Diffusion of Solutes in Dehydrated Vegetables', in J. Hawthorn and J. M. Leitch (eds.), *Recent Advances in Food Science*, vol. II, p. 46, Butterworths, London.

Eichner, K. and Karel, M.: 1972, 'The Influence of Water Content and Water Activity on the Sugar-Amino Browning Reaction in Model Systems under Various Conditions', *J. Agr. Food Chem.* **20**, 218.

Farkas, D. and Lazar, M.: 1969, 'Osmotic Dehydration of Apple Pieces', *Food Technol.* **23**, 689.

Fish, B. P.: 1958, 'Diffusion and Thermodynamics of Water in Potato Starch Gel', in *Fundamental Aspects of the Dehydration of Foodstuffs*, Society of Chemical Industry, London, p. 143.

Flink, J. and Gejl-Hansen, F.: 1972, 'Retention of Organic Volatiles in Freeze-Dried Carbohydrate Solutions: Microscopic Observations', *J. Ag. Food Chem.* **20**, 691.

Flink, J. and Karel, M.: 1970, 'Retention of Organic Volatiles in Freeze-Dried Solution of Carbohydrates', *J. Agr. Food Chem.* **18**, 295.

Flink, J. and Karel, M.: 1972, 'Mechanisms of Retention of Organic Volatiles in Freeze-Dried Systems', *J. Food Technol. (London)* **7**, 199.

Fritsch, R., Mohr, W., and Heiss, R.: 1971, 'Untersuchungen über die Aroma-Erhaltung bei der Trocknung von Lebensmitteln nach verschiedenen Verfahren', *Chem. Ing. Technik* **43**, 445.

Ghanem, A., Higuchi, W. I. Simonelli, A. P.: 1969, 'Interfacial Barriers in Interphase Transport: Retardation of the Transport of Diethylphthalate across the Hexadecane-Water Interface by an Adsorbed Gelatin Film', *J. Pharm. Sci.* **58**, 165.

Ito, K.: 1971, 'Freeze-Drying of Pharmaceuticals. Eutectic Temperature and Collapse of Solute Matrix upon Freeze-Drying of Three-Component Systems', *Chem. Pharm. Bull. (Japan)* **19**, 1095.

Jason, A. C.: 1958, 'A Study of Evaporation and Diffusion Processes in the Drying of Fish Muscle', in *Fundamental Aspects of the Dehydration of Foodstuffs*, Society of Chemical Industry, London, p. 103.

King, J.: 1970, 'Freeze Drying of Foodstuffs', *CRC Critical Reviews in Food Technology* **1**, 379.

MacKenzie, A. P.: 1966, 'Basic Principles of Freeze-Drying for Pharmaceuticals', *Bull. Parenteral Drug. Assoc.* **20**, 101.

Marcuse, R.: 1967, *The Influence of Oxygen Partial Pressure on Fat Oxidation*, S.I.K. (Swedish Institute for Food Preservation) Report No. 231.

McNulty, P. B.: 1972, *Factors Affecting Flavor Release and Uptake in Oil-in-Water Emulsions*, Ph.D. Thesis, Massachusetts Institute of Technology, Cambridge, Mass.

Michaels, A. S. and Bixler, H. J.: 1968, 'Membrane Permeation: Theory and Practice' in E. S. Perry (ed.), *Progress in Separation and Purification*, Wiley-Interscience, New York.

Morley, M. J.: 1971, 'Measurement of Oxygen Penetration into Meat Using an Oxygen Micro-Electrode', *J. Food Technol.* **6**, 371.

Mukerjee, P. and Mysels, K. J.: 1971, *Critical micelle concentration of aqueous surfactant systems*, Publication NSRDS-NBS 36, National Bureau of Standards, Wash. D.C.

Quast, D. and Karel, M.: 1968, Dry Layer Permeability and Freeze-Drying Rates in Concentrated Fluid Systems', *J. Food Sci.* **33**, 170.

Quast, D. and Karel, M.: 1971, 'Effects of Oxygen Diffusion of some Dry Foods', *J. Food Technol.* **6**, 95.

Plevan, R. E. and Quinn, J. A.: 1966, 'Effect of Monomolecular Films on Gas Absorption into a Quiescent Liquid', *AIChE J.* **12**, 894.

Rey, L. and Bastien, M. C.: 1962, 'Biophysical Aspects of Freeze-Drying: Importance of the Preliminary Freezing and Sublimation Periods', in F. R. Fisher (ed.), *Freeze-Drying of Foods*, National Academy of Sciences – National Research Council, Washington, D.C., p. 25.

Schultz, W. G. and Randall, J. M.: 1970, 'Liquid CO_2 for Selective Aroma Retention', *Food Technol.* **24**, 1282.

Sherwood, T. K. and Pigford, R. L.: 1952, *Absorption and Extraction*, McGill Book Co., Inc. New York, N.Y.

Sivetz, M. and Foote, H. E.: 1963, *Coffee Processing Technology*, vol. I, AVI Publ. Co., Westport, Conn.

Spiess, W. E. L., Wolf, W., Tirtahusodo, H., and Sole C., P.: 1969, 'The Influence of the Structure on the Mass Transfer in Freeze-Drying', in *Bull. de l'Institut International de Froid Annexe 1969-4.*, p. 155.

Thijssen, H. A. C.: 1970, 'Concentration Processes for Liquid Foods Containing Volatile Flavours and Aromas', *J. Food Technol.* **5**, 211.

Thijssen, H. A. C. and Rulkens, W. H.: 1968, 'Retention of Aromas in Drying Food Liquids', *Ingenieur* **80**, 45.

CHAPTER XII

FACTORS INFLUENCING THE INSTRUMENTAL AND SENSORY EVALUATION OF FOOD EMULSIONS

PHILIP SHERMAN

Dept. of Food Science, Queen Elizabeth College, University of London, London, W.8., England

1. Introduction

The texture profile concept introduced by Szczesniak (1963) and Brandt *et al.* (1963) demonstrates quite clearly that texture is not a single property but rather a collective term which defines the overall sensory impression originating from a number of different properties. In the case of food emulsions (e.g. margarine, butter, ice cream, milk, cream, salad dressings, dairy desserts) adjectives used to describe the relevant properties would include thin, watery, viscous, greasy, creamy, sticky, slimy etc.

Unfortunately, little is known about the factors associated with the sensory recognition and evaluation of textural properties with the possible exception of sliminess and viscosity. Szczesniak and Farkas (1962) showed that the viscosities of non-Newtonian hydrocolloid solutions which appeared extremely slimy in the mouth decreased to a much lesser extent with increasing shear rate, when examined in a viscometer, than did the viscosities of hydrocolloid solutions which appeared to be only slightly slimy or not slimy at all. The textural property which has been studied in greatest detail is viscosity. Consequently, this paper deals specifically with the evaluation of this property and, in addition, it shows how the internal structure of emulsions influences the evaluation. Although attention has been focussed primarily on oral evaluation of viscosity it should be pointed out at once that a preliminary evaluation of viscosity is made as soon as the product is observed and during manipulation (e.g. shaking or tilting the container, stirring the contents with a spoon) before introducing into the mouth. All of the impressions formed at the various stages of handling and using the product contribute to the final overall impression.

Pharmaceutical scientists have been interested in the non-oral aspects of viscosity evaluation since about 1950. In the absence of reliable experimental data various estimates were made of the shear rates associated with the spreading of preparations on the skin, pouring from a bottle, extrusion through the fine orifice of a syringe etc. All of these calculations were based on generalised hypothetical spreading velocities and deposited layer thicknesses so that the shear rates associated with a particular action appears to be constant irrespective of the flow properties of the preparation. Recent work (Berry and Grace, 1972) utilising the approach described later in this paper has shown that these conclusions were incorrect.

ChoKyun Rha (ed.), Theory, Determination and Control of Physical Properties of Food Materials, 251–266.
Copyright © 1975 by D. Reidel Publishing Company, Dordrecht-Holland. All Rights Reserved.

2. Shear Rate Dependence of the Flow Properties of Non-Newtonian Food Emulsions and its Implications

Food emulsions generally show non-Newtonian flow characteristics. The only exceptions to this statement would be products with very low concentrations of dispersed phase, and there are not many that fall within this category. Consequently, food emulsions do not have a true viscosity, and the value derived from instrumental evaluation depends on the operative shear rate with the measured viscosity decreasing as shear rate increases.

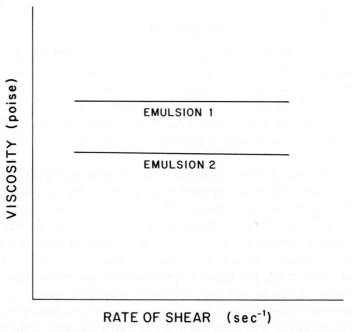

Fig. 1.   Viscosity-rate of shear plots for two Newtonian emulsions.

Some simple examples will serve to illustrate the complexity of the problem which this shear rate dependence presents to the experimentalist. A Newtonian system has a viscosity which is independent of shear rate. Thus, when comparing the viscosities of two Newtonian systems (Figure 1) it does not matter if they are measured at only one, arbitrarily selected, shear rate because the answer will be the same at all other shear rates viz. emulsion 1 has a higher viscosity than emulsion 2. If we now compare two non-Newtonian emulsions having viscosity-shear rate curves which do not intersect and cross over (Figure 2) we still observe that the viscosity of emulsion 1 is greater than the viscosity of emulsion 2 at all shear rates. However, there is one significant difference from the situation depicted in Figure 1, and this is that the difference between the two viscosities in Figure 2 is larger at low shear rates than it is at high shear rates. This is because, as we shall see later in this paper, the globules in emulsions

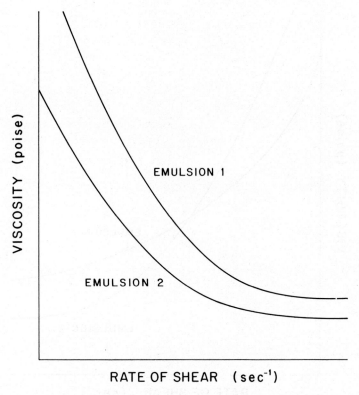

Fig. 2.   Non-intersecting viscosity-rate of shear plots for two non-Newtonian emulsions.

form three-dimensional interlinked structures in the stationary, zero shear, state and when shear is applied these structures break down. At low shear rates a higher degree of structure is present than at high shear rates, so that the structural differences between emulsions 1 and 2 are more pronounced at low shear rates.

As a final example let us consider the case of two non-Newtonian emulsions whose viscosity-shear rate curves do intersect at a point X and then cross over (Figure 3). The situation is now complex with three alternative answers possible depending on the shear rate ($\dot{\gamma}$) selected to analyse the curves. At the shear rate ($\dot{\gamma}_X$) corresponding to the point of intersection X, the two emulsions have identical viscosities. If the analysis is made at a shear rate lower than $\dot{\gamma}_X$ we find that emulsion 2 has a higher viscosity than emulsion 1, whereas at a shear rate higher than $\dot{\gamma}_X$ the reverse is true and emulsion 1 has the higher viscosity. We are then confronted with the question "which of these three possible analyses is the correct one?", and to give the correct answer we have to know what the shear conditions are during the particular application of these food emulsions in which we are interested. For the purpose of this paper this means that we need to know what shear conditions are associated with the consumer's evaluation of viscosity in the mouth, and also with other modes of evaluation before the food is introduced into the mouth.

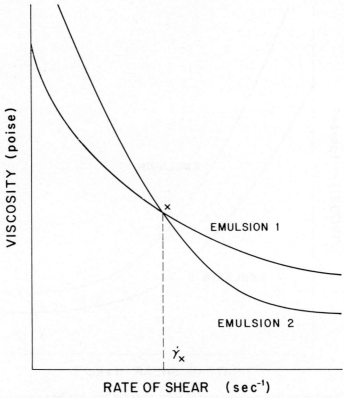

Fig. 3.   Intersecting viscosity-rate of shear plots for two non-Newtonian emulsions.

3. Determination of Shear Rates Associated with Viscosity Evaluation in the Mouth

Wood (1968) devised a technique for assessing the shear conditions prevailing in the mouth during viscosity evaluation. Four non-Newtonian sauces (a dried sauce base reconstituted with different amounts of water) were used along with a Newtonian glucose syrup as a standard against which the other samples could be compared. Viscometric examination of these materials indicated that the shear rate–shear stress curves for all four non-Newtonian sauces intersected and crossed the linear plot for the glucose syrup, and that they did so at different points (Figure 4). Samples of the five materials were presented to members of a sensory evaluation panel and they were asked to taste them and indicate, on the score sheet provided, which, if any of the sauces appeared to have a viscosity similar to that of the glucose solution. The panelists' responses indicated that samples 2 and 3 were considered to most closely resemble the standard glucose solution in viscosity. Reference to Figure 4 indicates that the flow curves for samples 2 and 3 intersect the plot for the glucose solution at a shear rate between 45 and 50 s^{-1}, although the corresponding shear stresses are widely different at 506 and 970 dyne cm^{-2} respectively. Discussions with the panelists suggest that viscosity was judged in the mouth by allowing each sample to flow through the channel between the raised upper surface of the tongue and the roof of the mouth. These remarks, taken in conjunction with the viscometer and panelist's evaluation data, suggested to Wood (1968) that viscosity evaluation is associated with perception of the stress developed at an approximately constant shear rate of 50 s^{-1}.

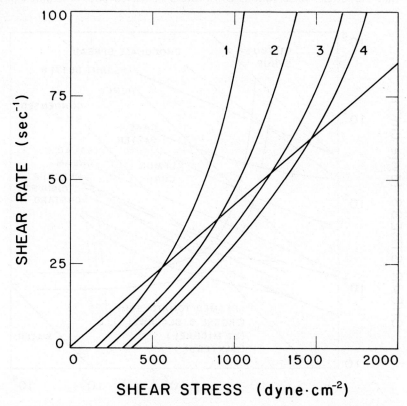

Samples 1, 2, 3 and 4 are the reconstituted non-Newtonian sources. The other sample is the Newtonian glucose syrup used as a standard.

Fig. 4. Instrumental shear stress–shear rate relationships for a Newtonian and non-Newtonian foods used to establish the shear rate-shear stress conditions associated with evaluation of viscosity in the mouth.

Furthermore, when the samples were compared in random pairs, and their viscosities were evaluated as multiples of the viscosity of the sample judged to be the most fluid, the data conformed to Stevens' (1957) psychophysical relationship.

$$\psi_R = kS^n, \qquad\qquad (1)$$

where ψ_R is the panelist's sensory response, S is the instrumentally measured shear stress developed at a shear rate of 50 s^{-1}, and k and n are constants with $n = 1.28$.

This means that viscometer determination of the shear stress developed at 50 s^{-1} shear rate by any food sample can be utilised in conjunction with Equation (1) to predict how the consumer would judge its viscosity in the mouth.

Subsequent work (Shama and Sherman, 1972) utilising the same approach but with foods covering a much wider range of flow properties than did Wood's samples has shown that the latter's observations constitute a restricted part of a much broader

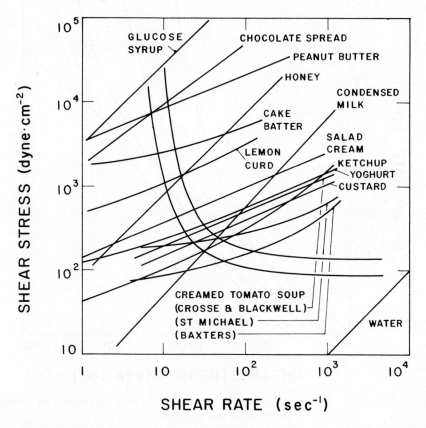

Fig. 5. Bounds for shear stress and shear rate associated with oral evaluation of viscosity.

picture. In fact, the shear rate operating in the mouth when viscosity is evaluated is not constant. It varies with the flow properties of the sample under investigation, a much higher shear rate being associated with viscosity evaluation of fluid foods than with evaluation of viscous and semi-solid foods. These findings are summarised in Figure 5, where the probable limits of shear stress and shear rate associated with oral evaluation of a wide range of viscosities are defined by the curved 'master' band. The widely different flow properties of a large number of food products, including food emulsions, are superimposed over this band in the form of shear rate-shear stress data. It is apparent that the double logarithmic shear rate ($\dot{\gamma}$)–shear stress (τ) plots for most of the foods examined give reasonable good straight lines over several decades of shear rate in accordance with the power law.

$$\dot{\gamma} = \frac{1}{\eta^*} (\tau - \tau_0)^{N_1} \tag{2}$$

or, in the form which is more correct dimensionally

$$\tau - \tau_0 = k (\dot{\gamma})^{N_2}, \tag{3}$$

where τ_0 is the yield stress, k is the consistency index, η^* is apparent viscosity, and N_1 and N_2 are constants. When $N_1 = 1$, the system is Newtonian, when $N_1 < 1$ it exhibits shear thinning and viscosity decreases as shear rate increases; when $N_1 > 1$ shear thickening is exhibited and viscosity increases with shear rate.

It appears that the nature of the stimulus associated with oral evaluation of viscosity also changes with the flow properties of the food. The stimulus for fluid foods appears to be the shear rate developed at an approximately constant shear stress of 100 dyne cm^{-2}, whereas for viscous and semi-solid foods it appears to be the shear stress developed at an approximately constant shear rate of 10 s^{-1}.

4. Determination of Shear Rates Associated with Viscosity Evaluation by Non-Oral Methods

As mentioned in Section 1 a preliminary impression of a product's viscosity is formed before it is inserted in the mouth, and this impression may be modified subsequently in the light of oral stimuli. It appears that the shear stresses and shear rates associated with non-oral methods of viscosity evaluation differ from those associated with oral evaluation. For example, the stimulus associated with viscosity evaluation by tilting a container in which there is a food emulsion is the shear rate ($0.1–40 \text{ s}^{-1}$) developed at a shear stress which varies ($60–600$ dyne cm^{-2}) with the flow characteristics of the food (Figure 6). Viscosity evaluation by stirring the contents of the container (Figure 6) is the shear stress ($10^2–10^4$) dyne cm^{-2}) developed at an approximately constant shear rate of $90–100 \text{ s}^{-1}$ (Shama *et al.*, 1972).

In effect, the shear rate–shear stress conditions associated with non-oral evaluation of viscosity by tilting the container parallel those associated with oral evaluation of the viscosity of fluid foods, while non-oral evaluation of viscosity parallels the oral evaluation of viscous and semi-solid foods.

5. Implications of the Variable Shear Rates Associated with Sensory Evaluation of Viscosity

5.1. FLOCCULATION AND EMULSION STRUCTURE

From what has been said in Sections 3 and 4 it is apparent that the shear rates and the shear stresses associated with viscosity evaluation by oral and non-oral methods vary within very wide limits according to the flow properties of the food being studied.

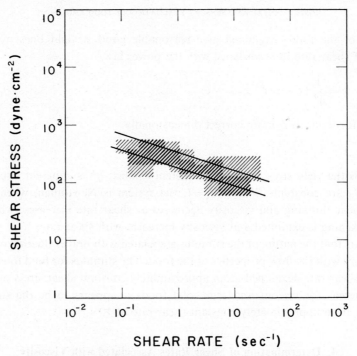

Fig. 6a.   Bounds for shear stress and shear rate associated with viscosity
evaluation by tilting the container.

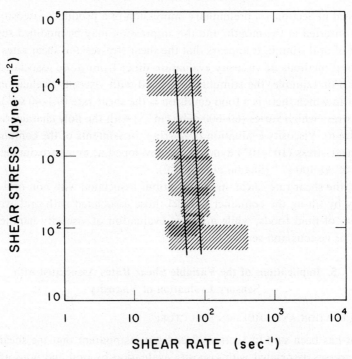

Fig. 6b.   Bounds for shear stress and shear rate associated with viscosity evaluation by stirring the
contents of a container.

With particular reference to food emulsions this means that when highly viscous or semi-solid products are evaluated orally, or by tilting the container, the assessment is made at a low shear rate such that the internal structure of the emulsions is only partially ruptured. Thus, the consumer's sensory response is associated with the shear stress developed within this partially broken down system. Sensory evaluation of the viscosities of these foods by stirring involves greater breakdown of their internal structures. In the case of fluid emulsions it is highly probable that their internal structures are almost wholly, if not wholly, broken down under the shear conditions associated with their evaluation.

Emulsions are prepared by mechanically dispersing a liquid (oil or water) as microscopic, or even sub-microscopic, size globules in another liquid (water or oil respectively) with which it is not miscible. The globules are stabilised by a third component, the emulsifying agent, which adsorbs (usually from the continuous phase) around each globule to form a protective layer so that when the globules come together (flocculate) they do not automatically coalesce together. During storage, following preparation of an emulsion, the globules flocculate and link up into aggregates. Initially, aggregates do not contain many globules, but at longer storage times the small aggregates link together to form larger aggregates and eventually all the globules may be connected as a three-dimensional network, somewhat like *a* very large bunch of grapes.

The globules in the aggregates are not in direct contact. They are separated by a distance (H_0) which is controlled by the net interaction force between them. This interaction (V) is defined by

$$V = V_R + V_A,$$ (4)

where V_R and V_A are the potential energies of repulsion and attraction respectively.

The globules have electrostatic charges on their surfaces and each surface is surrounded by a diffuse layer of counterions of equal and opposite charge. The magnitude of the surface charge depends on the nature and concentration of the various ions present, and the extent to which they are adsorbed on the globule surfaces. Globules repel each other when they are close enough for their diffuse double layers to overlap. Usually, the thickness $(1/\chi)$ of the diffuse layer is small with respect to the diameter (D) of the globules, and in such cases V_R can be calculated from

$$V_R = \frac{\varepsilon D \psi^2}{4} \{\ln [1 + \exp(-\chi H_0)]\},$$ (5)

where ε is the dielectric constant of the continuous phase of the emulsion, ψ is the potential in the outermost region of the diffuse layer (this is approximately equal to the zeta potential derived from electrophoretic mobility measurements). The value of $1/\chi$ is often taken as 10^{-7} cm.

An approximate, but reasonably reliable estimate of V_A is derived from

$$V_A = -\frac{AD}{24H_0},$$ (6)

where A is London-van-der Waals' constant. Published values of the latter range between 10^{-11}–10^{-14} erg depending upon the particular set of circumstances. Methods for calculating the value of A have been reviewed elsewhere. (Sherman, 1968).

When calculated values of V are plotted for different values of H_0 a curve of the general form shown in Figure 7 is obtained. The negative region of V defines the conditions under which the globules are attracted to one another. The curve has several distinctive features. Within the low H_0 region there is a potential energy barrier which globules must surmount if they are to come very close together and eventually make contact. It is usually assumed that if this barrier (V_{max}) has a value of 15–20 kT or more, where k is the Boltzmann constant and T is the temperature on the absolute scale, then only a proportion of the globules are able to pass over it. The number of globules that can do this decreases approximately exponentially as the potential energy

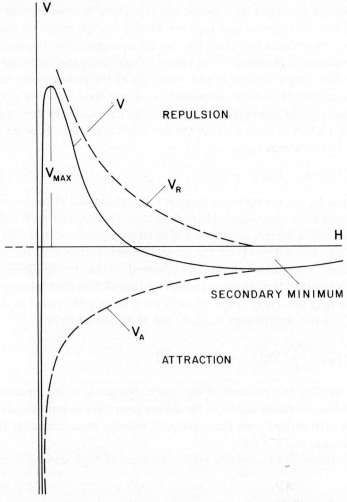

Fig. 7.   Influence of distance separating globules on their potential energy of interaction.

barrier increases beyond the critical value. If the location of the barrier on the H_0 axis is less than twice the thickness of the adsorbed layer of emulsifier around each globule then globules are prevented from making contact by steric barriers arising from the orientation and configuration of the emulsifier molecules at the oil-water interface. This latter effect is found in protein stabilised oil-in-water (O/W) emulsion and in water-in-oil (W/O) emulsions.

The value of V_R decays more rapidly with increasing H_0 than does V_A (Figure 7). Consequently, at some particular H_0 value V changes its sign from positive to negative and attraction forces prevail. The attraction force increases rather sharply at first with increasing H_0 to a maximum value and then it decays slowly with further increase in H_0. The H_0 region in which V exhibits maximum attraction is known as the secondary minimum.

In O/W emulsions the potential energy barrier to close approach of globules is usually large with respect to the secondary minimum of attraction. Consequently, after flocculation, most of the globules are separated by a distance corresponding to the location of the secondary minimum on the H_0 axis in Figure 7. The potential energy barrier in W/O emulsions is usually too small to be significant, and the distance separating drops following flocculation corresponds to approximately twice the thickness of the emulsifier layer adsorbed around the globules. This situation also applies for O/W emulsions stabilised by polymeric molecules such as milk proteins or gelatine. When the concentration of protein is very low it is possible that individual molecules bridge the gap between globules rather than undergo adsorption on their surfaces.

The three-dimensioned interlinked structure formed in both O/W and W/O emulsions after the globules have flocculated encloses liquid of the continuous phase within the voids between the globules. The volume of continuous phase held in this way depends on the size and number of voids.

Since globules tend to flocculate without passing over the potential energy barrier the initial rate of flocculation can be defined by Smoluchowski's (1916, 1917) theory of rapid flocculation. If there are N_0 globules cm^{-3} of freshly prepared emulsions, i.e. at time t_0, then at any other time t the number (N_1) of unassociated globules, the number (N_2) of aggregates containing two globules, the number (N_3) of aggregates containing three globules the number (N_i) of aggregates containing i globules, is given by

$$N_1 = \frac{N_0}{(1 + t/t_f)^2}, \qquad N_2 = \frac{N_0\,(t/t_f)}{(1 + t/t_f)^3},$$

$$N_3 = \frac{N_0\,(t/t_f)^2}{(1 + t/t_f)^4}, \qquad N_i = \frac{N_0\,(t/t_f)^{i-1}}{(1 + t/t_f)^{i+1}}, \tag{7}$$

where t_f, the time taken for the number of unassociated globules to reduce from N_0 to $N_0/2$, is given by

$$t_f = \frac{3\eta_0}{4kTN_0}, \tag{8}$$

where η_0 is the viscosity of the continuous phase. Thus, a wide range of aggregates containing different numbers of globules soon begin to form.

When η_0 is small, so that $t \gg t_f$, we have the situation

$$N_1 \approx N_2 \approx N_3 \ldots \approx N_i \approx \frac{N_0}{(1 + t/t_f)^2} \qquad (9)$$

and we have approximately equal numbers of aggregates containing $2, 3, \ldots, i$ globules, and of unassociated globules. In this situation only a relatively small proportion of the total number of globules cm^{-3} remain unassociated.

5.2. GLOBULE COALESCENCE

Following flocculation the globules coalesce, and this results in a decrease in N_0. The number (N_t) of globules cm^{-3} at any time t after the emulsion has been prepared is defined by

$$N_t = N_0 \exp(-Ct), \qquad (10)$$

where C is the rate of coalescence. At present we do not have precise information about the mechanism of calescence, but it is believed that important factors include the rate at which the films of continuous phase liquid between globules thin and drain away and the rheological properties of the emulsifier films adsorbed around the globules. The latter factor will determine the way in which the protective films rupture prior to actual coalescence.

The relative rates of flocculation and coalescence, and which of these two processes predominates, will have an important influence on the rheological properties of emulsions.

5.3. FACTORS INFLUENCING THE INSTRUMENTAL EVALUATION OF VISCOSITY

The chemical and physical properties of emulsions which influence the degree to which emulsions undergo structural alteration at low, intermediate and high rates of shear are summarised in Table I. Many of these factors and their implications have already been discussed. An important factor not discussed so far is solubilization in the presence of excess emulsifier. Most, if not all, food emulsions contain emulsifier in far greater concentration than that required to provide compact adsorbed layers around all globules. The excess emulsifier molecules form structures known as micelles (spherical or cylindrical, depending on the circumstances) within the continuous phase, and these micelles solubilize liquid from the dispersed phase. This changes the volume ratio of dispersed phase to continuous phase, and consequently the rheological properties of the emulsion are altered.

Some emulsifiers undergo partial transfer from the liquid phase in which they are initially dissolved, or dispersed, across the interface and into the other liquid phase. This may be associated with solubilization phenomena which influence the volume ratio of the two phases.

TABLE 1

Influence of shear rate on the phenomena associated with the rheological state of emulsions

Component of dispersed system	Variable through which effect exerted	Phenomenon involved	Phenomena operative at following rates of shear		
			Low	Intermediate	High
Dispersed phase	Volume concentration	(a) Hydrodynamic interaction between small globule aggregates		X	X
		(b) Hydrodynamic interaction between individual globules	X		
		(c) Properties of interlinked 3-dimensional globule networks	X	X	X
		(d) Flocculation and particle aggregation	X		
		(e) Interaction potential between individual globules	X		X
	Viscosity	(a) Deformation of globules by shear and/or due to internally developed stresses in aggregates after flocculation	X	X	X
		(b) Fluid circulation within globules	X		
	Mean particle size, and particle size distribution	(a) Deformation of globules by shear and/or due to internally developed stresses in aggregates after flocculation	X	X	X
		(b) Hydrodynamic interaction between small globule aggregates		X	X
		(c) Hydrodynamic interaction between individual globules			X
		(d) Interaction potential between individual globules	X		
		(e) Properties of interlinked 3-dimensional globule network	X		
Continuous phase	Chemical constitution	Solubilization phenomena involving the emulsifier and the continuous phase	X	X	X
	Viscosity	(a) Globule flocculation and aggregation	X	X	
		(b) Rheological properties of emulsions	X	X	X
	Chemical constitution	(a) Solubilization phenomena involving the emulsifier and dispersed phase	X	X	X
		(b) Potential energy of interaction between globules	X	X	
Emulsifier	Chemical constitution	(a) Solubilization phenomena involving the dispersed and continuous phases	X	X	X
		(b) Potential energy of interaction between globules	X	X	X
	Adsorption at oil-water-interface	(a) Effective particle size	X	X	X
		(b) Attraction forces between globules in aggregates	X	X	
		(c) Potential energy of interaction between individual globules			X
		(d) Globule deformation by shear and/or due to internally developed stresses in aggregates after flocculation	X	X	X
		(e) Mass transfer between the 2 liquid phases	X	X	X
		(f) Rate of globule coalescence	X	X	X

5.4. Equations Defining the Rheological State of Emulsions

5.4.1. *Dilute Emulsions*

The viscosity (η) of very dilute Newtonian emulsions can be defined under certain circumstances, by Einstein's (1906, 1911) equation

$$\eta = \eta_0 (1 + a\phi), \tag{11}$$

where ϕ is the volume fraction of globules, and a is a constant with a value of 2.5 provided the globules behave as rigid, spherical, particles, When globule diameter does not exceed a few microns this is the case. Other assumptions on which the validity of Equation (11) depends are that the globules do not interact with one another, and also that they are large relative to the size of the continuous phase molecules.

When the globules are deformable, and the adsorbed emulsifier film around each globule cannot prevent the transmission of the tangential component of an applied shear stress across the oil-water interface, the fluid within the globules undergoes internal circulation, and Equation (11) has to be modified (Taylor, 1932; Nawab and Mason, 1958) to

$$\eta = \eta_0 \left[1 + a \left(\frac{\eta_i + \frac{2}{5}\eta_0}{\eta_i + \eta_0} \right) \phi \right], \tag{12}$$

where η_i is the viscosity of the liquid forming the globules. The net effect is to reduce the value of a from 2.5.

5.4.2. *Concentrated Emulsions*

Attempts have been made to modify Equation (11) so that its range of validity can be extended to higher values of ϕ, but no agreement has been reached yet as to the precise form which such an equation should take. Many alternative forms have been proposed, and most of them have the general form

$$\eta = \eta_0 \exp \left[\frac{a\phi}{1 - k\phi} \right], \tag{13}$$

where k is a particle interaction coefficient that is inversely proportional to globule diameter. This means that if the mean globule size of an emulsion is reduced without altering its composition, the viscosity will nevertheless increase.

Expansion of Equation (13) gives

$$\eta = \eta_0 (1 + a\phi + b\phi^2 + c\phi^3 + \cdots), \tag{14}$$

where $b, c, \ldots$ are constants with values dependent on mean particle size.

In general all emulsions, apart from very dilute ones, show non-Newtonian flow properties, and Equations (13) and (14) then apply only at such high shear rates that all the internal structure due to globule flocculation has been destroyed, and the globules flow independently of one another. When the shear rate is lower than that required to achieve this condition a concentrated emulsion will exhibit a viscosity which is higher than is suggested by Equations (13) and (14). The emulsion behaves as if it had an apparent volume concentration of globules (ϕ_a) greater than the true value, ϕ, and this is defined by the relationship

$$\phi_a = \frac{\pi D^3}{6} [N_1 + f(N_t - N_1)] \tag{15}$$

where $f(=\phi_a/\phi)$ is the swelling factor (Mooney, 1946). The apparent increase in ϕ to ϕ_a is due to residual globular aggregates and the immobilisation of fluid from the continuous phase within these aggregates. The factor f decreases exponentially with increasing shear rate.

Under very low or zero shear conditions concentrated emulsions form globule aggregates in three-dimensions, and these structures exhibit viscoelastic properties i.e. they have the properties both of solids (elasticity) and of liquids (viscosity). The elastic shear modulus (G_s) of such structures is given by

$$G_s = \frac{\phi(1 - 1.828v) A}{36\pi D^3 H_0^3}, \tag{16}$$

where v is the volume fraction of continuous phase held in the voids between the globules (Sherman, 1970). As the shear conditions are increased G_s decreases in accordance with the degree to which the network structure is ruptured.

5.5. CONCLUDING REMARKS

Information is now available on the way in which the stress/shear rate behavior of emulsions influences their structure, and on the chemical and physical factors which influence these relationships. In addition, data is appearing on the stress-shear conditions associated with the sensory evaluation of viscosity in the mouth and by non-oral means. By combining these colloid, rheological and psychophysical observations it should be possible to predict the structural condition of food emulsions during sensory evaluation i.e. whether they are completely broken down, partially or only slightly broken down, and how this can be modified so as to influence the sensory response. For example, the mean particle size of both the fat and ice crystals in frozen ice cream can be reduced by lowering the freezer temperature by 2–4°C. As the mean particle size decreases the frozen ice cream becomes harder as assessed sensorily (Sherman, 1967) and instrumentally (Sherman, 1969). The consistency of protein stabilized O/W emulsions can be significantly altered by adding small concentrations of electrolytes, or alternatively by homogenizing at different pressures. By the first treatment the electrical charge on the surfaces of the globules is altered, while the second treatment alters the mean particle size and particle size distribution. Finally, the consistency of butter and margarine can be drastically changed by passing them through a perforated plate, or fine nozzle, to partially break down the internal fat network structure.

References

Barry, B. and Grace, A. J.: 1972, *J. Pharm. Sci.* **61**, 335.
Brandt, M. A., Skinner, E. Z., and Coleman, J. A.: 1963, *J. Food Sci.* **28**, 404.
Einstein, A.: 1906, *Ann. Phys.* **19**, 289.
Einstein, A.: 1911, *Ann. Phys.* **24**, 591.
Mooney, M.: 1946, *J. Colloid Sci.* **1**, 195.
Nawab, M. A. and Mason, S. G.: 1958, *Trans. Faraday Soc.* **54**, 1712.
Shama, F. and Sherman, P.: 1973, *J. Texture Studies* **4**, 111.
Shama, F., Parkinson, C., and Sherman, P.: 1973, *J. Texture Studies* **4**, 102.
Sherman, P.: 1967, *Food Technol.* **21**, 107.
Sherman, P.: 1968, in P. Sherman (ed.), *Emulsion Science*, Academic Press, London and New York, Chapter 4.
Sherman, P.: 1969, *J. Texture Studies* **1**, 43.
Sherman, P.: 1970, *Proc. 5th Intern. Congr. Rheol.* **2**, 327.
Smoluchowski, M. von: 1916, *Physik. Z.* **17**, 557, 585.
Smoluchowski, M. von: 1917, *Z. Physik Chem. (Leipzig)* **92**, 129, 155.
Stevens, S. S.: 1957, *Psychol. Rev.* **64**, 153.
Szczesniak, A. S.: 1963, *J. Food Sci.* **28**, 385.
Szczesniak, A. S. and Farkas, E.: 1962, *J. Food Sci.* **27**, 381.
Taylor, G. I.: 1932, *Proc. Roy. Soc.* **A138**, 41.
Wood, F. W.: 1968, in '*Rheology and Texture of Foodstuffs*' *S.C.I. Monograph* **27**, p. 40.

CHAPTER XIII

BASIC CONCEPTS OF COLORIMETRY

FREDERICK J. FRANCIS

Dept. of Food Science and Technology, University of Massachusetts, Amherst, Mass., U.S.A.

This chapter deals with color measurement and interpretation for opaque solids and transparent liquids. The problems associated with color measurement of turbid liquids and slurries are much more complicated and deserve treatment on their own.

The theoretical concepts involving color measurement of clear transparent liquids are identical with those of completely opaque solids except that the light is transmitted through the sample rather being reflected from the surface. These concepts have been described previously in *Food Technology* (Clydesdale, 1969; Francis, 1972).

1. The Concepts of a Colorimeter

A colorimeter is an instrument to reproduce optically and electronically the physiological sensation of the human eye. A colorimeter measures color as such and is not to be confused with the earlier use of the word colorimeter in chemical analysis. Instruments used for estimating the amount of a chemical are called color comparators and absorptimeters. Spectrophotometers can, of course, be used for both chemical analysis and calculation of color coordinates.

The design of a modern colorimeter can best be understood by an analogy to the way the human eye sees color. We have two anatomically distinct types of receptors in the human eye – the rods and cones. The rods are concerned with black and white vision in dim light and have no color function. There are three types of cones in the human retina, one sensitive to red, one to blue and the other to green. They are anatomically indistinguishable, and it was only three years ago that physical evidence was obtained that they were different even though Helmholtz had postulated 80 years ago that they had to be different. The human eye receives light reflected from an object to the retina and a signal from each type of cone is sent to the brain. The brain interprets the signals and assigns a 'color' to the object.

A simple colorimeter can be designed to duplicate the response of the human eye (Section 5). In Figure 1, three projectors with a red, green and blue filter, respectively, in front of the lens, shine a colored beam on a screen. Another projector with a filter of unknown color is projected on the same screen. If the operator can vary the amount of red, green and blue light reaching the screen, he can match almost any color. Then the unknown color can be described by the amount of red, green and blue required to match it (Figure 2). This principle has been used in several visual colorimeters to define the fundamental color solid. The colorimeter itself is too crude for every day

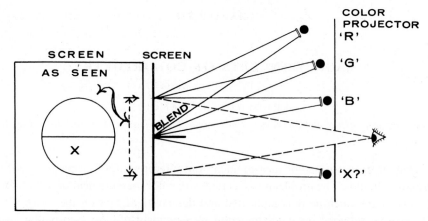

Fig. 1.   Color matching by the addition of lights. Screen is seen on edge and, to the left edge, in full view. Arrows indicate area on screen illuminated by lights used for matching (top half of circle) and light to be matched (bottom half, marked 'x'). Blending is accomplished by varying the intensity of light from each projector of a primary color; it occurs where the three beams overlap and the word 'BLEND' appears.

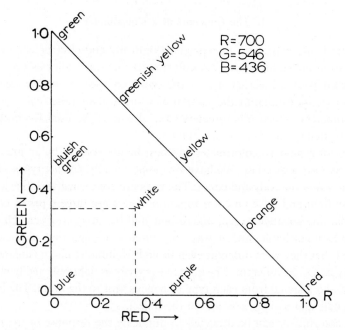

Fig. 2.   A right-angled *G R B* chromaticity diagram (Maxwell triangle).

use, but the data can be used to define a much more appropriate color solid. The triangle in Figure 2 is shown in the left hand portion of Figure 3 and a new set of stimuli called *X Y Z* are shown. The *G R B* primaries are physically realizable in the laboratory whereas the *X Y Z* primaries are not. The *X Y Z* points were chosen for mathematical convenience, and the fact that they cannot be made physically does not detract from their usefulness. Although it is not quite true, for ease of remembering, the X may be considered as degree of redness, Z blueness and Y greenness. The Y value also carries all the brightness factor. In the right hand side of Figure 3, the coordinate axes are shifted until the $X Y Z$ triangle is right angled, and of course, the $G R B$ is distorted. The $X Y Z$ diagram is accepted world wide as the fundamental

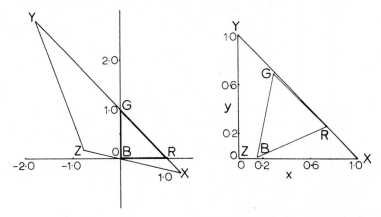

Fig. 3.   Transformation of the $G R B$ to the $X Y Z$ chromaticity diagrams. The right diagram is the $X Y Z$ obtuse triangle transformed into a right-angled triangle.

color solid. Every realizable color will have three coordinates which locate the point within the color solid. When one says "Measure the color," he is asking for the three coordinates which locate a point in space.

2. The Design of a Colorimeter

The problem in designing a colorimeter to duplicate the response of the human eye can be appreciated with the setup in Figure 1. We can shine a spectral color, say a blue of 400 nm, on the screen and ask the operator to match it with his red, green and blue controls. We can repeat the process for 410 nm and so on through the spectrum. The data obtained can be transformed from $G R B$ units to $X Y Z$ units and plotted as in Figure 4. The curves obtained will represent how the human eye sees the spectral colors. This information is all an optical engineer requires to design an instrument to duplicate the response of the human eye. A set of glass filters with transmission curves shaped like those in Figure 4 are shown in Figure 5 in a simple colorimeter. A photo-cell and meter can be used to take a reading of the light reflected from an object through each filter in turn and the readings are the $X Y Z$ values of the object. Every colorimeter uses this basic principle.

Newcomers to the science of colorimetry may be confused by the types of readout used by various models of colorimeters. For example, the Hunterlab instruments, the Color-Eye, the Colormaster and the Lovibond Tintometer, all use a different color solid, and hence, different coordinates, but they all have one thing in common. They

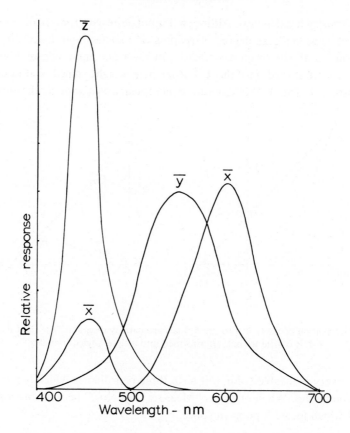

Fig. 4.   Standard observer curves based on the CIE, $X\,Y\,Z$ system.

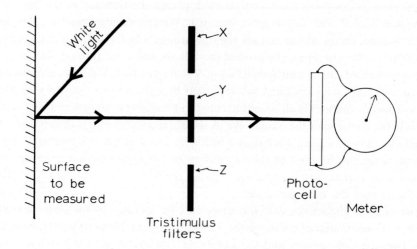

Fig. 5.   A simple tristimulus colorimeter.

all give three coordinates to locate a color in space. If one wants to convert data from one system to another, the conversion equations are only approximate, not exact.

The Judd-Hunter system as illustrated in Figure 6 is the color solid used by both the Gardner and the Hunterlab instruments and seems to be the most popular in the food field in America. Most of the color data on horticultural crops in America is in

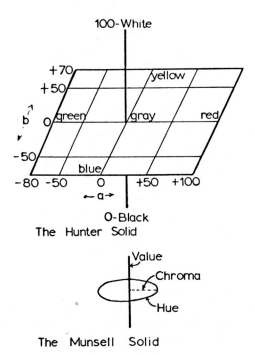

Fig. 6. The Hunter and Munsell Color Solids.

this system. It is relatively simple to understand with $+a$ for degree of redness, $-a$ for greenness, $+b$ for yellowness and $-b$ for blueness. The degree of lightness or darkness is represented by the vertical L or Rd scales. The algebraic quadrant concept is very familiar to us in view of early mathematical training, so it is easy to visualize a color in this system.

3. The Concepts of a Spectrophotometer

Another approach to color measurement is equally important, namely spectrophotometry. A conventional spectrophotometer can be used to obtain either transmission or reflection curves from which $X\,Y\,Z$ data can be calculated. We can illustrate this by Figure 7 which shows diagrammatically the calculation of $X\,Y\,Z$ data. R represents the transmission curve of a sample as obtained with a specific light source with energy distribution E. If we wish to characterize the sample in terms of say a standard light

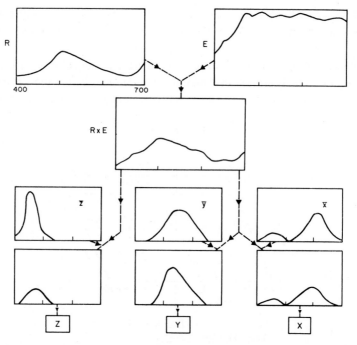

Fig. 7.   A diagrammatic illustration of the calculation of $X\,Y\,Z$ primaries.

source with energy distribution E, we multiply $R \times E$. The curve $R \times E$ is multiplied by the functions $\bar{x}$, $\bar{y}$, and $\bar{z}$ of the standard observer curves. The multiplication may take place at intervals across the visible spectrum, so that at any one wavelength $X = R\,E\,\bar{x}$, $Y = R\,E\,\bar{y}$ and $Z = R\,E\,\bar{z}$. This is actually equivalent to an integration between 380 and 750 nm as follows

$$X = \int_{380}^{750} R\,E\,\bar{x}\,dx$$

$$Y = \int_{380}^{750} R\,E\,\bar{y}\,dy$$

$$Z = \int_{380}^{750} R\,E\,\bar{z}\,dz$$

In order to obtain tristimulus coefficients or chromaticity coefficients from $X\,Y\,Z$ data, these values are merely expressed as fractions of their total such that

$$x = \frac{X}{X + Y + Z} \qquad y = \frac{Y}{X + Y + Z} \qquad z = \frac{Z}{X + Y + Z}$$

4. Interpretation of Color Data

In the 1940's and early 1950's nearly all instrumental color data for foods were calculated from spectrophotometric data. The calculation of X Y Z for one sample required at least 30 min, thus the approach was not often used. The calculation labor inspired the development of tristimulus colorimeters, and the 1960's saw the development of a number of different types. They had the advantage of ease and rapidity of measurement and relative simplicity of interpretation. However, in the late 1960's and 1970's rapid electronic computers have become much more accessible, and color measurement, at least in the more sophisticated applications, is returning to calculations directly from reflection or transmission curves. One of the latest developments is a spectrophotometer which obtains a reflection or transmission curve from a sample by means of a series of interferance filters spaced through the visible spectrum. The spectrophotometer is attached to a small computer which is attached to a teletypewriter. The computer can be programmed to print color data from any standard source of illumination, in any color scale, in direct units or in any desirable color-difference equations. Similar systems are available for conventional spectrophotometer. The above is a very sophisticated color system in use with some textile, paint and

Fig. 8. A Diano-Hardy spectrophotometer attached to a minicomputer and a teletypewriter for color measurement and calculations.

plastic applications. It is not being used insofar as I am aware in any food applications. However, there are many less pretentious applications in the food area, since most research laboratories have access to computers. With the calculation labor removed, this approach is very attractive. It will become even more so in the future for research and continuous to on-line measurement, but there is little doubt that conventional tristimulus colorimetry will constitute the bulk of food colorimetry for some time to come.

References

Clydesdale, F. M.: 1969, 'The Measurement of Color', *Food Technol.* **23**, 16–22.
Francis, F. J.: 1972, 'Colorimetry of Liquids', *Food Technol.* **26**, 39–42.

CHAPTER XIV

METHODS AND MEASUREMENTS OF FOOD COLOR

FERGUS M. CLYDESDALE

Dept. of Food Science and Technology, University of Massachusetts, Amherst, Mass. 01002, U.S.A.

Appearance involves all aspects of visual experience by which things are recognized. These include form and markings as well as color and gloss. Color is the most important optical attribute of product appearance, and the one which shall be discussed in detail, but it should be remembered that gloss, luster, turbidity, haze, distinctness-of-image, reflection, texture and other attributes also play an important role.

Consumers have strong preferences for products which have appearances which appeal to them. Certainly, where there is a choice, the products with the greatest visual appeal will be chosen first. In fact, preferences for visual appeal and visual uniformity of appearance are economically so important that quantitative identifications of appearance, not mere subjective descriptions, must be demanded by businesses dealing with consumer products.

There have been seven different procedures described by R. S. Hunter for the identification of a product color:

(1) by spectrophotometric curves (physical analysis);

(2) by visually equivalent additive mixtures of red, green and blue;

(3) by location in a three-dimensional color solid; for example: (a) CIE trilinear coefficients x, y, %Y; and (b) opponent colors with gray in the center (L, a, b, or $U^*V^*W^*$);

(4) by densities of three substractive-primary inks or dyes required to represent the color (color photography and color printing);

(5) by location with respect to visually systematic arrangement of color chips (Munsell);

(6) by location with respect to an array of chips of systematic substractive mixtures, of white, black and colored pigments of each separate hue (Ostwald). These substractive mixture dimensions of each color are familiar to the formulator or dyer who is experienced in obtaining product colors 'by eye';

(7) by ingredients required to obtain the color in a given product: (a) derived by color formulator's estimates; (b) derived by optical model of product, optical constants of ingredients and a computer.

This discussion will be basically concerned with objective techniques within the general area of tristimulus colorimetry. Therefore techniques and problems in measurement will be limited to method (3).

ChoKyun Rha (ed.), Theory, Determination and Control of Physical Properties of Food Materials, 275–289.

When considering instrumentation used in tristimulus colorimetry one must consider what happens when light falls upon the object and also the non-optical characteristics of the object.

R. S. Hunter has pointed out:

that a number of things happen when light falls upon an object. First if the object is opaque, light will be reflected by it. If it is transparent or translucent, light will pass through the object, or be partially reflected or transmitted by the object primarily in one of two manners; specularly or diffusely. Specularly reflected light is that which (as with shiny objects) is concentrated in the mirror direction of reflection, as a continuation of the incident beam. The straight-through transmission of the light by a transparent object can also be thought of as specular, although this admittedly is taking some liberties with the literal meaning of the word. Diffusely reflected or transmitted light on the other hand, is that which leaves the object uniformly in all directions. Similarly, a white translucent film transmits light, uniformly in all directions; that is, the source of light is not recognizable through it as the source.

The foregoing analysis provides the basis for an arbitrary classification of objects into four groups, based on whichever of the four manners of light projection is dominant:

(1) Opaque, non-metallic, such as a yellow ceramic vase.

(2) Opaque, metallic, such as a metal bell.

(3) Translucent, such as a plastic tumbler.

(4) Transparent, such as a bottle of clear yellow liquid.

Reflecting objects fall in groups 1 or 2; transmitting objects in groups 3 or 4. Objects which reflect or transmit light primarily in a specular manner fall into groups 2 and 4. Groups 1 and 3 are made up of objects that reflect or transmit light in a diffusing manner.

Having established a very rough framework of optical groups it is necessary to briefly consider non-optical characteristics which affect measurement.

One of the simplest problems which arises when measuring a sample is the manner in which you place the cuvette for the sample itself on the instrument. This depends on the type of sample, i.e. whether it is a liquid or a solid, and the type of measurement you want to make. For instance, if a sample is a liquid, obviously it must be in a container. The container (cuvette) may then be placed on top of the aperture of the instrument and the light from the source will then pass through the bottom of the cuvette. Another method would be to place the cuvette in a vertical position with reference to the instrument. In this case, the light enters the sample through the side of the cuvette. When it is advantageous to view the sample directly and not have the light pass through the cuvette, it is possible on some instruments to invert the viewing head and view the sample from above. These three types of viewing may be seen in Figure 1. Solids may be treated in the manner as liquids, but if they are compact and dry they can be placed directly on the viewing aperture. Another problem is caused by the irregularity of coloring in a food material. An example of an irregularly colored food is an apple which is partially red and partially green.

Depending on the size of each object in the sample, we can measure such a sample by either a spinning technique or a large area aperture. The concept of a spinning technique is illustrated in Figure 2. With large objects such as apples, a spinner to rotate each object would have to be used. If the sample is rotated fast enough, the resultant mixture of light reflected from the red and green portions would make up the signal to the phototube. With smaller objects, either a spinner or large area aperture could be used.

The large area aperture is merely an attachment for an instrument, along with appropriate optics, which allows a large number of samples having varying colors to be viewed simultaneously (e.g. cranberries), or a large portion of a sample to be viewed

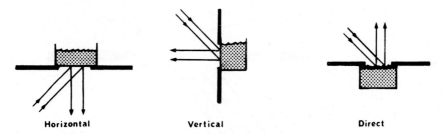

Fig. 1.   Horizontal, vertical and direct viewing of a sample.

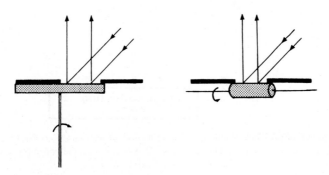

Fig. 2.   Examples of spinning techniques to measure irregularly colored foods.

(e.g. a slice of mellon). The spinner accomplished the same thing by exposing a maximum of sample through an aperture.

Such measurements give rise to a readout which is, in most cases, a composite of several colors. This is an empirical measurement and as such has to be correlated with visual judgment. That is, visual judgment will be used to predict what is desirable and then some function which can be derived from the instrument measurement can be used to make a decision about the product.

Use of the large area aperture is probably a more convenient method than the spinner because it is faster and there is less physical manipulation of the sample. However, a large unit cannot always be used when directional problems are apparent in a food material. For example directional problems exist in a cob of corn or a piece of meat. The manner in which the kernels or fibers are lined up have an affect not only on the amount of light reflected from the sample, but on the type of light reflected as well.

Irregular solid shaped food materials also may cause a problem. This problem is illustrated in Figure 3 and is often termed a "pillowing problem" (a term coined by R. S. Hunter). In a pillowing problem 2 sub-problems exist, one is geometrical and the other may be called a lens problem.

The geometrical problem exists because an instrument is calibrated for light to strike the sample at a specific distance from the source and from the phototube. However, as shown in Figure 3, if the sample is irregularly shaped and extends below

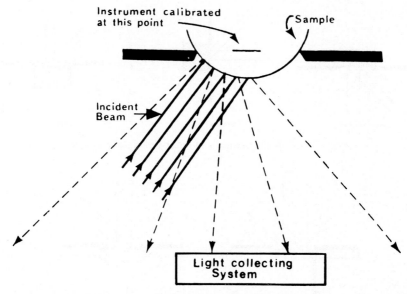

Fig. 3. The 'pillowing' problem. Some specular reflection as well as diffused reflection may enter the light collecting system. The sample surface is closer to the collecting system than the plane of calibration.

the aperture, then the calibration of the instrument is affected and the actual color of the readout obtained is not representative of the actual color of the sample. In this case, when a portion of the sample is closer to the photocell than the calibration, the actual reading will be lower in reflectance.

This could occur because when two beams incident on the sample at 45° are focused such that they will be perfectly superimposed in the plane of calibration, they will not be perfectly superimposed on the sample in a plane closer to the detector system. This could result in a larger area of the sample being illuminated; if the detector system did not have a large enough area to catch all the reflected light, a low reading would result. The same affect could occur in an instrument designed with an expanding incident beam and a detector system appropriate in size for a flat sample. A curved sample would exaggerate the expanding beam and result in a loss of reflected energy. The other problem is due to the fact that a curved sample surface could act as a lens and expand such a reflected beam. This might result in some specular reflection reaching the detector system (Figure 3). The answer to these problems, obviously, is to flatten the sample but this may be difficult to do. We should present an irregular solid like this on a spinner or perhaps measure a large area and use more sample. The directional affect mentioned previously is not a very serious problem in the food industry but it could occur. Usually if the directional problem is apparent, such effects may be minimized by using a spinner or a circular light source such as that developed by Hunterlab. Having mentioned briefly the four sub-groups in the object mode and also the gross non-optical attributes which affect the measuring technique of a sample it is now necessary to consider in more detail the optical characteristics which affect measurement.

When energy in the form of light encounters any object, the following may occur:
(1) reflection from the surface;
(2) refraction into the object;
(3) transmission through the object;
(4) diffusion;
(5) absorption within the object.

Other phenomena, less important since they are very specialized, may also affect the color. These include polarization, fluorescence, and interference, however for this discussion only the afore mentioned five factors will be considered.

When light strikes an object, reflection at the surface occurs. However, the type of reflection depends upon the surface which the light strikes. Where the boundary between two media consists of a series of small interfaces oriented at all possible angles to the normal, such as occurs when light strikes a rough surface, the distribution of the reflected energy follows the Lambert Cosine Law (Birth and Zachariah, 1971). Such an incident is shown in Figure 4. This might be thought of as a rough metallic surface where any radiation that enters the medium is quickly absorbed by free electrons, so essentially the only reflectance is regular reflectance. As shown in the figure, some of the rays strike the surface more than once before being reflected thus losing

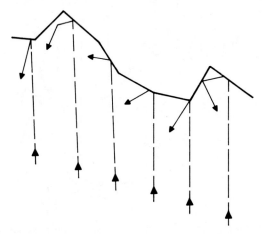

Fig. 4.   Light striking a rough surface (Birth and Zachariah, 1971).

some energy. As a result, a rough surface would be expected to have a lower total reflectance than a smooth surface. Figure 5 represents a very common situation in which 2 optical paths can be traced (Birth and Zachariah, 1971):

(1) The incident light encounters an initial smooth surface, and the regular reflection produces gloss or glare, commonly referred to as the specular component of reflectance. This type of reflection is not normally considered as a function of color.

(2) The light that is transmitted through the first interface undergoes absorption according to Beer's Law and Lambert's Law through some function of the distance and then is reflected at randomly oriented internal interfaces, and a fraction of this reflected radiation is transmitted back through the initial interface. The radiation that follows the second path constitutes diffuse reflection, which is normally considered a function of color. However it should be stressed that the absorption of the material affects both these types of reflectances differently and this should be kept in mind.

Figure 6 represents another condition where the smooth surface shown in Figure 5

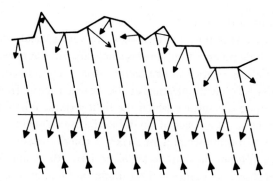

Fig. 5.   Light striking a surface which is initially smooth but has a rough layer underneath (Birth and Zachariah, 1971).

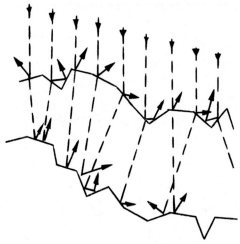

Fig. 6.   Light striking an initially rough surface with a rough surface also underneath (Birth and Zachariah, 1971).

is replaced by a rough surface. In this situation both the regular reflectance and diffuse reflectance are Lambertian; therefore the two forms of reflectance are not so easily separated.

Birth and Zachariah (1971) describe an excellent biological example of these forms of reflectance with an apple. As harvested, the apple has a minute wax coat which is not naturally smooth, so the regular reflectance is Lambertian, while the diffuse reflectance conveys the characteristic color to the eye. Since the eye sees the sum of both reflectances, a low color value is perceived as in an unpolished apple. Polishing the apple makes the wax coating smooth so the regular reflectance or gloss can be distinguished from the diffuse reflection. The regular reflectance imparts to the apple the bright glossy appearance of a polished fruit. However, the polished fruit seems to have a higher color value because the observer unconsciously orients the polished fruits so that the specular component of reflectance is not observed and one sees only that radiation which has undergone absorption by the pigment in the fruit skin.

Refraction measures the degree to which light is slowed down relative to its travel in air. At every boundary between two materials of different refractive index, light changes its speed and a small fraction of it is reflected. This usually is about 4% reflection for most common materials. This isn't a great deal of reflection, but when we are dealing with a particulate compound whose particles have a different refractive index than the surrounding medium, these encounters with their subsequent 4% reflection are repeated over and over again until you get a thorough diffusion of the light and total reflectances of up to 80–90%.

The effect of refractive index upon scattering in a particulate medium is illustrated graphically in Figure 7. From this graph it may be seen that when the refractive index

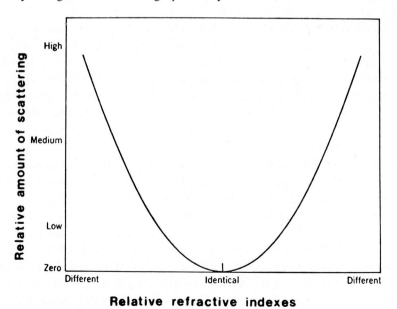

Relative refractive indexes

Fig. 7.   The affect of refractive index upon scattering (Billmeyer and Saltzman, 1966).

of the particles and the medium in which they are suspended are the same, there is virtually no scattering. On the other hand, as the difference in refractive index between the two increases, the scattering and thus the reflectance increases to a maximum.

The transmission of light is described by a transmitted ray that passes entirely through an object and is a continuation of the incident beam. This is an over simplification obviously, but for the purposes of this discussion it will suffice.

Diffusion is really a sub-section of reflection which we have discussed. However, its importance to color is such that further consideration is merited. Diffusion is a phenomenon which occurs when light comes in contact with a body which is non-homogeneous, i.e. which does not have a smooth surface. It is the term often used to identify what happens to a ray of light that is affected either by contract with a smooth surface, or by particles within the body of a granular or fibrous material. The first phenomenon is often called surface diffusion and the second, internal diffusion. As mentioned previously, surface diffusion is a function of color more than a function of gloss. Therefore, the color and shininess of a surface will depend upon the roughness of that surface. This is due to the action of light waves bouncing from a rough surface and either reinforcing or interferring with each other after reflection. When two rays are reflected from a rough surface at slightly different times due to irregularities in the surface, they will be out of phase. Consequently, the degree of apparent shininess depends not only on the roughness of the surface but also on the angle from which it is viewed. The two rays will be less out of phase when the surface is viewed at near grazing angles than they will be when viewed at near perpendicular angles. This is very apparent when we view a rough surface from a point directly above it, that is perpendicular to it. It will appear less shiny than if we viewed it from a point where our eyes were almost level with the surface.

Internal diffusion is a phenomenon extremely important in the color of an object. Indeed, internal diffusion gives rise to much more complete spreading of the light than surface diffusion. When light penetrates the surface of an object it may be subjected to reflection and refraction if particles are present. One of the requirements for internal diffusion, therefore, is the existence of many randomly oriented interfaces between materials of different refractive indices. As explained previously when this situation occurs, the many encounters caused by the refraction of the light rays give rise to very high levels of reflection.

Another phenomenon which affects internal diffusion is the particle size of the material within the medium. If particle size becomes smaller, diffuse reflection increases down to a limiting size. Graphically (Figure 8) the effect of particle size upon scattering (and thus internal diffusion) may be shown as a function of relative particle diameter. As the particle diameter decreases, the scattering, or diffusion, of the material increases inversely with the square of the particle diameter to a maximum. This maximum occurs when particle diameter reaches approximately 0.1 μ which is about $\frac{1}{4}$ the wavelength of light. At this diameter the particles become too small to reflect and defract in the usual manner and reflectance decreases with approximately the cube of particle diameter. As pointed out by Clydesdale and Francis (1969), the paint industry takes advantage of this fact by attempting to maintain pigment particle diameter at $\lambda/4$. The food industry might take note of the paint industry's procedures, particularly when dealing with prefabricated foods. For instance if one were preparing a granulated instant drink he might want to achieve maximum color from the food colors used, and thus reduce the expense. It might be possible in making this product to choose dispersible colorant particles of diameter, $\lambda/4$ and having a different refractive index than the substrate in which they are imbedded. That is, each particle in the powdered drink would be compounded in this manner. A start in this direction could

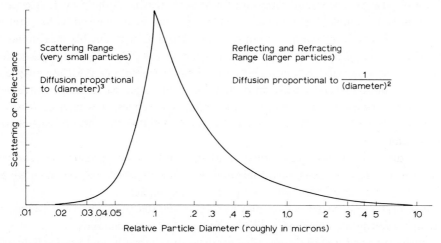

Fig. 8. Approximate relation between the diameter of spherical particles in suspension and the scattering of a beam of light by a thin layer of the suspension (constant weight of particulant material) reproduced by permission of Hunterlab Associates Laboratory, Incorporated.

be made by controlling the particle size and refractive index of lakes. Lakes for use as food colorants are made by absorbing the pigments upon substrates such as alumina hydrate. This produces an insoluble pigment which can be finely ground in order to achieve dispersibility in different media.

The next phenomenon which has been mentioned is absorption within the object. As stated previously, when light travels through an object, some of the light is electively absorbed and the rest is reflected or transmitted; the net affect is that the object looks colored. For instance, a green body would absorb all light except green light which it reflects.

A very important controlling feature in selective absorption is particle size. This might best be explained by an analogy. If one viewed large chunks of dark blue glass he would see a dark blue color. This is because light impinging on these chunks does not have encounters and reflectance is low. Therefore, the sample appears dark. The blue, of course, results because all other colors but blue are selectively absorbed as they pass through the glass. If the glass were ground to fine particles approximately $\lambda/4$ in diameter, many encounters would occur and total reflectance would be increased tremendously. Also, the light passing through these small particles would be so small that the blue colorant could not impose selective absorption on the light. The net result would be that the ground glass would appear bright and white, instead of dark blue. Therefore, if one were making a fabricated food, one would choose very small particles for high reflectance, but some compromise would have to be reached to achieve color. That is, the particles must be (not only) large enough to impose selective absorption on the light but also small enough to give a reasonable degree of reflectance. Another factor, of course, would be to use a strong colorant which would impose selective absorption even through short path lengths. Aside from the effects described previously, the particle size is important for another reason. If the particle size varies from sample to sample, the amount of reflectance and the amount of 'color' that we measure will vary from sample to sample, and constant, reproducible measurements cannot be made on any given series of the same sample. Therefore, if a particulate material such as cereals or potato chips are being measured, one must use a sieve or some other means whereby uniform particle size can be presented to the instrument every time a measurement is made. Another method suggested is to use presses to compress granular materials into a compact 'cake' such that the effect of particle size will be minimized.

In this discussion it has become apparent that there are many parameters to consider in measuring a solid material or turbid solution. Some of which may be taken advantage of to maximize colorant effect and total visual impact.

However, these same parameters may cause considerable problems when attempting to design appropriate color measuring techniques for such materials.

Translucent food materials which transmit, diffuse, absorb, and scatter light also may cause a problem in measurement.

The instructional manual for the Hunterlab color difference meter states, "with translucent liquids like fruit juices, the exposed area should be appreciably larger than that illuminated. This permits light entering the specimen and traveling latterly within it to emerge in the direction of the photoelectric pick-up." In order to clarify this quote, a diagrammatic approach is shown in Figure 9. In the first case (Figure 9a) a sample has been placed horizontally on an instrument such that the area of sample being illuminated is about the same size as the aperture of the instrument. This light enters the sample and is scattered with some of the light coming back into the instrument to

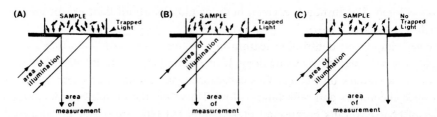

Fig. 9.   (a) Light trapping in a translucent sample due to the exposed sample area being equal to the illuminated area. (b) Light trapping is decreased as the exposed area increases with respect to the area of illumination. (c) Light trapping disappears as the exposed area is greatly increased with respect to the area of illumination.

be measured in a phototube. However, some of the light is trapped because it is scattered and cannot pass through the aperture.

In Figure 9b the aperture size is increased with respect to the light area illuminating the sample, and much less of the light is trapped.

In Figure 9c the aperture size has been increased with reference to the area of the light illuminating the sample, and thus very little of the light is trapped, and in fact nearly all the scattered light is reflected back for measurement.

Therefore, by increasing the aperture size with relation to the area of illumination, we not only achieve a better measure of the total energy but also overcome another problem, namely the selective absorption of the trapped light. For example, with red tomato juice, it is mostly red light that is reflected, so that the reading we get if the trapping affect occurs will be different not only in terms of the amount of total measured energy, but also in terms of hue. That is, the hue measurement which we obtain will not really show the total degree of redness in such a measurement.

This problem has been overcome to some extent in the case of citrus and tomato juices since R. S. Hunter has designed specific instruments and color scales for these products. The instruments are known as the Citrus Colorimeter and the Tomato Colorimeter.

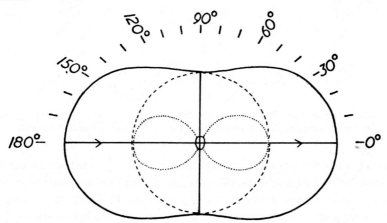

Fig. 10. Radial distribution of scattered light from small particles above the scattering center o. The broken and dotted curves represent the vertically and horizontally plane polarized components, respectively, and their sum is given by the outer continuous curves. (Thorne)

As well as these effects, scattering causes many problems in the measurement of certain food commodities. Beer is an excellent example of a food which creates such problems due to the fact that the size of particles, which can vary widely in such a sample, affects both the amount and the radial distribution of light scattering.

Scattered light is resolvable into horizontally and vertically planed-polarized components. If the scattering particles are small in relation to the wavelength of the light, the radial distribution of these components is as shown in the vector diagram in Figure 10, where the direction of the incident light is shown by the arrowheads. The vertically polarized component is of uniform intensity in all directions and is represented by the broken curve, a circle drawn about the scattering center. The horizontally polarized component, on the other hand, varies as $\cos^2\theta$, where theta is the scattering angle, and is represented by the dotted Figure 8 curve. The net light scattered in any direction is the sum of these two, and its radial distribution is given by the outer continuous curve. Figure 10 shows that the scattered light is not uniform in all directions, although it is symmetrical about the 90° direction, therefore, the scattering due to such particles could be adequately measured on a nephelometer. If the scattering particles are a size comparable with the wavelength of light, a further complication arises, namely mutual interference of the scattered light rays. The effect of such interference is shown

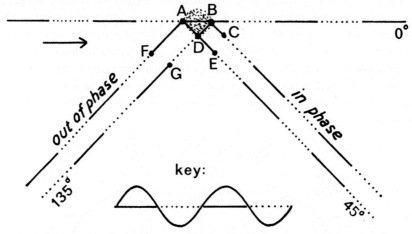

Fig. 11.   Interference of scattered light rays from large particles giving more intense light scattering in the forward direction than in the backward direction. (Modified Thorne)

in Figure 11 where for simplicity the light waves are represented not by the conventional wave, but by alternating continuous and dotted lines each denoting pulses of half a wavelength. This is shown by the key in Figure 11 where the continuous line represents a wave crest and the dotted line a wave trough. The light scattering is shown from two points, A and B, of a large particle separated from each other by an appreciable fraction of a wavelength, and in two directions, 45 and 135°, to the direction of the incident light (arrow).

The light scattered in the 45° direction is essentially going forward with one ray following the path ADE (a crest) and the other ray passing through the particle, following path ABC (a crest). Since both rays are going in the forward direction they're essentially in phase at points C and E where troughs begin. Thus, light in this direction undergoes constructive interference and is scattered. On the other hand light scattered at 135° is essentially in the backward direction, with one ray following the path AF (a crest) and the other path ABD (a trough). In this case path ABD goes forward then backward such that at points G and F a crest and a trough begin, respectively. This creates destructive interference and light is therefore not scattered in this direction.

The scattering from large particles is therefore dissymmetrical, i.e. more light is scattered forwards than backwards. For such particles the symmetrical vector curve of Figure 10 becomes distorted producing skew forms as shown in Figure 12 where X and Y represent scattering diagrams of large, but different sized particles. The dissymmetry of light scattering is expressed by a coefficient, z, which is the ratio of the scattered light intensities at two angles symmetrically disposed above 90°, i.e. 45 and 135°. For the small particles in Figure 10, there is no dissymmetry and z equals 1, while for the larger particles in Figure 12, z equals 1.7 and 5.0, respectively. For still larger particles, the dissymmetry can attain much higher values.

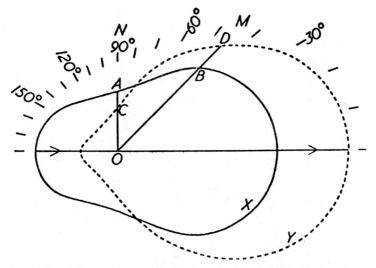

Fig. 12.   Dissymmetrical light scattering (producing inconsistent nephelometer readings at different angles). X and Y represent two different particle diameters. (Modified Thorne)

Problems involved in measuring materials such as beer with large varying particles on a nephelometer now become evident. Not only does the light intensity change but the readout will depend on the radial distribution. In Figure 12 it may be seen that a reading at 90° will produce values A and C for different particle sizes. Likewise, for a given particle size the readout will vary with the viewing angle, therefore, readings from different instruments will vary (Clydesdale and Francis, 1970a, 1970b, 1971a). There has been a great amount of work done in an attempt to resolve such problems, however there has been no final solution and work is still continuing in an attempt to produce adequate color measurement. A review of this area has been written by Clydesdale and Francis (1971a).

Problems induced by scattering can also occur with solid media. Little and Mackinney (1969) proposed an interesting hypothesis concerning the relationship of scattering to absorption in samples of tuna fish. They found that homogenized samples invariably show spectral reflectance curves almost completely devoid of the maxima and minima characteristics of the ferrohemochrome pigment even when the sample was prepared in an inert atmosphere. However, the spectral characteristics of the heme system became apparent when the pigment was reduced with sodium dithionite as shown in Figure 14. These authors point out that they could not definitely give a reason for the obliteration of the spectral characteristics of the heme pigment. However they suggest that the explanation may lie in the physical relationships of the pigment system to the muscle fiber matrix with which it is associated. The cellular matrix

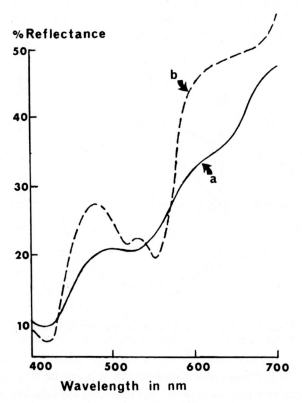

Fig. 13.   Reflectant spectra of homogenized canned tuna. A-untreated, B-after treatment with sodium dithionite. (Little)

can be considered as a spectrally non-selective light-scattering medium with scattering coefficients S. The pigment system selectively absorbs light, with absorption coefficients, K_λ. Then the value of the ratio S_λ/K_λ becomes critical. With S_λ/K_λ large, the scattering affect of the cellular matrix may mask the pigment affect, K_λ. As K_λ increases on treatment with dithionite, the ratio S_λ/K_λ decreases, thus reducing the masking affect of the background. Support of this hypothesis may be seen in Figure 14. This is a reflectance spectra of β-carotene applied to filter paper. A is the reflectance spectrum of the filter paper and $F-B$ show decreasing concentrations of β-carotene. It may be seen that as more pigment is applied thus increasing K_λ with respect to S_λ the spectral characteristics of β-carotene become more pronounced.

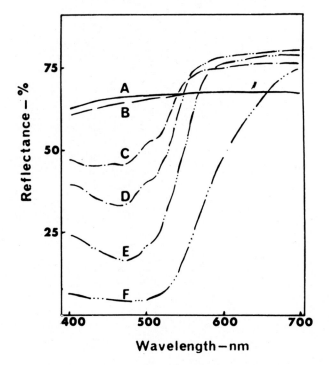

Fig. 14. Reflectance versus wavelength for Beta-carotene absorbed on filter paper. Curve *A* is the filter paper, *B* to *F* show increased Beta-carotene.

This hypothesis, if true, implies that when manufacturing a food material one major objective would be to minimize background scatter in order to maximize color. A factor which should be considered in this area of technical sophistication and in the increased production and use of synthetic and manufactured foods.

It is hoped that this paper has shown the importance of the optical and non-optical physical parameters of a sample in the measurement of color. Although it has been merely a cursory review of the subject, readers perhaps may be stimulated to inquire in other references.

References

Birth, G. S. and Zachariah, G. L.: 1971, 'Spectrophotometry of Agricultural Products', Paper #71–328, presented at the Asae Meeting, St. Joseph, Michigan.

Clydesdale, F. M. and Francis, F. J.: 1969, 'Color Measurement of Foods: 13. Sample Presentation – Physical Attributes which Influence Measurement', *Food Product Development* 3, 23–28.

Clydesdale, F. M. and Francis, F. J.: 1970a, 'Color Measurement of Foods: 22. Sugars', *Food Product Development* 4, 51–56, 96.

Clydesdale, F. M. and Francis, F. J.: 1970b, 'Color Measurement of Foods: 23. Beer. Part I', *Food Product Development* 4, 67, 71, 73, 76.

Clydesdale, F. M. and Francis, F. J.: 1971a, 'Color Measurement of Foods: 24. Beer. Part II', *Food Product Development* 5, 84, 88.

Little, A. C. and Mackinney, G.: 1969, 'The Sample as a Problem', *Food Technol.* 23, 25–38.

CHAPTER XV

APPLICATION OF COLOR THEORY TO COMMODITY AREAS

FREDERICK J. FRANCIS

Dept. of Food Science and Technology, University of Massachusetts, Amherst, Mass., U.S.A.

The measurement of color in food products has been applied in commodity areas where there is an economic need. This will be illustrated by several examples.

1. Tomatoes

Color grades for raw tomatoes and tomato products have been the object of more color research than any other agricultural commodity. Processors of tomatoes have known for a long time that optimum flavor is associated with optimum color of the raw fruit. There was no practical way to measure flavor but they could measure the color. In the early 1920's McGillivray in California developed a visual system of measuring the color of processed tomatoes and tomato juice based on the color of spinning discs. This method is still the official U.S. grading system for processed products but it was not suitable for raw tomatoes.

Processors of tomatoes desired to purchase raw tomatoes, from their contract growers, on a graded basis with monetary incentives for higher quality. The grading was done, and still is in most areas, by the visual opinion of the graders. Objective systems of color measurement were desired and their concepts were developed in the 1950's. They were based on skin color, cut surface color and juice color. It soon became apparent that the outer skin color of a tomato did not accurately reflect the color of juice made from the tomatoes, primarily because the skin contains both red and yellow pigments. Methods involving the color of the cut surface of a tomato and the color of the juice were developed and are used today.

The method involving cut surface was developed primarily through the efforts of the Magnuson Company in California and the California Department of Agriculture. A method was desired which would serve as an electronic check on the judgment of the visual graders. A tomato could be cut in half, the two halves measured for color, after which the halves could be put together and the tomato used to demonstrate the grade to the growers. An instrument to measure the color of the two halves of a tomato is illustrated in Figure 1. The instrument actually measures the reflectance at 546 (green) and 640 nm (red) and computes the ratio between the two. The method of calculating the score is shown in Figure 2. This instrument was adopted as the official grading instrument for the State of California in 1953 and the Province of Ontario,

Canada in 1957. With this system, the grading is done visually and the instrument serves as a check on the visual decisions.

A different approach was used in the eastern parts of the U.S. Since most tomatoes are used to produce homogenized products such as juice or puree, processors were

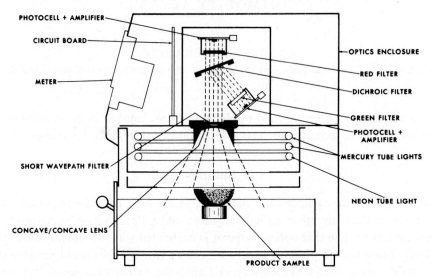

Fig. 1.   The Agtron E, an instrument designed to measure the color of the cut surface of a tomato.

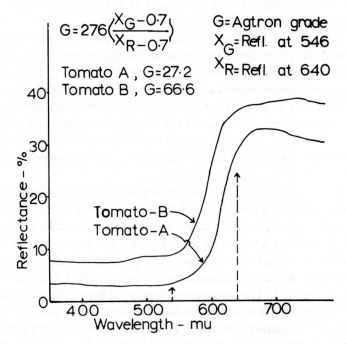

$$G = 276\left(\frac{X_G - 0.7}{X_R - 0.7}\right)$$

G = Agtron grade
X_G = Refl. at 546
X_R = Refl. at 640

Tomato A , G = 27.2
Tomato B , G = 66.6

Fig. 2.   Calculation of the color score of a cut tomato from two points on a reflection curve.

interested in a method which would be used to measure the color of the raw juice to predict the color of the processed products. Research at a number of universities and the U.S.D.A. laboratories defined the area in color space of the raw juice obtained from tomatoes of many varieties and many areas. The effect of processing variables and relationships between color of raw juice and color of the processed product was well established. Several basic colorimeters were used to establish this data but most of the data were in the Judd-Hunter system. This system employs a three dimensional color solid with L=lightness, $+a$=redness, $-a$=greenness, $+b$=yellowness and $-b$=blueness. Using the Hunter Color and Color Difference Meter, the position of the various samples in L, a, b color space was defined. However the users did not want a three-dimensional read-out and insisted on a single number scale. R. S. Hunter, of Hunterlab Associates, was asked to design a Tomato Colorimeter which would take the three dimensional data for raw tomato juice and condense it into a linear scale to relate to the visual judging system. The mathematical relationship between the instrument read-outs and the visual judgment had been established previously primarily by the work of Yeatman at the U.S.D.A. laboratory. The relationship actually adopted was Tomato Color=2000 cos Q/L or in C.I.E. terms=$21.6/(Y)^{1/2} - (3.0/(Y)^{1/2})$ $(Y-Z/X-Z)$. The Tomato Colorimeter was put into commercial production and is used today for measurement of color of raw tomato juice. For processed juice, there is a more accurate relationship between visual judgments and instrument read-out and the following equation is recommended:

$$\text{Processed juice color} = b\,L/a.$$

A conventional Hunterlab colorimeter is available which can be converted to a Tomato Colorimeter by a switch. If desired, it would be a relatively simple matter to convert the read-out to the optimum equation for processed juice, but this convenience is not available commercially as yet.

The same type of approach can be used to convert conventional color scales such as the Judd-Hunter L a b to scales which relate linearly to pigment content. This has been done by Eagerman *et al.* at the University of Massachusetts for a series of water soluble red dyes and anthocyanin pigments in fruit juices. The same instrument can be used for conventional appearance scales, or additional scales related to pigment content.

2. Citrus Juices

The color of frozen concentrated orange juice, and the citrus products is an important quality attribute. Apparently with orange juice from Florida, the public prefers the deep orange color of juice from the late season Valencia oranges over that produced from the early varieties. Orange juice is a difficult product to measure accurately since the juice both absorbs and reflects light. The readings obtained by reflection from a sample placed on a colorimeter are necessarily empirical but nevertheless useful. The area in L a b color space occupied by orange juice samples was defined primarily by the researchers at the University of Florida and the Florida Department of Citrus.

Again a three-dimensional specification of color was complicated to administer and R. S. Hunter was asked to develop an instrument to duplicate the response of the visual graders. He developed the Citrus Colorimeter (Figure 3) and it is in commercial production. This instrument takes the reflected light from a test tube of orange or other citrus juice and a read-out is provided in terms of two new color scales called CR (for citrus red) and CY (for citrus yellow). This instrument appears to have had good commercial acceptance.

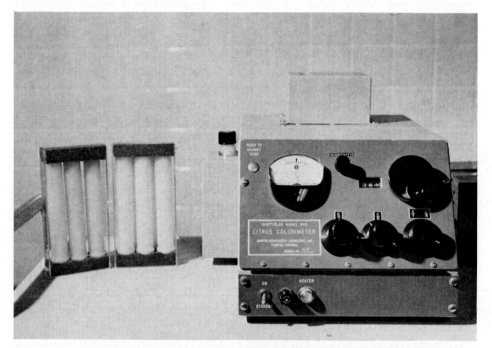

Fig. 3.   The Citrus Colorimeter for measurement of color of citrus juices.

3. Cranberries

The two examples above are fairly similar and the concept has been employed in other commodities. For example, cranberries are used primarily to produce juice and sauce, and the limiting factor is usually the pigment content. Processors pay a premium for well colored berries for juice production and a method of measuring the skin color of the raw berries and relating it to the pigment content was desired. Such a method could be used for incentive payments to growers. The basic work defining the areas in color space for cranberries and the relationships to pigment content and color of products was done at the University of Massachusetts. An instrument to measure the color of the berry skin was developed at the British Columbia Research Institute and it is based on a read-out ratio of the red reflectance over the total reflectance.

4. Apples

The skin color of apples is an important criteria of market quality, particularly for the red varieties. Methods of measuring individual apples by rotation, then over the exposure head of a colorimeter are available, but they are slow and involve sampling and samples presentation errors. The B.C. Research Corporation has developed an ingenious portable Apple Colorimeter which can be used to measure the color of apple skin by placing the apple on the instrument. The light emerging from the apple is measured rather than simple reflectance and this eliminates the problems with specular reflectance.

5. Sugar Solutions

The problems involved with color of sugar solutions are completely different from the examples described above. In the refining process with sugar, the colored pigments have to be removed. The refiners want to know the extent and nature of the colored impurities in order to estimate the cost of removal.

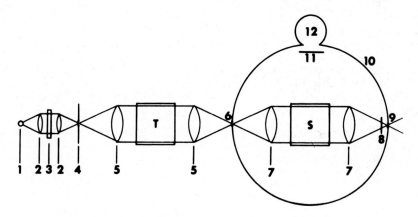

Fig. 4.   The Sphere Colorimeter for measurement of color of sugar solutions.

Some very sophisticated methods have been developed in the sugar industry but they usually are related to two concepts. There are two types of impurities of interest in unrefined sugar, the 'absorbers' and the 'scatterers'. Actually the compounds which absorb light also scatter light to some degree and vice versa. Bernhard developed a 'Sphere Colorimeter' for use with sugar solutions which would give an index of both types of compounds. It is shown diagrammatically in Figure 4. With a sample cell in the T position, the instrument would measure essentially the absorbers. With the same cell in the S position, the degree of scattering would be measured. This instrument would appear to be of sound design but may be too sophisticated for everyday use. It has not received widespread adoption by industry.

6. General Applications

The above are examples of specialized instruments designed for a specific purpose. Many other examples are available in which conventional all-purpose colorimeters can be used. Each commodity has its own color interpretations for consumer appeal, effect of processing, relationship to pigment content, etc. Much research has been done on meats, tuna, salmon, eggs, poultry, oilseeds, flour, cereals, preserves, etc. The color problems can usually be handled with a conventional colorimeter but the interpretation will be unique for each commodity.

CHAPTER XVI

ANALYSIS AND PROCESSING OF COLORIMETRIC
DATA FOR FOOD MATERIALS

FERGUS M. CLYDESDALE

Dept. of Food Science and Technology, University of Massachusetts, Amherst, Mass. 01002, U.S.A.

The measurement of any food material, be it transparent, translucent or opaque, may be carried out on existing equipment, as long as appropriate precautions are taken, and the optical characteristics of the food material have been defined.

However, once the measurement has been taken with all such precautions, the investigator finds himself or herself with a vast array of data which might seem almost useless. Therefore, the problem of handling or reducing data becomes all important at this stage.

Since color measurement basically consists of the specification of a point in three-dimensional space, it is logical to assume that the value of each of the three coordinates will change, as the color is changed.

Under normal circumstances, however, such as in quality control, it would be advantageous to have only one or two variables, rather than three. If only one or two of the variables are used, there is normally going to be a corresponding decrease in the accuracy of the measurement made. Therefore, it is the operator's choice to decide, if the extent of the increase in accuracy justifies the computations involved with the use of three variables.

It should be reemphasized that any function of color used for a particular product should first be correlated with a subjective panel of visual judgments.

Before beginning a discussion of data reduction, it would be advantageous to first briefly describe the general area of color scales.

If one wished to purchase a colorimeter for research on quality control, he would probably survey the available array of instruments and choose one to suit his needs. The data obtained could be reported in terms of the read-out system for that particular type of instrument and color specifications could be set accordingly. In such a system, he need never be concerned with other read-outs or color scales. However, problems arise, when dealing with other systems from other read-outs. In this case, specifications would have to be converted from one color system to another.

The types of colorimeters that have been employed widely, to date, in food applications in America are the Hunterlab instruments, the Gardner series, the Color-Eye, the Colormaster and the Tintometer. Conversion of data from one system to another, via the CIE, $X Y Z$ system, is reasonably simple with the use of computers and the conversion equations supplied by the manufacturer. The Agtron has also been used widely in food applications, but conversion to the $X Y Z$ system is not possible.

The C.I.E. $X Y Z$ system of specifying color emerged in the 1920–1930 era as the most important system. The color solid involved is usually represented as the familiar horse-shoe shaped curve on an x, y axis with the brightness factor ($\% Y$) arising from the point of white light. In Figure 1 the contour lines which enclose all real colors are shown for various values of Y for C.I.E. source C. The lighter the color, the more restricted is the range of chromaticity. These contour lines are sometimes called the 'MacAdam limits' after MacAdam (1935) who first calculated them. They would look different under tungsten light (Source A). The chromaticity gamut approaches zero slowly as the Y value approaches 100 and very quickly as the Y value approaches zero. The MacAdam limits are very useful for visualizing the $X Y Z$ solid.

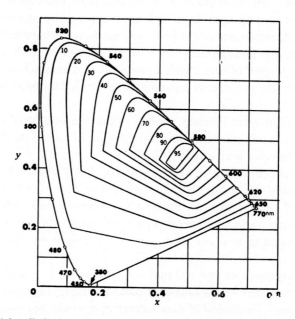

Fig. 1.   The MacAdam limits for colors viewed in C.I.E. source C (daylight). The figure beside the contour lines represents the luminance ($\% Y$). The lighter the color, the more restricted is the color gamut.

The $X Y Z$ system has been remarkably successful, but it was developed primarily for convenience in color specification. It was not designed for color measurement. It was apparent to those who set up the C.I.E. system that it was not a visually uniform system. This means that two samples when judged to have the same visual difference in one part of the solid, as another pair in another part of the solid, will not have the same distance between them on an x, y plot. This non-uniformity was investigated in detail by MacAdam in 1942. He had an observer make over 25 000 judgments of least perceptible chromaticity differences. Fortunately, he was a typical observer, as later observations with more subjects confirmed the data (MacAdam, 1965). The data from his observer came to be known as the MacAdam ellipses and are illustrated in

Figure 2. Every radius on every ellipse has the same visual magnitude, but obviously, they differ in length on the x, y plot. Wright (1941) and Stiles (1946) also produced extensive studies on the perceptibility of chromatic differences and together with MacAdam data provided a means of testing the visual uniformity of any coordinate transformation.

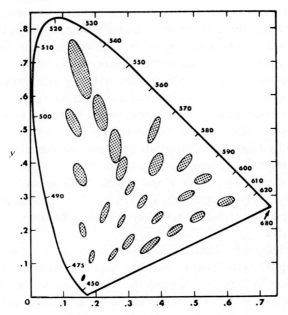

Fig. 2.   The MacAdam ellipses plotted on an C.I.E. x, y chromaticity diagram. The diameter of the ellipses is 10 × the value of the just noticeable differences of chromaticity from 25 typical standard colors.

Actually, the study of the non-uniformity of the C.I.E. scales was known when Judd (1935) attempted to develop a scale which would produce visually uniform chromaticity differences. He developed the r, g, b system using triangular coordinates. MacAdam (1937) presented a simpler system using rectangular coordinates called u and v. Breckenridge and Shaub (1939) converted Judd's scales to rectangular uniform-chromaticity-scales (R.U.C.S.) coordinates. The three are quite similar.

Judd's U.C.S. scales were converted to the Hunter alpha-beta system in 1942. This system placed the illuminant at the center of the chromaticity diagram, and therefore, the origin is located at the point representing MgO which was then the reference white for surface colors. This coordinate system was specifically designed to measure surface colors and was used in Hunter's multipurpose reflectometer. It became the forerunner of the Rd or L a b scales used today.

The MacAdam u v system developed in 1937 has been singled out for special attention because the C.I.E. in 1959 adopted it as the provisional standard U C S system. It is referred to as the 1960 CIE-UCS diagram. Subsequent detailed testing o$_s$ the spacing of this diagram showed that it agreed very well with experimental data obtained by the Optical Society of America Committee on Uniform Color Scales. This data was obtained under conditions simulating those used in industrial inspection. Present consensus of opinion is that the 1960 CIE-UCS diagram is as good as any available.

Another transformation was made at the same time as the Alpha-Beta system by Hunter (1942a) where he proposed the first color solid based on a uniform surface-color scale. He labelled the axes L' a' and B'. This solid is important in that it is the scale associated with the current definition of an NBS unit of color difference. Scofield simplified the equations (1943) and introduced the L a b system.

Workers up to this time had been concerned primarily with chromaticity of colors, but it is obvious that lightness was important also. This was the time that the Munsell system, which had been published in 1929 even before the C.I.E. had agreed on the Standard Observer scales, was studied intensively for color spacing. The Munsell system was used by Adams (1942) to study problems of color vision. He developed a new color space termed the chromatic value scales. He assumed that the eye measured chromaticity by comparing X with Y and Y with Z. He applied the Munsell Value scale to X, Y and Z proposed two scales $Vx - Vy$ and $Vz - Vy$. This system has enjoyed considerable popularity. Adams' second system, the Chromatic Valence scale, has been used very little.

Hunter (1958) followed Adams' lead by using $X - Y$ and $Y - Z$ signals in his Color Difference Meter. These differences could be obtained electronically and, after applying a correction for lightness, became the a and b scales. The original luminous attribute (lightness) was Rd. This was replaced by L, a more uniform measure of lightness. These L a b scales are the ones currently in use with the Hunterlab series of instruments.

A number of other color spaces have been proposed, i.e. Moon and Spencer's 'omega' space, Saunderson and Milner's 'zeta' space and the Adams-Nickerson Scales. The Adams chromatic value system was modified by Glasser and Troy (1952) for use in the Colormaster colorimeter. This system, called the Modified Adams scale, can be designated L_m a_m b_m. Glasser and coworkers (1958) further modified the Adams scale and produced the Glasser Cube Root scale (L_c a_c b_c). Wyzecki (1963) introduced the U^* V^* W^* scale. Friele introduced a new set of primaries, r, g, b, based on a transformation of the C.I.E. system. MacAdam (1966) modified Friele's equations to fit his own data and introduced the $\bar{p}$ $\bar{q}$ $\bar{s}$ primaries. Chickering (1967) optimized the Friele-MacAdam primaries and introduced the Friele-MacAdam-Chickering (F.M.C.) system.

All the preceding systems involved attempts to develop a more visually uniform color system or to adapt the calculations to electronic instruments. There is some doubt that a completely visually uniform space can ever be attained. MacAdam concluded, after a careful analysis of his chromaticity data, that "no linear or even nonlinear transformation of the (x, y)-chromaticity diagram could convert his ellipses into perfect circles of equal size." A curved surface would be required which would look like a badly rumpled felt hat.

Several new systems have also been proposed, but these will be discussed later.

From this description, it becomes obvious that there are a large number of scales which produce at least a tri-dimensional read-out.

The problem of data reduction mentioned previously may be applied to any of these scales in a similar manner.

One of the simplest reductions conceptually would be the prediction of a color difference between two samples by means of a single number. This means that a difference (without any direction being implied) between two points in a three dimensional color solid is being described.

This in itself is simple enough, but the difficulty arises when the non-uniformity of color space is taken into account. The modifying factors which complicate the formulas for calculating color differences are usually added to account for non-uniform color space. Other complications arise when factors are added to weight each of the three parameters in color space to comply better with human visual judgments. A third set of modifying factors are introduced when the various color difference formulas are weighted to give approximately equal results.

A trend is emerging in the calculation of color differences. Years ago when the early workers were struggling with attempts to develop uniform color scales, it became apparent that mathematically complex adjustments were required in color difference formulas to make one unit of color difference visually equal in all portions of color space. The adjustments were too complex to make routine calculation of color differences practical. Consequently, graphical methods and simplifying assumptions were made in order to reduce the computational labor. These methods necessarily were less accurate but were justified in view of the relatively wide range encountered in human visual judgments. In recent years with the use of computers becoming much more widespread, the importance of the computational labor was reduced considerably. Consequently, some of the recent formulas for calculating color differences make the use of a computer almost mandatory.

The development of color difference formulas has followed three main paths. The first was based primarily on the Munsell system and was pioneered by the work of Nickerson, Balinkin and Godlove. The second, based on the CIE system and its transformations, was developed by Judd, Hunter, Adams, Glasser, Wyzecki and many others. The third was based on the MacAdam color scales, as interpreted by MacAdam, Brown, Davidson, Hanlon, Simon, Friele and many others. All systems give slightly different results, but two major units have evolved. One is the NBS unit of color difference (sometimes called the judd, after Dean B. Judd), designed to approximate a commercially acceptable color match. The other is the MacAdam unit which approximates the least perceptible color difference. One NBS unit is roughly three times as large as a MacAdam unit.

Further information on specific formulations for color differences may be found in an article by Francis and Clydesdale (1969).

Color differences are extremely useful, but at times should be used with some other parameter to indicate the direction of the difference.

Other sets of reduced functions are often used to describe a hue shift from sample to sample. Depending on the system in use, these may be as simple as X/Y, X/Z, Y/Z, a/b, etc.

The use of the a/b ratio as an index of color change is well established in the food field, possibly because it offers a very convenient way of reducing two color parameters to one. It has been used as a method of describing color changes in a number of products and also a part of the military specification of color for dehydrated sweet potato flakes. The actual meaning of the a/b ratio can be illustrated by reference to work with carrots.

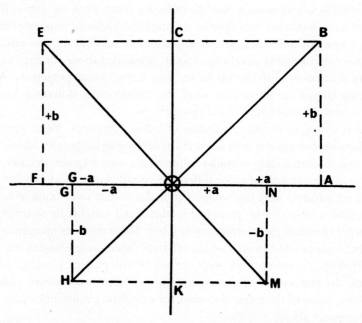

Fig. 3.   Hunter a b plot showing how the a/b ratio and the $\tan^{-1} a/b$ values vary with the color.

Figure 3 is an abbreviated Hunter a b plot showing the usual limits of the color of carrot slices and puree. Most carrot colors will fall within the angle $D\ O\ E$, the lines $D\ O$ and $E\ O$ representing the yellow ($a/b=0.2$) and red ($a/b=1.0$) limits respectively. It is clear that the a/b ratio is a function of hue. We can expand this concept by reference to Figure 3. The point B represents an orange color with an a/b ratio equal to unity. The fraction $+a/+b$ is actually the tangent of angle $O\ B\ A$, which in this case would be 45 deg. As the hue becomes more yellow and moves towards point C, the tangent of $O\ B\ A$ ($OBA=BOC$) approaches zero, as well as the ratio. In the second quadrant, the ratio a/b would represent the yellow-green colors and would go from zero at C, to infinity at F, to zero at K and back to infinity at A.

It is obvious from Figure 3 that the a/b ratio is a tangential function, not a linear one. If the ratio is in the vicinity of one, where the angle is approximately 45 deg, the departures from linearity are not too great. It was in this area with tomatoes that the

a/b usage becomes firmly entrenched. However, if one attempts to use the a/b ratio in a portion of the color solid where it approaches infinity, the concept obviously breaks down. In such cases, it is preferable to use the actual angle that the point makes with the vertical axis, rather than the ratio itself. A rough rule of thumb would be that if the a/b ratio is within 0.2 and 2.0, it probably is satisfactory by itself. If the ratio is outside these limits, the angle should be used, not the ratio. The angle is conveniently written as $\tan^{-1} a/b$. Note that the superscript $^{-1}$ is used in the trigonometrical sense and is read as, "the angle whose tangent is a/b." It does not mean the reciprocal $1/\tan a/b$.

The use of the function $\tan^{-1} a/b$ was first suggested by Francis (1952b) for use with apples. It was believed to be original at that time, but it became apparent later that it was very similar to the hue angle concept suggested by Hunter (1942b).

Chroma, although not often used singularly, may also be defined as a reduced function in the form $\sqrt{a^2+b^2}$. This is simply the hypotenuse of the right-angled triangle OBA (Figure 3).

A singular function which incorporates all three tristimulus functions may be seen in the formula $a/\sqrt{(a^2+b^2)}/L$ developed by Yeatman et al. (1960) for grading tomato juice.

Although such conversions or reductions are extremely useful, care must be used in interpretation. In reality, one of the major practical concerns of conversions or reduced data systems is to enable one to set up tolerances for a given product.

Certainly, visual tolerances using systems such as the Munsell are important. However, they are expensive, and color chips may be damaged or lost.

Francis and Clydesdale (1969b) have pointed out that three dimensional visual tolerances have not been used to any extent in the food industry primarily because very few foods look like a Munsell chip. Most foods are somewhat translucent, and this complicates human judgment, when a comparison is made with an opaque chip.

There are many other ways to set tolerances. For example, we can measure any color instrumentally in terms of tristimulus colorimetry, as described in previous chapters. Suppose we do specify a desired color by locating it in the $X Y Z$ system. If we wish to set tolerances on the color, we need only give values for $\pm X$, $\pm Y$ and $\pm Z$. We can do a three-dimensional plot on a two-dimensional sheet of paper, but it is difficult to visualize. It is much easier to make 2 or 3 plots as in Figure 4. We can define an area on each plot and say that these areas define the outer limits of acceptable color for a given product.

This concept is easy to understand and can be used with any system employing $X Y Z$ data. It is not in general use, because the use of a color reading $\pm$ a tolerance implies that the defined area is a rectangle. This is undesirable as shown in Figure 5 in which a desired color is plotted in three planes using Hunter $L a b$ data. The color g is further away from s than any of the acceptable colors c, d, e and f. Color g would be acceptable but would have a higher value for a color difference from s than any of the other four. Obviously, the defined area should be an ellipse, not a rectangle as shown in Figure 5. It is a simple matter to draw an ellipse instead of a rectangle, when the major and minor axes are known, and the same terminology can be used. The color would be described as the volume enclosed by the ellipsoid defined by the required color plus or minus the radii in three dimensions. In Figure 5, the tolerances would be L (25±0.8), a (23±0.4), and b (11±0.5). This particular color would be a

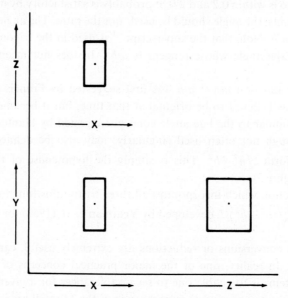

Fig. 4.   Data in the $X Y Z$ system plotted with a tolerance in three dimensions.

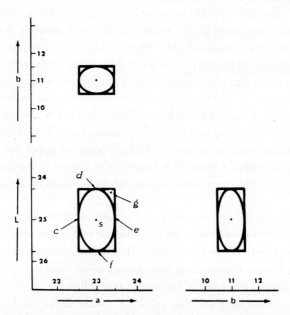

Fig. 5.   Data in the $L a b$ system showing a desired color s, and acceptable limits c, d, e, f. The volume within the ellipsoid contains the acceptable colors.

red similar to that of canned tomato juice. It has yellow and red upper and lower limits as well as limits for lightness or darkness. This system would be simple and workable.

Plotting ellipses to define an ellipsoid for a tolerance is more adaptable to data systems such as the Hunter solid for which approximately equidistant visual steps are used over the whole color solid. In other systems, it might be preferable to convert the data, but this is not essential.

Another interesting concept which involves the use of scale conversion and data reduction, to set tolerances for use in quality control, has been proposed by Little (1969).

This method relates a chromaticity shift in tuna meat to the corresponding CIE Y (brightness) value. This technique would appear to have important implications for quality control methods.

Ground tuna meat was prepared and placed in custom-made Lucite sample holders as described previously. The holders were designed such that colorimetric or spectro-photometric measurement of the sample could be made through the 5.5 cm diameter Lucite face. Each cell holder was then opened and 0.5 ml of freshly prepared ,5% sodium dithionite solution was applied to the sample surface. The airtight cover was replaced and reflectance readings taken 30 min later. A Colormaster Differential Colorimeter was used to record the reflectance readings and the data were converted to CIE, x, y values.

The shift in x, y values on reduction with dithionite is a measure of the amount of reducible pigment present. The Y value also is related to the amount of pigment present. The 67 tuna samples used in Little's study were plotted as chromaticity shifts (difference in x and y between controls and samples treated with dithionite) in Figure 6. The rectangle (labelled area 1 in Figure 6) enclosing the area, x -8 to $+6$, y 0 to -8 ($\times 10^{-3}$) included all the white samples. These, of course, are all albacore samples. No albacore samples plotted outside this area. The samples in Group 2, light, were concentrated in the rectangle labelled area II in Figure 6. The samples in Group III, dark, fell mainly in area III, but the light and dark samples could not be separated unequivocally by the chromaticity shift. There was a marked tendency for the chroma-ticity shift to increase as the Y value decreased. This of course, is to be expected, since a darker sample would have more pigment and have a larger shift. The degree of shift and the Y value correlated highly with visual scores assigned to duplicate cans from the same lot of samples. One sample, in Group III, was given a high visual score, yet it had a small chromaticity shift, indicating the possibility of browning.

The advantage of using a measure of Y together with a chromaticity shift is that the latter provides an indication of 'available' pigment which should correlate with the Y value. If it does not, it may be an indication that improper processing, such as browning, scorching or stackburn, or even oxidative degradation had taken place. A small chromaticity shift combined with a low Y value would be evidence that the dark color was the result of something other than the content of normal pigment.

A study of the chromaticity shift was included in the collaboration study described previously. The Agtron was very successful in separating the samples by the Y value but failed completely in the chromaticity shift. This is because of the unfortunate coincidence that the 436 and 546 nm lines isolated for the Agtron blue and green readings coincide with points of minimal difference between the oxidized Fe^{+++} and the reduced Fe^{++} spectral curves (Figure 6).

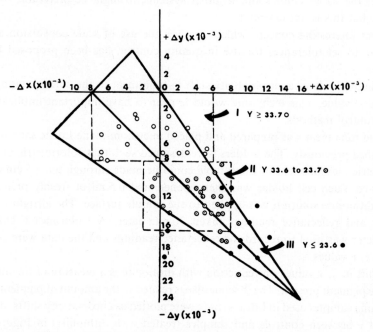

Fig. 6. Relationship of chromaticity shift to Y value (Little).

The Kubelka-Munk colorant layer theory has also been applied in the measurement of translucent food systems as a data reduction tool. Through the use of the appropriate mathematical equations, tristimulus values or reflectance readings may be reduced to a scattering component, S, and an absorption component, K. The ratio K/S is a combination of the two which takes into consideration the optical properties of a translucent material and relates to total visual impact. This is an essential requirement for media which scatter, absorb and transmit light. Visual impact is the final arbiter in the setting of tolerances, and the functions which do not describe this are rather meaningless.

Tolerances, obviously, have many uses, but it is essential that both buyer and seller agree on and understand the tolerances. As well, in some cases, it would be an ideal situation if similar equipment was used for all systems.

In the general area of color scales, there is a great amount of work going on in an attempt to find particular scales for particular applications.

R. S. Hunter has proposed the use of two new quantities a' and b' for metallic objects and clear solutions which give a scale expansion towards the darker colors. More recently, Eagerman et al. (1972) investigated thirteen scales, both proposed and currently in use, to determine their usefulness for transmission colorimetry of dark colored beverages. The colored solutions tested consisted of serial dilutions of known concentrations of F. D. and C. Red #1, F. D. and C. Green #1 and Cyanidin-3-glucoside obtained from blackberries. It was found in all cases that an 'area of confusion' developed below a certain degree of lightness. That is, the scale reading did not remain linear when the solutions attained a dark color which ranged between an L value of approximately 19 and 54.

In order to evaluate the possible importance of these results in industrial quality control, various dark colored commercial juices were purchased from a local super-market and measured in two different cell thicknesses, 3 mm and 10 mm. The two sizes were used, since a 3 mm cell would provide a lighter sample, due to physical thickness; but the 10 mm cell would simulate a size more likely to be used in quality control, due to ease of handling and filling. The juices used were: Welch's grape juice and grape drink, Za-Rex fruit punch syrup and raspberry syrup, Ocean Spray cranberry juice cocktail and Sunsweet prune juice.

Table I shows the raw data for all the juices obtained from the GERS and D25.

TABLE I

Tristimules readings for some commercially available juices showing that the luminosity values fall within the area of confusion of most color scales tested

Juice	Cell thickness	X, Y, Z data			Hunter L, a, b, data		
		X	Y	Z	L	a	b
Welch's grape	3 mm	19.89	9.64	3.03	29.34	57.08	13.99
juice	10 mm	5.85	2.67	0.81	14.50	31.40	8.80
Welch's grape	3 mm	43.01	27.57	31.76	51.13	56.22	0.22
drink	10 mm	14.01	6.92	3.36	24.85	48.60	10.30
Za-rex fruit	3 mm	22.02	10.44	1.33	29.80	61.86	17.70
punch syrup	10 mm	5.84	2.67	0.85	15.20	39.80	10.20
Za-rex raspberry	3 mm	30.97	15.42	3.61	37.18	70.90	19.07
syrup	10 mm	15.57	7.04	0.78	24.30	54.40	15.10
Ocean spray cranberry	3 mm	72.78	62.78	62.86	78.65	25.77	7.38
juice cocktail	10 mm	44.98	29.11	16.43	53.37	50.60	17.80
Sunsweet prune	3 mm	34.03	28.92	3.43	52.70	17.91	30.11
juice	10 mm	6.54	3.79	0.78	17.65	26.10	10.63

The 10 mm cells have more than three times the sample thickness of 3 mm cells, and therefore, the light beam must pass through three times the amount of pigment before reaching the photocell. From this logic the color quality parameter (a and b) for the juices in the thicker cells should be greater than those for the thinner cell; however, in every case, except for cranberry juice cocktail, the a and b values for the juice in the thicker cell were smaller than those in the thinner cell. This indicates the inadequacy of several of the scales in properly assessing colorant concentration. Measurements of cranberry juice correlated with colorant concentration, because the L values were above that required for an area of confusion. All other juices had at

TABLE II

Formulas and degree of linearity for color quality parameters which are linear
to colorant concentration

Colorant or pigment	Parameter (a^* or b^*)	Formula	Slope	Coefficient of determination
F.D.&C. Red #1	a^*	$\dfrac{170(1.02X - Y)}{Y^{2.33} - 2.5}$	2.50	0.9995
	a^*	$\dfrac{170(1.02X - Y)}{Y^{2.35} - 2.5}$	2.43	0.9994
	a^*	$\dfrac{100X}{Y^{2.35} - 2.5}$	2.36	0.9994
	a^*	$\dfrac{170(1.02X - Y)}{Y^{2.35} - 3.0}$	2.45	0.9993
F.D.&C. red #2	a^*	$\dfrac{100X}{Y^{2.30} - 0.5}$	7.61	0.99986
	a^*	$\dfrac{170(1.02X - Y)}{Y^{2.30} - 0.5}$	7.57	0.99982
	a^*	$\dfrac{170(1.02X - Y)}{Y^{2.25} - 0.5}$	7.94	0.9996
	a^*	$\dfrac{170(1.02X - Y)}{Y^{2.15} - 1.0}$	9.32	0.9993
Cyanidin-3 #-glucoside	a^*	$\dfrac{100X}{Y^{1.60} - 1.5}$	18.53	0.9996
	a^*	$\dfrac{170(1.02X - Y)}{Y^{1.60} - 0.2}$	22.25	0.9992
	a^*	$\dfrac{170(1.02X - Y)}{Y^{1.65} - 2.0}$	20.55	0.9990
F.D.&C. yellow #6	a^*	$\dfrac{170(1.02X - Y)}{Y^{3.10}}$	0.0127	0.9992
F.D.&C. green #3	b^*	$\dfrac{120(Y - 0.847Z)}{Y^{1.65}}$	-6.47	0.9996
F.D.&C. blue #1	b^*	$\dfrac{120(Y - 0.847)}{Y^{1.90}}$	-147.6	0.9994

least one reading below the area of confusion for the scales tested indicating that such
scales would be inadequate for color quality control of these products.

For this reason, several new scales were developed in an attempt to lower the degree
of darkness where an area of confusion exists.

It was noted that the scales proposed by R. S. Hunter for transmission, differed
from those proposed for reflectance only in the denominator (Y, a function of lightness
or darkness) of the conversion formula. That is the denominator was increased in
order to increase the lightness function and thus expand the color scales at darker
values. On this basis, the denominator was varied with the different systems, as shown
in Table II until a suitable new color scale was found. In each case, the area of confu-

sion does not occur until a Y value of approximately 2.5 is reached. This provides a suitable measurement for all the commercial juices tested. As well, it was found that the new scales provided a function linear to pigment concentration, as seen by the coefficient of determination in Table II. This may allow the use of tristimulus colorimetry to predict pigment content rather than laborious chemical techniques.

References

Adams, E. Q.: 1942, 'X-z Planes in the 1931 ICI System of Colorimetry',' *J. Opt. Soc. A.* **32**, 168.

Breckenridge, R. C. and Shaub, W. R. :1939, 'Rectangular Uniform-Chromaticity-Scale Coordinates', *J. Opt. Soc. Am.* **29**, 370.

Chickering, K. D.: 1967, 'Optimization of the MacAdam Modified 1965 Friele Color Difference', *J. Opt. Soc. Am.* **57**, 537.

Eagerman, B. A., Clydesdale, F. M., and Francis, F. J.: 1972a, 'Comparison of Color Scales for Dark Colored Beverages', *J. Food Sci.* **38**, 1051–1055.

Eagerman, B. A., Clydesdale, F. M., and Francis, F. J.: 1972b, 'Development of New Transmission Color Scales for Dark Colored Beverages', *J. Food Sci.* **38**, 1056–1059.

Francis, F. J. and Clydesdale, F. M.: 1969a, 'Color Measurement of Foods: XV. Color Differences', *Food Prod. Devel.* **3**, 38.

Francis, F. J. and Clydesdale, F. M.: 1969b, 'Color Measurement of Foods: XVI. Color Tolerances', *Food Prod. Devel.* **3**, 44.

Francis, F. J.: 1952a, 'A Method of Measuring the Skin Color of Apples', *Proc. Am. Soc. Hort. Sci.* **60**, 213.

Francis, F. J.: 1952b, 'A Method of Measuring the Skin Color of Apples', *Proc. Am. Soc. Hort. Sci.* **81**, 409.

Glasser, L. G., McKinney, A. H., Reilly, D. C., and Schnelle, P. D.: 1958, 'Cube-Root Coordinator System', *J. Opt. Soc. Am.* **48**, 736.

Glasser, L. G. and Troy, D. J.: 1952, 'A New High Sensitivity Differential Colorimeter', *J. Opt. Soc. Am.* **42**, 652.

Hunter, R. S.: 1958, 'Photoelectric Color Difference Meter', *J. Opt. Soc. Am.* **48**, 985.

Hunter, R. S.: 1942a, 'Photoelectric Tristimulus Colorimetry with Three Filters', U.S. Nat. Stand. Circular C-429, Wash., D.C.

Hunter, R. S.: 1942b, 'Photoelectric Tristimulus Colorimetry with Three Filters', National Bur. Std. Circ. C-429.

Judd, D. B.: 1935, 'A Maxwell Triangle Yielding Uniform Chromaticity Scales', *J. Opt. Soc. Am.* **25**, 24.

Little, A. C.: 1969, 'Reflectance Characteristics of Canned Tuna. 1. Development of an Objective Method for Evaluating Color on an Industry Wide Basis', *Food Technol.* **23**, 1301.

MacAdam, D. L.: 1935, 'Maximum Visual Efficiency of Colored Materials', *J. Opt. Soc. Am.* **25**, 361.

MacAdam, D. L.: 1937, 'Projective Transformations of the I.C.I. Color Specifications', *J. Opt. Soc. Am.* **27**, 294.

MacAdam, D. L.: 1966, 'Smoothed Versions of Friele's 1965 Approximations for Color Metric Equivalents', *J. Opt. Soc. Am.* **56**, 1784.

MacAdam, D. L.: 1965, 'Official Digest', *Fed. Soc. Paint Tech.* **37**, 1487.

Scofield, F.: 1943, 'A Method for Determination of Color Differences', National Paint, Varnish, Lacquer Assoc. Sci. Soc. Circ. 664.

Stiles, W. S.: 1946, 'A Modified Helmholtz Line Element in Brightness – Color Space', *Proc. Phys. Soc. London*, **58**, 41.

Wright, W. D.: 1941, 'The Sensitivity of the Eye to Small Color Differences', *Proc. Phys. Soc. London*. **53**, 93.

Wyzecki, G.: 1963, 'Proposal for a New Color Difference Formula', *J. Opt. Soc. Am.* **53**, 1318,

Yeatman, J. N., Sidwell, B. A. P., and Norris, K. H.: 1960, 'Derivation of a New Formula for Computing Raw Tomato Juice Color from Objective Color Measurement', *Food Technol.* **14**, 16.

CHAPTER XVII

THERMAL PROPERTIES OF FOOD MATERIALS

CHOKYUN RHA

Dept. of Nutrition and Food Science, Massachusetts Institute of Technology, Cambridge, Mass., U.S.A.

1. Introduction

Food processing, preparation and storage in one way or the other requires addition or removal of energy from food materials or maintaining the temperature at a given level.

One of the most common methods of food processing, canning, is often referred to as thermal processing, literally indicating that the process is accomplished by a heat treatment. In every method of drying and evaporation, the energy for vaporization of water has to be supplied and in many cases prior to those effects taking place, the temperature of the product has to be brought up to the vaporization point under the prevailing condition. Even unprocessed or semiprocessed food requires storage under refrigeration or freezing to prolong the shelf life, in which case a decrease in the temperature of the product is brought about.

In fact, these processing techniques resort to temperature manipulation to obtain

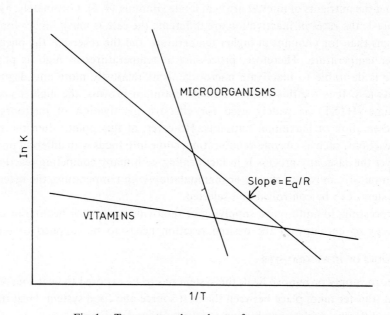

Fig. 1. Temperature dependency of rate constants.

the desired effect. This is because the rate of reaction depends on the temperature, as indicated by reaction theory. The rate process theory is applied earlier in visualizing the theory of viscosity in Chapter II, and the same extrapolation may be applied here. Assuming that activation energy is independent of temperature in the range of our interest, according to Arrhenius [1]

$$\ln K = \frac{E_a}{RT} + \ln A,$$ (1)

Where K is the rate constant, E_a is the activation energy, R is gas constant, T is absolute temperature and A is a constant, a frequency factor. The Arrhenius equation shows that rate constant K depends on the absolute temperature T, or the log of the rate of reaction is inversely proportional to the absolute temperature (Figure 1).

There are many different reactions taking place in a system as complex as food. In food processing, it is desirable to bring the system into a condition that encourages desired reactions to proceed faster than undesired ones. Some of the reactions which contribute to lowering quality, such as degradation of nutrients and growth of micro-organisms of health significance, are undesirable and should be kept at a minimum, while some of the reactions which are responsible for maintaining the quality, such as inactivation of enzymes, microbes and toxic factors, should be encouraged. For instance, in the case of thermal processing, two important and independent reactions may be considered among the numerous reactions taking place. The most important reaction to consider is the inactivation of microorganisms and this should be done with the minimum loss of nutrient. In general, the organism of concern is the public health significant or spoilage-causing spore formers such as genuses *Clostridium* or *Bacillus* [2, 3] and the nutrients of interest are heat labile vitamins [4, 5]. Fortunately, as shown in Figure 1, the rates of inactivation are different: the rate is much higher for micro-organisms than for vitamins at higher temperature and the reverse is the phenomena at lower temperature. Therefore, processing at temperatures as high as physically possible is desirable to inactivate microorganisms relatively more and degrade the nutrients less. It is for this reason that the optimum process, the high temperature short time (HTST), is widely used for effective inactivation of microorganisms and preservation of maximum nutrients. However, at this point, often the rates of other reactions, such as enzyme denaturation, come into focus with different problems. Whatever the case, any process is in fact dealing with many competing reactions and by addition or removal of energy or manipulation with temperature, the reactions in food systems can be controlled and selected.

In processing, in addition to supplying or removing energy for heating or cooling, the energy or enthalpy of the desired reaction needs to be supplied or removed.

1.1. MODES OF HEAT TRANSFER

In order to supply or remove heat, the system has to be exposed to heat sources; then the heat transfer takes place between the heat source and food system. Heat transfers by three different modes: conduction, convection, and radiation.

Conduction Heating. In conduction heating, the energy is transmitted by direct molecular communication and contact.

Convection Heating. Convection is a process of energy transport by combined action of heat conduction, energy storage, and mixing motion.

Radiation Heating. Radiation heat transfer takes place when heat flows from a high temperature body to a lower temperature body when they are separated in space, even though there is a vacuum between them.

Usually, more than one of the above mechanisms of heat transfer is involved in food processing. In order to supply or remove heat, it is necessary to subject the system to a temperature gradient. The flow is proportional to the driving force and inversely proportional to resistance:

$$\text{Flow} \ \alpha \ \frac{\text{Driving force}}{\text{Resistance}}.$$

As the driving force is the velocity gradient in the theory of viscosity (Chapter II), the driving force in heat transfer is the temperature gradient. Heat flow or heat flux, q, is

$$q \ \alpha \ \frac{dT}{dx},$$

where T is temperature and x is the thickness subjected to heat transfer.

1.2. CONDUCTION

When food is placed in direct contact with a solid heating medium, conduction is the prevalent mechanism of heat transfer. In conduction heating, the coefficient of thermal conductivity K is defined by Fourier's Law:

$$q = - KA \frac{\partial T}{\partial x}, \tag{2}$$

where q is conductive rate of heat transfer, A is the area for heat transfer, T is temperature and x is the length. The coefficient of thermal conductivity is a proportionality constant between the heat flux and temperature gradient.

Phenomena of thermal conduction is explained by body interactions and multiple collisions involving intermolecular forces and structures, again like what is considered in the theories of viscosity in Chapter II. At the molecular level, when the temperature gradient exists in a liquid which is assumed to have a lattice structure, the excess energy between the layer is transferred down the gradient by two mechanisms, a convective contribution due to the motion of molecules from cell to cell and by a vibrational mechanism in which molecules vibrate within the cell. Values for the thermal conductivity of the materials commonly used as heating media in food processing are taken from literature [6, 7, 8, 9] and presented in Table I.

TABLE I[a]

Thermal conductivity of heating media used in food processing

	Temperature (°F)	Thermal conductivity (BTU/ft °F h)
Ice	−13	1.40
	32	1.28
Water (liquid)	32	0.343
	100	0.363
	200	0.393
Air	32	0.0140
	212	0.0183
	392	0.0226
Olive oil	68	0.097
	212	0.095
Sodium Chloride brine		
(25%)	86	0.33
(12.5%)	86	0.34
Nitrogen	−148	0.0095
	32	0.0140
	212	0.0180
Sulfar dioxide	32	0.0050
	212	0.0069
Water Vapor (saturated)	32	0.0132
	200	0.0159
	400	0.0199
	600	0.0256

[a] Compiled from References:

[6] Kreith, Frank: 1963, *Principles of Heat Transfer*, International Textbook Company, Scranton, Pa., 7th Printing.

[7] McAdams, William H.: 1954, *Heat Transmission*, McGraw-Hill Book Company, Inc., New York, 3rd ed.

[8] Perry, John H. (ed.): 1950, *Chemical Engineers Handbook*, McGraw-Hill Book Company, Inc., 3rd ed.

1.3. CONVECTION HEATING

When food is placed directly into a fluid heating medium, convection heat transfer takes place. In convective heat transfer, heat transfer is accomplished by actual physical movement of the heating medium, which is usually fluid. The equation that expresses the convective heat transfer is similar to that used for conduction heating, except that the convective heat transfer coefficient replaces the conductivity,

$$q = h_c A (T_s - T_\infty),$$ (3)

where

q = convective rate of heat transfer,

h_c = convective heat transfer coefficient,

A = area for heat transfer,

T_s = surface temperature,

T_∞ = temperature of heating medium.

The movement of the fluid heating medium which is responsible for heat transfer is illustrated in Figure 2. If one looks into this movement of the fluid a laminar layer exists near the surface even when the main stream flows in well-developed turbulent motion [6, 7]. In this laminar boundry layer, physically, most of the temperature gradient occurs and the resistance to heat transfer is encountered mostly within this layer; thus the boundary layer contributes significantly to the convective heat transfer

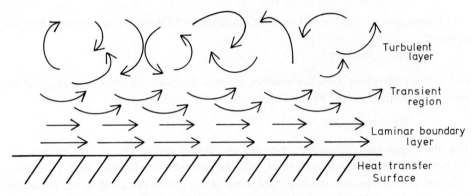

Turbulent layer

Transient region

Laminar boundary layer

Heat transfer Surface

Fig. 2.   Movement of heating medium in convection heating.

coefficient h_c. Usually, theoretically, thermal-boundry layer thickness is defined as the distance from the surface at which the temperature difference between the wall and the fluid reaches 99% of the free-stream value [6, 7].

There are few methods for obtaining values for h_c. The traditional and simplest method is to find out the value of h_c from the empirical relationship established between the dimensionless groups which resulted from the dimensional analysis of the variables involved in convective heat transfer. The dimensionless groups are:

$$\text{Nusselt number} = \frac{h_c D}{k} \tag{4}$$

$$\text{Reynolds number} = \frac{D V \varrho}{\mu} \tag{5}$$

$$\text{Prandtl number} = \frac{C_p \mu}{k} \tag{6}$$

$$\text{Grashof number} = \varrho^2 g \beta (T_s - T_\infty) L^3 / \mu^2 \tag{7}$$

where

h_c = convective heat transfer coefficient;
D = length parameter;
ϱ = density of the fluid;
k = thermal conductivity of the fluid;
μ = viscosity of the fluid;
β = temperature coefficient of volume expansion;
V = velocity of fluid;
C_p = specific heat at constant pressure
T_∞ = temperature of heating medium at the distance far;
    removed from surface of heat transfer; and
L = characteristic length parameter.

In forced convection, Nusselt number is a function of Reynold number and Prandtl number,

$$Nu = f(Re, Pr)$$

and in free convection, Grashof number replaces Reynolds number, making Nusselt number dependent on the Prandtl number and Grashof number

$$Nu = f(Gr, Pr).$$

The above general relationships between the dimensionless groups are numerically defined for various conditions [6, 7, 8]. The related experimental data have been collected extensively over the past, and some of the pertinent examples are presented below. For instance, in forced convective heat transfer, it is established that when the heating medium cross flows over a cyclinder [6]

$$Nu = 0.82 \, (Pr)^{0.3} \, (Re)^{0.4} \tag{8}$$

and when the heating medium flows over a plate (6)

$$Nu = 0.664 \, (Pr)^{0.33} \, (Re)^{0.5}. \tag{9}$$

For free convective heat transfer in heated square plates facing upward or cooled plates facing downward,

$$Nu = 0.14 \, (Gr \, Pr)^{1/3} \tag{10}$$

in turbulent range, Gr from 2×10^7 to 3×10^{10}, and

$$Nu = 0.54 \, (Gr \, Pr)^{1/4} \tag{11}$$

in the laminar range, Gr from 10^5 to 2×10^7, where L is length of the side of the square. For heated plates facing downward and cooled plates facing upward, the equation

$$Nu = 0.27 \, (Gr \, Pr)^{0.4} \tag{12}$$

is applied (6, 7).

The order of magnitude of the convective heat transfer coefficients is given in Table II [6, 7, 8].

TABLE II[a]

Order of magnitude of convective heat transfer coefficients

Air, free convection	1–5	BTU/h ft^2 °F
Superheated steam or air, forced conv.	5–50	BTU/h ft^2 °F
Oil, forced convection	10–300	BTU/h ft^2 °F
Water, forced convection	50–2000	BTU/h ft^2 °F
Water, boiling	300–10000	BTU/h ft^2 °F
Steam, drop-wise condensing	5000–20000	BTU/h ft^2 °F
Steam, film-type condensing	1000–3000	BTU/h ft^2 °F
Organic vapor condensing	200–400	BTU/h ft^2 °F

[a] Compiled from References:

[6] Kreith, Frank: 1963, *Principles of Heat Transfer*, International Textbook Company, Scranton, Pa., 7th Printing.
[7] McAdams, William H.: 1954, *Heat Transmission*, McGraw-Hill Book Company, Inc., New York, N.Y. 3rd ed.
[8] Perry, John H., Ed.: 1950, *Chemical Engineers Handbook*, McGraw-Hill Book Company, Inc., New York, N.Y., 3rd ed.

1.4. RADIATION HEAT TRANSFER

In radiation heat transfer, electromagnetic waves travel in a straight line at the speed of light. The chance of these electromagnetic waves hitting the object depends on the position, size and shape of the object. When radiation energy strikes a body, it is partly absorbed and partly reflected, except in black bodies in which all the energy is absorbed.

The heat transferred by radiation between body 1 to body 2 when one is enclosed in another is the difference between the energy emitted by body 1,

$$E_1 = \frac{q_1}{A} = e_1 \sigma T_1^4 \tag{13}$$

and by body 2,

$$E_2 = \frac{q_2}{A} = e_2 \sigma T_2^4, \tag{14}$$

where

q_1 and q_2 = heat flux
T = absolute temperature
e_1 and e_2 = emissivity
σ = Stefan-Boltzmann constant
= 1730×10^{-12} BTU/ft^2 h °R^4.

Taking into account the geometric factor,

$$q_{1-2} = A\sigma f \left[e_1 T_1^4 - e_2 T_2^4 \right],$$
$$= A\sigma f \left[e_1 T_1^4 - e_2 T_2^4 \right], \tag{15}$$

but if $e_1 \approx e_2$

$$= A\sigma f e \left[T_1^4 - T_2^4 \right].$$

The emissivity of some materials of interest in food processing is given in Table III. The published data on the emissivity and absorptivity of food materials is very limited.

TABLE III[a]

Emissivity

Material	Temperature (°F)	Emissivity	Reference
Water	32–212	0.95–0.963	[8]
Ice	32	0.97	[6]
Beef, lean (69.5% H_2O)	70–90	0.74–0.73	[12]
Beef fat	90–95	0.78–0.77	[12]
Freeze Dried Beef	70	0.75	[12]
Paper	66	0.924	[9]
Aluminum (highly polished)	440–1070	0.039–0.057	[6]
Copper (polished)	242	0.023	[6]
Red Brick (rough)	70	0.93	[8]

[a] Compiled from References:

[6] Kreith, Frank: 1963, *Principles of Heat Transfer*, International Textbook Company, Scranton, Pa., 7th Printing.
[7] McAdams, William H.: 1950, *Heat Transmission*, McGraw-Hill Book Company, Inc., New York, 3rd ed.
[8] Perry, John H. (ed.): 1950, *Chemical Engineers Handbook*, McGraw-Hill Book Company, Inc., 3rd ed.
[9] Charm, Stanley E.: 1963, *Fundamentals of Food Engineering*, The Avi Publishing Company, Inc., Westport, Conn.

As in the conduction and convection heat transfer, radiation may be rewritten also as a linear function of the temperature difference between the two bodies,

$$q_r = h_r(T_1 - T_2) = A\sigma f e [T_1^4 - T_2^4], \tag{16}$$

now

$$h_r = fe\sigma A \frac{[T_1^4 - T_2^4]}{T_1 - T_2},$$

$$= feAF_t, \tag{17}$$

where F_t is a temperature factor. The numerical values for the temperature factor can be found from the temperatures of radiating and absorbing surfaces [6]. This concept of a radiation heat transfer coefficient makes it convenient to compare or combine the heat transfer of different modes, since during the processing of food, heat transfer is accomplished usually by some combination of the above three mechanisms.

2. Thermal Properties of Food Materials

The above discussion deals with heat transmission into food systems of interest. The heat transfer by these three modes depends on the thermal properties of the heating medium such as thermal conductivity, heat capacity, density, viscosity, coefficient of thermal expansion, emissivity, and absorptivity.

Once the heat is transferred into the system, the change in the temperature is governed similarly by properties of the food, heat capacity or specific heat and density, viscosity, and thermal expansion, as well as by its thermal conductivity and enthalpy of phase change, if involved. Temperatures of food products during heating and cooling are not generally, in practice, predicted or calculated for the process. The main reason for this is because of diversity of the conditions in which the thermal data is obtained and unavailability due to ineffective coordination and tabulation of the existing data, as well as lack of data. A number of good articles on thermal properties have been published, containing pertinent quantitative information [12, 9,18, 13].

2.1. SPECIFIC HEAT AND ENTHALPY

An important thermal property, specific heat, determines the quantity of heat to be supplied or removed in order to bring the food material to the desired temperature. The specific heat is:

$$C_p = \frac{Q}{W}(\Delta T), \tag{18}$$

where

  C_p = specific heat or heat capacity;
  Q = amount of heat input;
  W = weight of the material;
  ΔT = resultant temperature change in the material.

Theoretically, the heat capacity of a composite material, such as food, can be determined by the sum of the heat capacities of each fractional ingredient. Two major components of food are water and fat, and the rest may be grouped as solid. Accordingly, the following equation may be used to estimate the specific heat of the food:

$$C_p = 1.0\, X_m + 0.5\, X_F + 0.3\, X_s, \tag{19}$$

where X_m, X_F and X_s are the weight fraction of water, fat and solid respectively [9]. This equation is applicable above freezing and below boiling points. An even more simple expression was found by Riedel [10,11], who extensively studied the specific heat of fresh beef, veal, chicken, venison, haddock, cod, and perch with the water contents above 25%, and found the specific heat of these products can be expressed by:

$$C_p = 0.4 + 0.006\, W, \tag{20}$$

where W is percent weight water content. This equation applies also for fruit and vegetables when the water content is greater than 50%.

The above two equations are useful in estimating the heat capacity when experimental data is not available. A very similar relation between the specific heat and percent moisture content,

$$C_p = 0.3337 + 0.0077 \, W \tag{21}$$

was found with grain sorghum [25]. Although this equation was derived from an experiment with one particular commodity, it probably would give a good approximation for other grains.

Experimental values of the specific heat of foodstuffs given in Table IV are adopted from an extensive compilation of data, 'Plant Handbook Data,' published in *Food Engineering* [20].

Specific heat or heat content, interchangeably used here, is the quantity required to raise by one degree the temperature of a unit mass compared with that required for a unit mass of water. This applies under the condition where there is no change in the phase or no reaction involved. For such conditions, the specific heat of a material does not irregularly deviate or change appreciably over the temperature range. Therefore, the specific heat can be represented theoretically by a particular value, for instance, as in Table IV, one value for specific heat above freezing and another below freezing.

However, in practice, phase change takes place over a temperature range instead of sharply at one temperature. At the temperature range where the change in the phase or some reaction takes place, the heat required for this change, i.e., enthalpy of fusion, vaporization, sublimation or reaction, needs to be supplied or removed along with the energy required for further effective change in temperature. In other words, in that region, there is no corresponding change in sensible temperature with the energy input. Therefore, when phase change or chemical reaction are involved in the process, they will also affect the temperature of the system.

It is convenient to group the enthalpy of phase change or reaction and the heat capacity together as an apparent heat capacity, because as discussed above, usually such change does not take place sharply at any one particular temperature, especially in materials, such as food, containing a number of soluble solids. Instead the increasing

TABLE IV

Specific heat of food stuffs [a]

	Water content (%)	Average freez. pt. (°F)	Specific heat	
			Above freezing	Below freezing
Vegatables				
Artichokes	90.0		0.93	
Globe	83.7	29.6	0.87	0.45
Jerusalem	79.5	27.5	0.83	0.44
Asparagus	93.0	30.4	0.94	0.48
Avocados	65.4	30.0	0.72	0.40
Beans				
Green or snap	88.9–90.0	30.2	0.91–0.92	0.47
Lima beans	65.5–66.5	30.9	0.73–0.40	0.40
Fresh	90.0		0.92	
String beans	88.9		0.91	0.47
Dried beans	12.5		0.30	0.24

Table IV (Continued)

	Water content (%)	Average freez. pt. (°F)	Specific heat	
			Above freezing	Below freezing
Beets				
Topped	87.6	29.2	0.90	0.46
Broccoli, sprouting	89.9	30.3	0.92	0.47
Brussels Sprouts	84.9	30.2	0.88	0.46
Cabbage, late	92.4	30.5	0.94	0.47
white, fresh	90.0–92.0		0.93	
Carrots				
Bunch	86.0–90.0		0.93	
Topped	88.2	28.8	0.90	0.46
Boiled	92.0		0.90	
Cauliflower	91.7	30.2	0.93	0.47
Celeriac	88.3	30.2	0.91	0.46
Celery	93.7	30.9	0.95	0.48
Corn, sweet	73.9	30.8	0.79	2.42
green	75.5		0.80	0.43
dried	10.5		0.28	0.23
Cucumbers	96.1–97.0	30.5	0.97–0.98	0.49
Eggplants	92.7	30.4	0.94	0.48
Endive	93.3	31.1	0.94	0.48
Garlic, dry	74.2	28.0	0.79	0.42
Kale	86.6	30.7	0.89	0.46
Kohlrabi	90.1	30.0	0.92	0.47
Leeks, green	88.2–92.0	30.4	0.90–0.95	0.46
Lettuce	94.8	31.2	0.96	0.48
Lentils	12.0		0.44	
Mushrooms, fresh	90.0–91.1	30.0	0.93–0.94	0.47
dried	30.0		0.56	
Okra	89.8	28.6	0.92	0.46
Onions	87.5	30.1	0.90	0.46
Parsley	65.0–95.0		0.76–0.97	
Parsnips	78.6	29.8	0.84	0.46
Peas				
Dried	9.5		0.28	0.22
Green	74.3–76.0	30.1	0.79–0.81	0.42
Air dried	14.0		0.44	
Peppers				
Sweet	92.4	30.5	0.94	0.47
Chili (dry)	12.0	30.9	0.30	0.24
Popcorn, unpopped	13.5		0.31	0.24
Potatoes	75.0	0.84		
Late crop	77.8	29.8	0.82	0.43
Sweet potatoes	68.5	29.2	0.75	0.40
Pumpkins	90.5	29.9	0.92	0.47
Radishes				
Spring, bunched	93.6	30.1	0.95	0.48
Spring, prepacked	93.6	30.1	0.95	0.48
Winter	93.6		0.95	0.48
Rhubarb	94.9	29.9	0.96	0.48
Rutabagas	89.1	29.7	0.91	0.47
Sorrel	92.0		0.96	
Salsify	79.1	29.6	0.83	0.44

Table IV (Continued)

	Water content (%)	Average freez. pt. (°F)	Specific heat	
			Above freezing	Below freezing
Spinach	85.0–92.7	31.3	0.90–0.94	0.48
	93.0		0.94	
Squash				
Acorn		30.0		
Summer	95.0	30.4	0.96	
Winter	88.6	29.8	0.91	
Tomatoes				
Mature green	85.0–94.7	30.4	0.89–0.95	0.48
Ripe	94.1	30.4	0.95	0.48
Turnips	90.9	29.8	0.93	0.47
Fruit				
Apples	75.0–85.0	28.2	0.87–0.96	0.45
	85.0		0.89	
Apricots	85.4	29.6	0.88	0.46
Bananas	74.8	29.6	0.80	0.42
Dates				
Dried	20.0	−4.2	0.36	0.26
Fresh	78.0	27.1	0.82	0.43
Dried Fruits			0.30–0.32	
Figs				
Dried	24.0		0.39	0.27
Fresh	78.0	27.1	0.82	0.43
Grapefruit	88.8	28.6	0.91	0.46
Lemons	89.3	29.0	0.92	0.46
Limes	86.0	28.2	0.89	0.46
Mangoes	81.4	29.4	0.85	0.44
Melons				
Cantaloupe, Persian	92.7	29.9	0.94	0.48
Honey Dew, Honey Ball	92.6	29.8	0.94	0.48
Casaba	92.7	29.9	0.94	0.48
Watermelons	92.1	30.6	0.97	0.48
Nectarines	82.9	29.0	0.90	0.49
Oranges	87.2	28.0	0.90	0.46
Papayas	90.8	30.1	0.82	0.47
Peaches	86.9	29.6	0.90	0.46
Pitted	90.0		0.91	
Pears	0.45	82.7	27.7	0.86
Persimmons	78.2	27.5	0.84	0.43
Pineapples				
Mature green		29.1		
Ripe	85.3	29.7	0.88	0.45
Plums (and fresh Prunes)	81.0–85.7	28.7	0.87–0.88	0.45
Dried	28.0–35.0		0.53–0.59	
Pomegranates		26.5		
Quinces	85.3	28.1	0.88	0.45
Tangerines	87.3	29.5	0.90	0.46
Berries				
Blackberries	84.8	29.4	0.88	0.46
Blueberries	82.3	28.6	0.86	0.45

Table IV (Continued)

	Water content (%)	Average freez. pt. (°F)	Specific heat	
			Above freezing	Below freezing
Cherries	83.0	27.7	0.87	0.45
Cranberries	87.4	30.0	0.90	0.46
Currants	87.4	30.2	0.88	0.45
Dewberries		29.2		
Gooseberries	88.9	30.0	0.90	0.46
Grapes				
American type	81.9	29.4	0.86	0.44
European type	81.6	27.1	0.86	0.44
Logan Blackberries	82.9	29.5	0.86	0.45
Raspberries				
Black	80.6	29.4	0.84	0.44
Red	84.1	30.3	0.87	0.45
Strawberries				
Fresh	89.9–91.0	30.2	0.92–0.96	
Frozen	72.0			0.42
Dairy products, eggs				
Butter	15.0–16.0		0.33–0.64	0.30
Frozen	16.0			0.25
Cheese, non-fat	50.0		0.64	0.35
Cheeses	30.0–38.0	28.0	0.44–0.50	0.29–0.31
Cream				
Sweetened	75.0		0.85	0.50
40% fat	73.0	28.0	0.85	0.40
15% fat			0.92	
30% fat			0.93	
60% fat			0.99	
Sour cream	57.0–73.0		0.70	0.30
Cream Cheese	80.0		0.70	0.45
Ice Cream	58.0–66.0	27.0	0.78	0.45
Milk				
Whole	87.5	31.0	0.92–0.93	0.49
Skim	91.0		0.95	0.60
Powdered	12.5		0.31–0.93	0.49
Dried			0.23	
Whey			0.97	
Eggs				
Shell	67.0–76.0	28.0	0.41–0.74	0.40
Shell, farm cooler	67.0–70.0	28.0	0.74–0.76	0.40–0.41
Frozen	73.0	28.0		0.42
Dried, whole	5.0		0.25	0.21
Dried, yolk	3.0		0.22	0.21
Spray dried albumen	up to 6%		0.25	
Fermented albumen	3.0–15.0		0.22–0.32	
Meat, poultry				
Bacon				
Cured, farm style	13.0–29.0		0.30–0.43	0.24–0.29
Beef				
Fresh	62.0–67.0	28.0–29.0	0.70–0.84	0.38–0.43
Bones			0.40–0.60	

Table IV (Continued)

	Water content (%)	Average freez. pt. (°F)	Specific heat	
			Above freezing	Below freezing
Fat beef	50.0		0.60	0.35
Lean beef	70.0–76.0		0.76	0.41
Dried	5.0–15.0		0.22–0.34	0.19–0.26
Mincemeat			0.84	
Hams and Shoulders				
Fresh	47.0–54.0	28.0–29.0	0.58–0.63	0.34–0.36
Cured	40.0–45.0	0.52–0.56	0.52–0.56	0.32–0.33
Kidneys			0.86	
Lamb				
Fresh	60.0–70.0	28.0–29.0	0.68–0.76	0.38–0.51
Lard			0.38	
Mutton	90.0		0.93	
Pork				
Fresh	35.0–42.0	28.0–29.0	0.48–0.54	0.30–0.32
Fresh, fat	39.0		0.62	
Fresh, non-fat	57.0		0.73	
Smoked	57.0		0.60	0.32
Sausages				
Fresh	65.0	26.0	0.89	0.56
Miscellaneous				
Bread				
White	44.0–45.0		0.65–0.68	
Brown	48.5		0.68	
Dough			0.45–0.52	
Flour	12.0–13.5		0.38–0.45	0.28
Grains	15.0–20.0		0.45–0.48	
Gelatin			0.31	
Macaroni	13.0		0.44–0.45	
Nuts	3.0– 6.0		0.22–0.25	0.21–0.22
Dried	3.0–10.0		0.21–0.29	0.19–0.24
Oils (Vegetable)	75.0–90.0		0.47–50.0	0.35
Oleomargarine	15.5		0.32	0.25
Olives, fresh	75.2	28.5	0.80	0.42
Pearl Barley			0.67–0.68	
Porridge (Buckwheat)			0.77–0.90	
Raisins	24.5		0.47	
Rice	10.5–13.5		0.42–0.44	
Salt			0.27–0.32	
Sugar			0.20	0.20
Maple Sugar	5.0		0.24	0.21
Maple Syrup	36.0		0.49	0.31
Yeast	70.9		0.77	0.41
Chocolate, ground				0.63
Cacao, ground				0.63
Cocobutter				0.60

[a] Adapted from:
Editors of *Food Engineering*: 1962, 'Special Report: Plant Handbook Data', *Food Engineering* **34**, 89–104.

fraction undergoes the change as temperature continues to increase or decrease further after the initiation of phase change, and some portions never go through the change, as in the case of a small amount of unfrozen water present in frozen food. Figure 3 shows the usual behavior of apparent specific heat over a temperature range. Experimentally, it is difficult to determine what portion of the system has actually undergone phase transition during this period. Using the apparent heat capacity eliminates the need for the separation between the energy required for the reaction and the temperature change and is more practical.

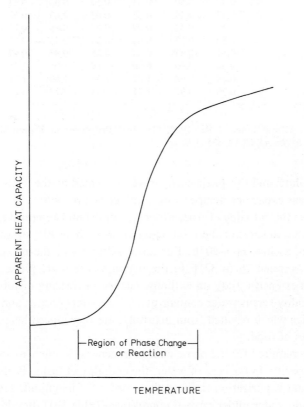

Fig. 3.   Apparent heat capacity versus temperature.

Apparent heat capacities of sucrose solutions [26] are given in Table V for a number of concentrations over the temperature range where fractional freezing takes place. At $-40°F$, there is only a very limited amount of melting taking place, and the apparent heat capacities are almost the same at all concentrations. As the temperature rises, the apparent heat capacities increase, indicating the corresponding increase in the portion undergoing fusion. This increase in the apparent heat capacity starts with higher concentration solutions, followed by those with lower concentrations sequentially until fusion is completed. The freezing point is depressed with increased con-

TABLE V[a]

The apparent heat capacity of sucrose solutions

Temperature range	Percent sucroce concentration							
	2	4	10	16	20	25	30	35
− 40 to − 35 F	0.43	0.44	0.46	0.44	0.43	0.44	0.44	0.44
− 35 to − 30 F	0.43	0.44	0.46	0.46	0.44	0.45	0.45	0.44
− 30 to − 35 F	0.45	0.46	0.48	0.51	0.52	0.51	0.53	0.56
− 25 to − 20 F	0.46	0.48	0.52	0.55	0.55	0.57	0.62	0.64
− 20 to − 15 F	0.46	0.48	0.52	0.57	0.58	0.59	0.63	0.65
− 15 to − 10 F	0.47	0.50	0.54	0.59	0.60	0.63	0.67	0.72
− 10 to − 5 F	0.47	0.49	0.55	0.62	0.63	0.67	0.70	0.75
− 5 to 0 F	0.48	0.52	0.59	0.67	0.68	0.73	0.79	0.84
0 to 5 F	0.50	0.53	0.64	0.74	0.76	0.82	0.93	0.97
5 to 10 F	0.51	0.55	0.68	0.85	0.85	0.97	1.08	1.17
10 to 15 F	0.54	0.60	0.80	0.99	1.05	1.21	1.38	1.53
15 to 20 F	0.58	0.68	1.02	1.36	1.50	1.71	2.02	2.28
20 to 25 F	0.70	0.92	1.71	2.43	2.79	3.26	3.76	4.16

[a] Adopted from:
[26] Keppeler, R. A. and Boose, J. R.: 1970, 'Thermal Properties of Frozen Sucrose Solutions', *Transactions of the ASAE* **13**, 335–339.

centration of solute, and this phenomena is well indicated by the relationship between the apparent heat capacities, temperatures, and concentration.

Table VI gives the enthalpy of frozen foods, adopted and compiled from Dickerson [12]. It shows that in some of the food materials as high as 10% of water is unfrozen at temperatures as low as − 20°F. For most of the food, the maximum enthalpy change occurs between 26 to 32°F in the region where more phase change occurs. Riedel made an extensive study on enthalpy change in freezing of food and the diagrams, for enthalpy versus water content at various temperatures and percentage of total water as ice which resulted from his study, are reproduced here (Figures 4–10) for several types of food.

The enthalpy required for the phase change in general is much greater than specific heat or heat capacity. In the case of water, the enthalpy of fusion is about two orders of magnitude and vaporization is about three orders of magnitude greater than heat capacity of liquid water under normal conditions (Table VII). Beside heat of fusion and vaporization, other reactions (such as denaturation or coagulation of protein, gelatinization of starch, and dissolvation of solute) involve energy and may be included in apparent heat capacity. Heat of denaturation of protein may vary from several hundred calories to several hundred kilo calories per mole depending on the protein and the condition. Irreversible denaturation of trypsin is 67600 cal/mole [32], while elsewhere the enthalpy for denaturation is given as 30000 cal/mole. Denaturation of myoglobin is usually considered a few hundred kilocalories per mole. The enthalpy for the denaturation of single cell protein is found to be about 800 cal/mole under the condition of the experiment [33, 34]. The protein denaturation related to food processing is characteristically endothermic, and thus the heat of

TABLE VI

Enthalpy of frozen foods[a]

Food	Water content (% by weight)	Mean heat capacity 40° to 90°F (BTU/lb °F)		Temperature (°F)												
				−40	−20	−10	0	10	20	22	24	26	28	30	32	40
Apple juice	87.2	0.92	A	0	10	16	24	34	51	57	66	79	105	151	153	161
			B	–	5	7	9	14	24	28	34	44	63	100	–	–
Apple juice concentrate	49.8	0.72	A	0	18	30	48	73	85	–	–	–	–	92	–	99
			B	–	33	45	63	93	–	–	–	–	–	100	–	–
Apple sauce	82.8	0.89	A	0	11	17	25	36	56	61	71	84	114	145	147	155
			B	–	5	7	11	17	28	33	41	52	76	100	–	–
Asparagus (peeled)	92.6	0.95	A	0	8	14	19	26	37	40	44	51	63	101	162	170
			B	–	–	–	4	6	10	12	16	20	28	55	100	–
Carrots	87.5	0.93	A	0	10	15	22	31	45	50	57	68	88	152	154	161
			B	–	–	5	7	11	19	22	27	35	50	100	–	–
Grape juice	84.7	0.91	A	0	10	17	25	34	56	63	73	88	119	147	149	156
			B	–	6	8	12	17	29	35	42	54	78	100	–	–
Onions	85.5	0.91	A	0	10	16	24	34	52	57	66	79	105	149	151	159
			B	–	5	7	9	15	24	28	35	45	65	100	–	–
Orange juice	89.0	0.93	A	0	10	16	23	31	48	53	60	72	94	154	156	164
			B	–	–	6	8	12	20	23	29	37	53	100	–	–
Peaches without stones	85.1	0.90	A	0	10	16	24	34	53	59	67	81	108	148	150	157
			B	–	5	7	10	15	26	30	37	48	69	100	–	–
Pears, Bartlett	83.8	0.89	A	0	10	17	25	35	53	59	69	83	111	146	148	155
			B	–	6	8	10	15	27	31	38	49	72	100	–	–
Plums without stones	80.3	0.87	A	0	12	19	28	40	64	73	85	113	139	141	–	149
			B	–	8	11	16	22	38	46	55	71	100	–	–	–
Raspberries	82.7	0.89	A	0	10	16	22	31	46	52	59	71	92	146	148	155
			B	–	8	11	16	22	31	39	46	56	77	100	–	–
Spinach	90.2	0.93	A	0	8	14	19	26	35	38	42	48	59	93	158	167
			B	–	–	–	5	5	10	11	14	18	25	50	100	–
Strawberries	89.3	0.94	A	0	9	15	21	29	41	45	51	60	77	127	158	165
			B	–	–	6	8	15	21	25	28	40	59	79	100	–
Sweet cherries without stones	77.0	0.86	A	0	12	20	29	42	67	76	89	110	134	136	138	146
			B	–	9	12	17	25	43	50	62	80	100	–	–	–
Tomato pulp	92.9	0.96	A	0	10	14	20	27	39	42	47	54	68	112	163	171
			B	–	–	–	–	6	12	14	18	22	31	62	100	–

Table IV (Continued)

Food	Water content (% by weight)	Mean heat capacity 40° to 90°F (BTU/lb °F)		Temperature (°F)												
				−40	−20	−10	0	10	20	22	24	26	28	30	32	40
Egg white	80.0	0.87	A	0	9	14	19	25	35	39	44	51	65	110	138	145
			B	—	—	—	—	12	18	19	22	26	36	70	—	—
Egg white	86.5	0.91	A	0	9	14	19	25	33	36	40	45	55	87	151	158
			B	—	—	—	—	—	12	13	14	17	22	48	—	—
Egg white	90.0	0.93	A	0	9	14	19	25	32	34	37	42	50	74	158	165
			B	—	—	—	—	—	—	—	10	12	17	36	—	—
Egg yolk	40.0	0.68	A	0	9	14	20	26	35	38	41	46	53	76	82	89
			B	20	—	—	24	27	—	34	38	43	54	89	—	—
Egg yolk	50.0	0.74	A	0	9	14	19	25	33	35	38	42	47	65	98	105
			B	—	—	—	—	—	—	20	23	27	32	66	—	—
Whole egg with shell	66.4	0.79	A	0	9	13	18	23	31	34	37	41	49	73	121	127
Egg shell	2.6	0.24	A	0	4	6	8	10	—	13	—	14	—	15	16	18
Meat																
Beef, lean, fresh	74.5	0.84	A	0	9	15	21	27	38	42	48	57	74	119	131	137
			B	10	10	11	12	15	22	24	28	37	48	92	100	—
Beef, lean, dried	26.1	0.59	A	0	9	14	20	28	33	—	36	—	38	—	40	45
			B	96	96	96	98	100	—	—	—	—	—	—	—	—
Cod	80.3	0.88	A	0	10	15	21	28	39	43	48	56	73	123	139	146
			B	10	10	10	12	14	20	22	26	32	45	88	100	—
Haddock	83.6	0.89	A	0	9	15	21	28	39	43	48	56	73	127	145	152
			B	8	8	9	10	12	17	19	23	29	42	86	100	—
Perch	79.1	0.86	A	0	9	14	20	27	38	42	46	53	68	117	137	144
			B	10	10	11	12	14	19	21	24	30	41	83	100	—
Chicken, young rooster	76	0.85	A	0	9	14	20	24	39	43	49	57	72	122	132	139
Chicken, dried young rooster	23.5	0.56	A	0	9	14	19	24	29	30	31	32	—	35	36	41
Veal, fresh	76.5	0.85	A	0	10	15	21	28	39	43	48	57	72	123	133	140
Venison	73	0.84	A	0	9	14	20	28	41	46	52	61	78	119	127	133

A Enthalpy BTU/lb
B % water unfrozen
[a] Adapted and compiled from:

[12] Dickerson, W., Jr.: 1968, Chap. 2: 'Thermal Properties of Foods in Freezing', *The Freezing Preservation of Foods*, Avi Publishing Co., P.O. Box 670, Westport, Conn., 4th ed., Vol. 2.

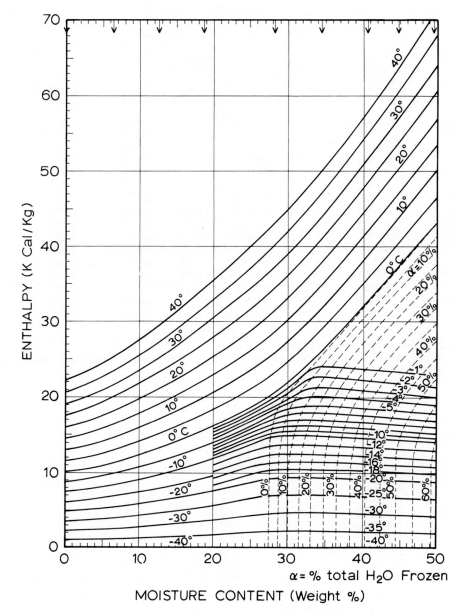

Fig. 4. Enthalpy diagram for white bread.

denaturation needs to be supplied and apparent heat capacity will increase in the region where the change takes place.

Heat of dissolution of some materials frequently present in food is given in Table VIII. The heat of dissolution which also must be supplied or removed during the processing contributes to the apparent heat capacity, although not predominantly.

TABLE VII

Enthalpy for phase transition of water[a]

Temperature (°F)	Absolute pressure (psia)	Density (lb/ft³)		Latent heat, (BTU/lb)		
		Ice	Vapor	Fusion	Vaporization	Sublimation
300	67.013		0.154655		910.1	
250	29.825		0.072354		945.5	
212	14.696		0.037313		970.3	
200	11.526		0.029723		977.9	
150	3.718		0.010302		1008.2	
100	0.9492		0.002854		1037.2	
50	0.17811		0.000587		1065.6	
32	0.0885	57.2	0.000303	143.35	1075.8	1219.1
30	0.0808	57.2	0.000277	143.35	1074.9	1219.3
20	0.0505	57.3	0.000177	149.31	1070.6	1219.9
10	0.0309	57.3	0.000111	154.17	1066.2	1220.4
0	0.0185	57.4	0.000068	158.93	1061.8	1220.7
−10	0.0108	57.4	0.000041	163.59	1057.4	1221.0
−20	0.0062	57.5	0.000024	168.16	1053.0	1221.1
−30	0.0035	57.5	0.000014	172.63	1048.6	1221.2
40	0.0019	57.6	0.000008	177.00	1044.2	1221.2

[a] Adapted and compiled from Reference:
[35] Keenan, Joseph H. and Keyes, Frederick G.: 1963, *Thermodynamic Properties of Steam*, John Wiley & Sons, Inc., New York. 1st ed., 35th printing.

TABLE VIII[a]

Heat of solution in water

	(cal/g mole) [b]
Acetic	− 2251
Dextrin	+ 268
Ethyl alcohol	+ 3200
Lactose	− 3705
Sodium chloride	− 1164
Sucrose	− 1319
Vanilic acid	− 5160

[a] Compiled from References:
[8] Perry, John H. (ed.): 1950, *Chemical Engineers Handbook*, McGraw-Hill Book Company, Inc., New York, 3rd ed.
[1] Moore, Walter J.: 1955, *Physical Chemistry*, Prentice-Hall, Inc., Englewood Cliffs, N.J., 2nd ed.
[b] at infinite dilution and approximately at room temperature
+ denotes heat evolved
− denotes heat absolved

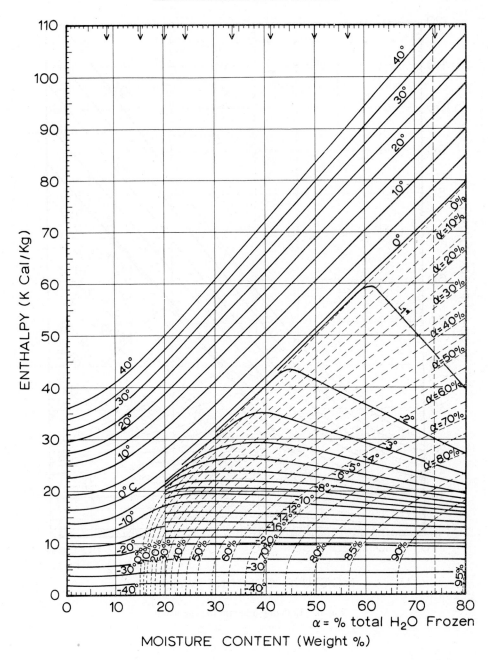

Fig. 5.   Enthalpy diagram for whole egg.

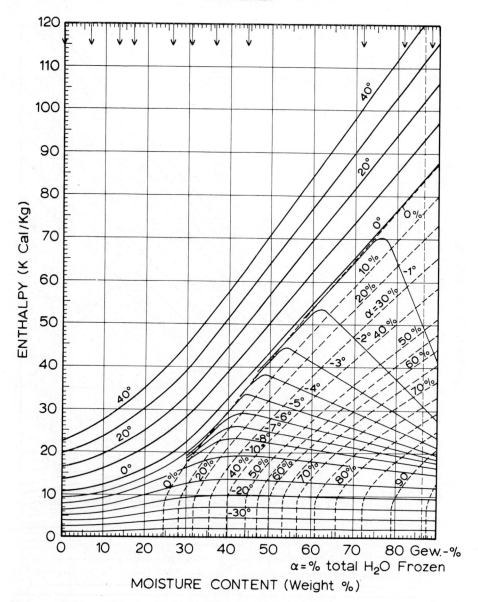

Fig. 6.   Enthalpy diagram for egg white.

2.2. THERMAL CONDUCTIVITY

The thermal conductivity of food depends on the thermal conductivity of the food components and it can be calculated theoretically from the thermal conductivity and fractional composition of individual components.

Fortunately, most fresh food contains a large percentage of water; therefore, the

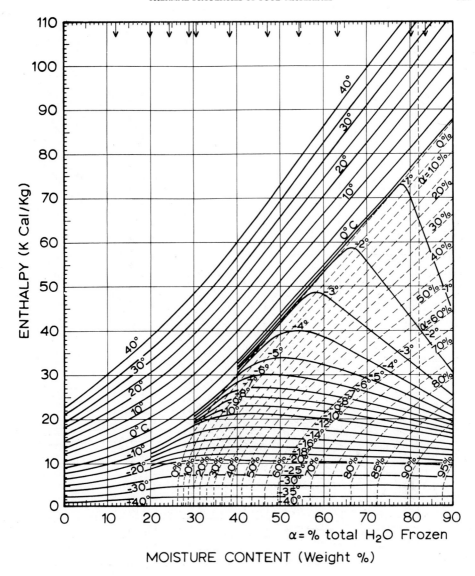

Fig. 7. Enthalpy diagram for lean sea fish muscle.

thermal conductivity of water at a similar condition may be used as a very rough approximation or first approximation of the thermal conductivity of the food materials when no data is available. This approximation can be made because thermal conductivity of most of the major food components is in the same order of magnitude not too different from water. One exception is air, which is often present or incorporated in food. Thermal conductivity of air is less than one order of the magnitude of that of liquid water (Table IX and X). Therefore, careful consideration should be

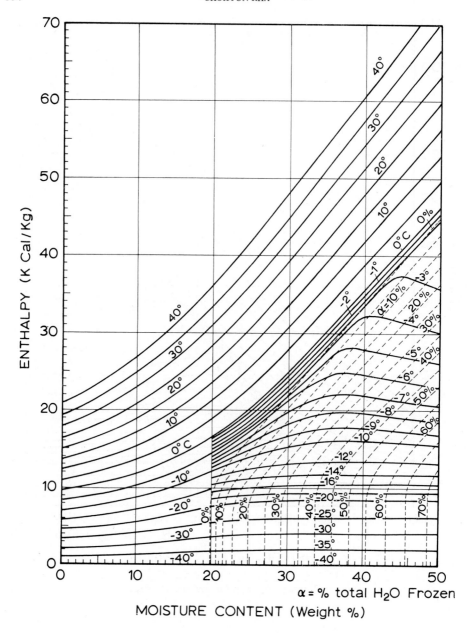

Fig. 8.   Enthalpy diagram for potato starch.

given to air content or void spaces in food materials. (Thermal conductivity and other thermal properties of air are given in Table IX). The thermal conductivity of water in liquid, vapor and solid states is given in Table X. The thermal conductivity of liquid water is greater than that of water vapor and the thermal conductivity of ice

TABLE IX

Thermal conductivity, specific heat, density and diffusivity of air[a]

Temperature (°F) (at atomic pressure)	Thermal conductivity (BTU/h ft °F)	Specific heat (BTU/lb °F)	Density (lb/ft³)	Thermal diffusivity (ft²/h)
0	0.0133	0.239	0.086	0.00130
32	0.0140	0.240	0.081	0.00145
100	0.0154	0.240	0.071	0.00180
200	0.0174	0.241	0.060	0.00239
300	0.0193	0.243	0.052	0.00306
400	0.0212	0.245	0.046	0.00378

[a] Adapted from Reference:
[6] Kreith, Frank: 1963, *Principles of Heat Transfer*, International Textbook Company, Scranton, Pa., 7th Printing.

TABLE X

Thermal conductivity, specific heat, density and thermal diffusivity of water[a]

Temperature (°F) (at atomic pressure)	Thermal conductivity (BTU/h ft °F)	Specific heat (BTU/lb °F)	Density (lb/ft³)	Thermal diffusivity (ft²/h)
Liquid				
32	0.320	1.008	62.4	0.00509
40	0.326	1.005	62.4	0.00520
50	0.333	1.002	62.4	0.00533
70	0.347	0.999	62.3	0.00558
100	0.362	0.999	62.0	0.00585
150	0.381	1.001	61.2	0.00622
200	0.392	1.006	60.1	0.00648
Vapor (at saturation)				
32 (0.0884 psia)	0.0092	0.466	0.000303	65.2
50 (0.1781 psia)	0.0096	0.458	0.000587	35.7
100 (0.949 psia)	0.0108	0.441	0.002854	8.57
150 (3.718 psia)	0.0121	0.432	0.0103	2.72
200 (11.526 psia)	0.0135	0.423	0.0297	1.07
Solid				
32	1.28	0.492	57.2	0.0455
20	1.31	0.483	57.3	0.0473
10	1.34	0.475	57.4	0.0491
0	1.37	0.467	57.4	0.0511
− 10	1.39	0.459	57.4	0.0528
− 50	1.57	0.426	57.6	0.0670
− 100	1.78	0.379	57.7	0.0814
− 150	2.02	0.330	57.8	0.1059

[a] Compiled from References:
[35] Keenan, Joseph H. and Keyes, Frederick G.: 1963, *Thermodynamic Properties of Steam, Including Data for the Liquid and Solid Phases*, John Wiley & Sons, Inc., New York, 1st ed. 35th Printing.
[12] Dickerson, R. W.: 1968, Chap. 2: 'Thermal Properties of Foods in Freezing', *The Freezing Preservation of Foods*, Avi Publishing Co., P.O. Box 670, Westport, Conn. 4th ed. Vol. 2.

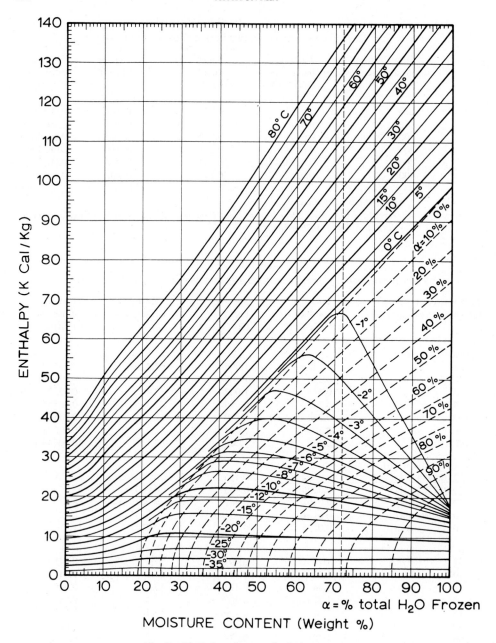

Fig. 9. Enthalpy diagram for bakers yeast.

is even greater than water. This would be expected, considering that molecules in ice are in closest proximity, but in the vapor state the least association of molecules exists.

Considering that the ingredients form layers in food, as in Figure 11, the thermal

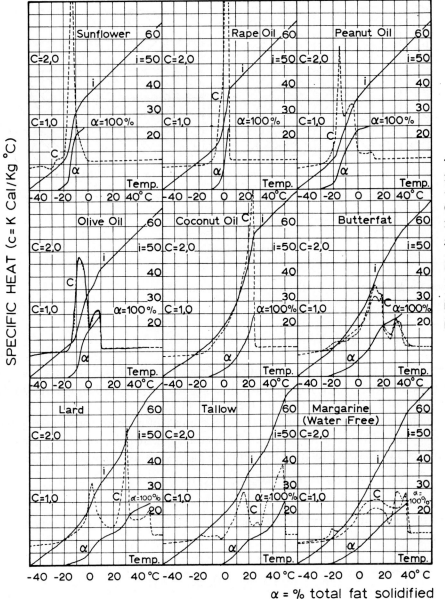

Fig. 10. Enthalpy and specific heat of fusion of fats and oils.

conductivity of the system can be calculated from the heat transfer equation when the layers are parallel to heat flow (Figure 11a),

$$K_{ave} = \sum_1^n K_i \Omega_i. \tag{22}$$

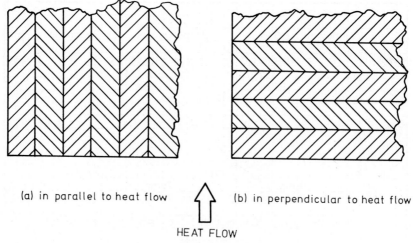

(a) in parallel to heat flow   (b) in perpendicular to heat flow

HEAT FLOW

Fig. 11.   Composite ingredient layers of food.

When the layers are perpendicular to the heat flow (Figure 11b),

$$K_{ave} = \frac{1}{\sum\limits_{1}^{n} \Omega_i/K_i},$$ (23)

where K_i is the thermal conductivity of each constituent and Ω_i is the volume fraction of each constituent or $W_i \varrho_T / W_T \varrho_i$ where W and ϱ are weight and density respectively and subscript T denotes the total for the system [37].

In addition to these two simple methods, a number of methods for estimating the thermal conductivity of a mixed system have been suggested [38, 39, 40, 41]. Among them, a few methods are presented here. For two-phase systems, Maxwell [38] suggested,

$$K_m = \frac{K_c [K_d + 2K_c - 2\Omega_d (K_c - K_d)]}{K_d + 2K_c + \Omega_d (K_c - K_d)}.$$ (24)

For a two-phase mixture with known particle shape [39], thermoconductivity of the mixed system may be expressed as

$$K_m = \frac{K_c [K_d + (S - 1) K_c - (S - 1) \Omega_d (K_c - K_d)]}{K_d + (S - 1) K_c + \Omega_d (K_c - K_d)}$$ (25)

and for a multiphase mixture with known particle shape, it is [39]

$$K_m = K_c \left[\frac{1 - \sum\limits_{i}^{n} \dfrac{\Omega_i (S - 1) (K_c - K_i)}{K_i + (S_i - 1) K_i}}{1 + \sum\limits_{i}^{n} \dfrac{\Omega_i (K_c - K_i)}{K_i + (S_i - 1) K_i}} \right],$$ (26)

where $S = 3/\psi$ (where ψ is sphericity and $S = 3$ for sphere) and subscript m refers to mixture, c to continuous phase and d to dispersed phase.

When the thermal conductivity of major food components is accurately determined and composition of food material is known, theoretically, some of the above equations may be useful in predicting the thermal conductivity of the food. In practice,

TABLE XI[a]

Thermal conductivity of common food components, grain materials and fruits products

Product	Temperature (°F)	Thermal conductivity (BTU/h ft °F)	References
Common food components			
Ethylene glycol	32	0.153	[8]
Glycerol (100%)	86	0.078	[8]
	212	0.167	[8]
Glycerol (20%)	68	0.278	[8]
Stearic Acid	212	0.0786	[8]
Acetic Acid (50%	68	0.099	[8]
Ethyl alcohol		0.105	[8]
Grain material			
Whole soybeans (11.2% H_2O)	50–150	0.056–0.077	[36]
Crushed soybeans (10.0% ^2H_2O)	50–150	0.048–0.073	[36]
Soybean powder (9.4% H_2O)	50–150	0.038–0.058	[36]
Fruit product			
Apple Sauce	72.5	0.400	[9]
Banana puree	60	0.320	[9]

[a] Compiled from References:

[8] Perry, John H. (ed.): 1950, *Chemical Engineers Handbook*, McGraw-Hill Book Company, New York, N.Y. 3rd ed.

[9] Charm, Stanley E.: 1963, *Fundamentals of Food Engineering*, The Avi Publishing Company, Inc., Westport, Conn.

[36] Jasansky, A. and Bilansky, W. K.: 1973, 'Thermal Conductivity of Whole and Ground Soybeans', *Transactions of the ASAE* **16**, 100–103. Published by the American Society of Agricultural Engineers, St. Joseph, Michigan.

variations in the biological material do not allow the precise reproduceability that is expected in other materials, accurate experimental confirmation of predicted values is difficult, and entertaining complex mathematical expressions in an attempt to increase the accuracy and precision is often meaningless.

Table XI lists the thermal conductivity of some common food components, grain materials and fruit products; Table XII, that of a number of meats; Table XIII, the thermal conductivity of frozen food, and Table XIV, the thermal conductivity of some freeze-dried food.

According to Table XII, most of the meats have thermal conductivity between 0.23 to 0.28 BTU/h ft °F above freezing, which is about 30% lower than water, but higher values 0.29 and 0.30 BTU/h ft °F have also been reported [11, 13]. The thermal conductivity of frozen meats (Table XIII) is much lower than that of ice and decreases with the increase in the temperature as in ice. The decrease is linear between the temperature range of $-10°$ to $20°F$. At above $20°F$ up to freezing, the change in thermal conductivity is unexpectedly fast and unfrozen or fresh food has thermal conductivity only about one-third or one-fourth that of frozen food.

TABLE XII[a]

Thermal conductivity of meats

		Temperature (°F)	Conductivity (BTU/h °F ft)
Beef (0.8% fat, 78.9% water)	A	8.7	0.769
lean, Inside Round,		18.4	0.718
Canner and Cutter Grade		44.5	0.275
		74.4	0.278
		97.3	0.270
		143.7	0.281
Beef (1.4% fat, 78.7% moisture)	B	11.3	0.806
lean, Inside Round,		19.6	0.757
Canner and Cutter Grade		46.2	0.249
		63.4	0.248
		89.8	0.251
		141.6	0.256
Chicken		40–80	
Broiler muscle			0.238
Broiler muscle and overlying skin			0.211
Broiler skin			0.018
Hen muscle			0.255
Hen muscle and overlying skin			0.229
Hen skin			0.014
Pork (6.7% fat, 75.9% moisture)	A	6.2	0.750
lean, leg		13.0	0.737
Premium Grade		17.4	0.738
		42.8	0.282
		46.9	0.286
		70.6	0.293
		112.7	0.306
		138.8	0.312
Pork (7.8% fat, 75.1% moisture)	B	8.2	0.823
lean, leg		12.8	0.800
Premium Grade		21.3	0.745
		23.6	0.730
		38.8	0.256
		43.0	0.261
		66.2	0.262

TABLE XII (Continued)

		Temperature (°F)	Conductivity (BTU/h °F ft)
		109.3	0.280
		141.3	0.283
Veal (2.1% fat, 75.0% moisture) lean, leg Premium Grade	A	14.8	0.750
		18.0	0.738
		24.3	0.708
		42.6	0.275
		50.0	0.259
		108.2	0.276
		125.6	0.282
		144.4	0.283
Veal (2.1% fat, 75.0% moisture) lean, leg Premium Grade	B	8.2	0.835
		13.8	0.814
		20.9	0.773
		23.6	0.762
		40.4	0.255
		47.2	0.260
		74.6	0.258
		106.4	0.254
		138.9	0.261
Lamb (8.7% fat, 71.8% moisture) lean, leg Premium Grade	A	5.6	0.650
		11.8	0.623
		17.0	0.607
		23.9	0.590
		41.8	0.260
		51.0	0.259
		93.6	0.271
		115.6	0.271
		142.0	0.276
Lamb (9.6% fat, 71.0 moisture) lean, leg Premium Grade	B	6.0	0.735
		10.1	0.730
		19.3	0.689
		24.8	0.679
		42.0	0.240
		50.8	0.226
		87.4	0.236
		119.3	0.243
		142.6	0.244

A Fiber perpendicular to direction of heat flow
B Fiber parallel to direction of heat flow

[a] Adapted from References:
[45] Leitman, J.: 1967, 'Thermal Conductivity of Nondehydrated Meats by Current Investigation', M.S. Thesis, Georgia Institute of Technology, Atlanta, Ga.
[46] Walters, Robert E. and May, K. N.: 1963, 'Thermal Conductivity and Density of Chicken Breast Muscle and Skin', *Food Technol.* **17**, 808.

TABLE XIII[a]

Thermal conductivity of frozen food

Product	Temperature (°F)	Thermal conductivity, (BTU/h ft °F)	References
Butterfat			
0.3% water, 0.1% air	–	0.39	[21]
16.9% water, 0.9% water, 6.5% air	–	0.47	[21]
14.5% water, 0.1% air	–	0.48	[21]
15.3% water, 10.0% air	–	0.42	[21]
Eggs			
Egg, whole, thawed and refrozen	17	0.56	[17]
Egg, whole, concentrated by a factor of 2	17	0.32	[17]
Fish			
Codfish (heat flow perpendicular to fiber)	– 10	0.90	[13]
	0	0.85	[13]
	15	0.81	[13]
Fish (cod and haddock)	– 20	1.1	[16]
	0	1.0	[16]
	15	0.93	[16]
Cod fillets, packed tightly (apparent density, 62.4 lb/ft³)	– 3	0.68	[17]
Herring, gutted whole, packed tightly (apparent density, 56.8 lb/ft³)	– 2	0.46	[17]
Salmon (heat flow perpendicular to fiber)	– 10	0.72	[13]
	0	0.68	[13]
	15	0.63	[13]
Fruit			
Gooseberries, dry, mixed sizes	5	0.16	[17]
Gooseberries, wet, mixed sizes	2	0.19	[17]
Plums, small, firm, and dry, 3 cm long, 2 cm diameter	4	0.17	[17]
Plums, large, 5 cm long, 4 cm diameter	2	0.14	[17]
Strawberries, mixed sizes, packed tightly	4	0.65	[17]
Strawberries, large sizes, 12 to 16 gm, wet, unsquashed	2	0.32	[17]
Strawberries, mixed sizes, firm	3	0.32	[17]
Strawberries, small sizes, 1½ to 3 gm, firm, dry	0	0.34	[13]
Poultry			
Turkey, breast, 74% water, 2.1% fat (heat flow parallel to fiber)	– 10	0.94	[13]
	0	0.91	[13]
	15	0.83	[13]
Turkey, breast, 74% water, 2.1% fat (heat flow perpendicular to fiber)	– 10	0.76	[13]
	0	0.73	[13]
	15	0.68	[13]
Turkey, leg, 74% water, 3.4% fat (heat flow perpendicular to fiber)	– 10	0.85	[13]
	0	0.81	[13]

Table XIII (Continued)

Product	Temperature (°F)	Thermal conductivity (BTU/h ft °F)	References
Meat			
Beef, sirloin, 75% water, 10.9% fat	0	0.84	[13]
(heat flow parallel to fiber)			
Beef, loin, 70% water (heat flow	0	0.63	[14]
parallel to fiber)	15	0.60	[14]
Beef, round, 77% water, 2.4% fat	0	0.80	[15]
(heat flow parallel to fiber)	13	0.75	[15]
	18	0.70	[15]
	22	0.69	[15]
Beef, flank, 74% water, 3.4% fat	0	0.66	[13]
(heat flow perpendicular to fiber)	15	0.62	[13]
Beef, round, 76% water, 3.0% fat	9	0.65	[15]
(heat flow perpendicular to fiber)	17	0.63	[15]
	19	0.62	[15]
	25	0.58	[15]
Beef fat, 9% water, 89% fat	10 to 20	0.17	[13]
Pork, lean leg, 72% water, 6.1%	−10	0.92	[13]
fat (heat flow parallel to fiber)	0	0.88	[13]
	15	0.83	[13]
Pork, lean leg, 72% water, 6.1%	−10	0.80	[13]
fat (heat flow perpendicular to fiber)	0	0.76	[13]
	15	0.72	[13]
Pork fat, 6% water, 93% fat	−10 to 35	0.12	[13]
Vegetables			
Beans, runner, sliced by hand and scalded	2	0.46	[17]
Beans, runner, sliced by machine and scalded	9	0.53	[17]
Broccoli, heads cut into 8 to 12 pieces			
and scalded	20	0.22	[17]
Carrots, whole, scraped, and scalded	1	0.36	[17]
Carrots, scraped, sliced, and scalded	2	0.39	[17]
Carrots, puree	18	0.73	[17]
Peas, shelled and scalded	3	0.29	[17]
Potatoes, mashed, and packed tightly	9	0.63	[17]
Potatoes, mashed, extruded as strings,			
loosely packed	17	0.24	[17]

ª Compiled from References:

[13] Lentz, C. P.: 1961, 'Thermal Conductivity of Meats, Fats, Gelatin, Gels, and Ice', *Food Technol.* **15**, 243–247.

[14] Miller, H. L. and Sunderland, J. E.: 1963, 'Thermal Conductivity of Beef', *Food Technol.* **17**, 124–126.

[15] Hill, J. E.: 1966, 'The Thermal Conductivity of Beef', M.S. Thesis, Georgia Institute of Technology, Atlanta, Ga.
nology, Atlanta, Ga.

[17] Smith, J. G., Ede, A. J., and Gane, R.: 1952, 'The Thermal Conductivity of Frozen Foodstuffs', *Mod. Refrig.* **55**, 254–257.

[21] Lagoni, H. and Merton, D.: 1956, 'Wärmeleitsfähigkeitsmessungen on Butter', *Kiel. Milchwirtsch. Forschungsber.* **8**, 85.

Thermal conductivity, like heat capacity, changes rapidly over the phase transition region, exhibiting discontinuity due to differences in the thermal conductivities at different phases as discussed above (Figure 12).

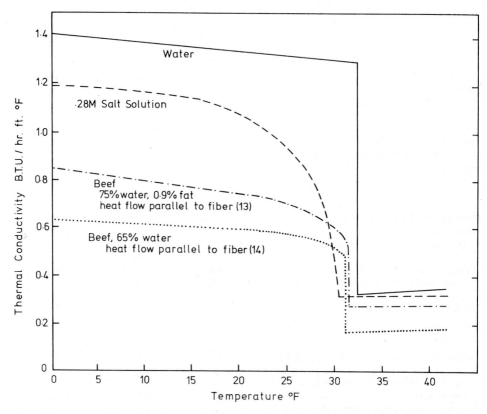

Fig. 12.   Thermal conductivity versus temperature.

In a frozen state, the orientation of fiber has been reported to influence the thermal conductivity of meat. Heat flow along the fiber shows the thermal conductivity 15% to 30% greater than when heat flow is perpendicular to the fiber [13], although such variation may not be accounted considering the physical dimension of the fiber and thermal conductivities of individual meat components.

The low thermal conductivity of freeze-dried foods (Table XIV) exhibits the phenomenon that thermal conductivity is lowered by void space or air.

2.3. THERMAL DIFFUSIVITY

In food processing, the heat transfer involved is more often unsteady state, that is, the temperature of the system changes with time. In such case, the response of the system upon exposure to the temperature gradient is of importance, i.e., what temperature

TABLE XIV[a]

Thermal conductivity of freeze-dried foods

	Gas in freeze-dried material	Temperature (°F)	Thermal conductivity (BTU/h ft °F) at absolute pressures in Hg					
			0.01 mm	0.1 mm	1 mm	10 mm	100 mm	760 mm
Haddock	–	20–120	–	0.011	–	–	–	–
Perch	–	20–120	–	0.013	–	–	–	–
Beef, round commercial	Air	0–110	0.022	0.024	0.028	0.035	0.037	0.037
Beef	Carbon dioxide	80–110	–	0.023	0.026	0.030	0.032	0.032
Beef	Helium	80–110	0.022	0.024	0.036	0.066	0.094	0.098
Beef	Nitrogen	80–110	0.022	0.023	0.028	0.035	0.038	0.038
Apple	Air	0–110	0.009	0.010	0.015	0.022	0.024	0.024
Apple and pear	Carbon dioxide	80–110	–	0.013	0.016	0.021	0.023	0.023
Apple and pear	Helium	80–110	–	0.015	0.023	0.049	0.090	0.098
Apple and pear	Nitrogen	80–110	0.013	0.014	0.018	0.025	0.029	0.029
Peach	Air	0–110	0.009	0.010	0.015	0.023	0.025	0.025

[a] Adapted from References:

[42] Harper, J. C.: 1962, 'Transport Properties of Gases in Porous Media at Reduced Pressures with Reference to Freeze-Drying', *Am. Inst. Chem. Eng. J.* **8**, 298–302.

[43] Harper, J. C. and El Sahrigi, A. F.: 1964, 'Thermal Conductivities of Gas-Filled Porous Solids', *Ind. Eng. Chem. Fundamentals* **3**, 318–324.

[44] Lusk, G., Karel, M., and Goldblith, S. A.: 1964, 'Thermal Conductivity of Some Freeze-Dried Fish', *Food Technol.* **18**, 121–124.

change in the system results as well as how fast the system transfers heat. Thermal diffusivity, which is the ratio of thermal conductivity to the product of heat capacity and density, defines this particular thermal property. This property of the material characteristically affects the unsteady state heat conduction, and the time rate of temperature change depends on its numerical value.

In other words, in such unsteady state, the amount of heat transferred into the system is equal to the change in the temperature of the system. Therefore, the energy balance or unsteady state heat transfer defines the thermal diffusivity as:

$$\frac{\partial T}{\partial \theta} = \frac{K}{C_p \varrho} \left(\frac{\partial^2 T}{\partial X^2} + \frac{\partial^2 T}{\partial Y^2} + \frac{\partial^2 T}{\partial Z^2} \right). \tag{27}$$

where T = temperature, θ = time, $K/C_p\varrho$ = thermal diffusivity where K is thermal conductivity, C_p is heat capacity and ϱ is density, and X, Y, Z are length in three dimensions.

Thermal diffusivity, along with other thermal properties of air, water and some foods is given in Table IX, X and XV respectively. Table XVI presents the thermal diffusivity of sucrose solutions at various concentrations below 25°F.

TABLE XV[a]

Thermal properties of foods

	Water content (% by weight)	Thermal conductivity at 60°F (BTU/h ft °F)	Apparent density (lb/ft³)	Mean specific heat to 90°F (BTU/lb °F)	Thermal diffusivity (ft²/h)
Apples		0.24	54.8	0.9	0.0049
Apple Juice	87.2	0.32	65.6	0.92	0.0053
Apple Juice Concentrate	49.8	0.25	76.6	0.72	0.0045
Beef (lean Sirloin)		0.29	72	0.84	0.0048
Beet, sugar					0.0049
Bilberry juice	89.5	0.32	65	0.93	0.0053
Carrots	87.5	–	–	0.93	–
Cherry juice	86.7	0.32	65.7	0.92	0.0053
Cod fish		0.31	62	0.88	0.0057
Grapefruit		0.23	55.2	0.9	0.0047
Grape juice	84.7	0.31	66.3	0.91	0.0051
Oranges		0.24	54.8	0.9	0.0049
Orange juice	89.0	0.32	65.1	0.93	0.0053
Potato Salad		0.28	63	0.79	0.0056
Raspberry juice	88.5	0.32	65.3	0.93	0.0053
Strawberry juice	91.7	0.33	64.5	0.95	0.0054

[a] Compiled from References:

[12] Dickerson, R. W.: 1968, 'Thermal Properties of Foods', in D. K. Tressler, W. B. Van Ardel, and M. J. Copley (eds.), *The Freezing Preservation of Foods*, Avi Publishing Co., Westport, Conn., 4th ed., Vol. 2.

[47] Dickerson, R. W. and Read, R. B.: 1973, 'Cooling Rates of Foods', *J. Milk Food Technol.* **36**, 167–171.

2.4. OTHER THERMAL CHARACTERISTICS

The thermal properties discussed above control the heat transfer or temperature change of the system. On the other hand, there are properties which are affected by the temperature of the system and consequently affect the heat transfer or temperature change. Such properties are density or coefficient of thermal expansion. For liquids and solids, which constitute most of the food components, the change in density with temperature is not significant. For instance, for liquid water the coefficient of thermal expansion ranges from -0.0004 to 0.0007 per °F between 32 to 350°F [6]. Therefore, density does not affect appreciably heat transfer or temperature change. However, this is not true for vapors or gases which may be contained in foods, or more often, used as heating media. Gases have a higher coefficient of thermal expansion. For air, this coefficient is 0.002 to 0.001 per °F between 0 to 400°F, and steam has a similar value between 212 to 400°F [6].

Density affects conduction heating by being inversely proportional to thermal diffusivity. The coefficient of thermal expansion affects free convective heat transfer by contributing to the Grashof number.

TABLE XVI

Thermal diffusivity (ft^2/h) for sucrose solutions

Temperature interval	Percent sucrose concentrations								
	2	4	10	14	16	20	25	30	35
−40 to −35 F	0.43	0.44	0.46	0.45	0.44	0.43	0.44	0.44	0.44
−35 to −30 F	0.43	0.44	0.46	0.46	0.46	0.44	0.45	0.45	0.44
−30 to −25 F	0.45	0.46	0.48	0.51	0.51	0.52	0.51	0.53	0.56
−25 to −20 F	0.46	0.48	0.52	0.54	0.55	0.55	0.57	0.62	0.64
−20 to −15 F	0.46	0.48	0.52	0.55	0.57	0.58	0.59	0.63	0.65
−15 to −10 F	0.47	0.50	0.54	0.58	0.58	0.60	0.63	0.67	0.72
−10 to − 5 F	0.47	0.49	0.55	0.61	0.62	0.63	0.67	0.70	0.75
− 5 to 0 F	0.48	0.52	0.59	0.65	0.67	0.67	0.73	0.79	0.84
0 to 5 F	0.50	0.53	0.64	0.72	0.74	0.76	0.82	0.93	0.97
5 to 10 F	0.50	0.55	0.68	0.80	0.85	0.85	0.97	1.07	1.17
10 to 15 F	0.54	0.60	0.80	0.95	0.99	1.05	1.21	1.38	1.53
15 to 20 F	0.58	0.68	1.02	1.28	1.36	1.50	1.70	2.02	2.28
20 to 25 F	0.70	0.92	1.71	2.23	2.43	2.79	3.26	3.76	4.16

[a] Adopted from Reference:
[26] Keppeler, R. A. and Boose, J. R.: 1970, 'Thermal Properties of Frozen Sucrose Solutions', *Transactions of the ASAE* **13**, 335–339.

2.5. Temperature effect on food properties

Some of the reasons for applying heat or removing heat are discussed in the Introduction and the effect of the temperature on some properties of food are discussed in the other chapters. In addition, there are a few important temperature effects on food properties which need to be mentioned, if not elaborated in detail. Carbohydrate in the presence of water swells and gelatinizes when heated. In general, fat melts when heated and solidifies when cooled in a relatively narrow range of temperatures. Protein coagulates or degrades when heated and some protein solutions gel upon cooling. During the heating, food is affected in many ways, and develops flavor, texture and color which are desirable.

2.6. Thermal properties related to food processing

It is not common present practice to predict heat transfer in food processing from calculations using numerical values of thermal properties or to design a process solely based on such calculations. Nevertheless, a heat transfer calculation is still the best way to demonstrate the importance of thermal properties in food processing and to construct overall perspectives. Of course, if such a quantitative or analytical approach was practised more often, it would increase efficiency considerably and benefit the food industry.

In practice heat is transferred by not one mode but some combination of more than one mode. In many cases, however, the rough order of magnitude analysis would indicate which mode is dominating, controlling or limiting and in such case mechanisms contributing insignificantly can be neglected.

When a piece of steak is broiled in the oven, the heat is transferred to the roast by both radiation and convection. The overall heat transfer coefficient $\bar{h}$ for the roast surface to heat flow by convection and radiation is

$$\bar{h} = \bar{h}_c + \bar{h}_r.$$

In this case, the hot surface faces downward in natural convection. Therefore, convective heat transfer $\bar{h}_c$ may be calculated with Equation (8). Considering oven temperatures or air temperatures of around 350°F, and initial surface temperatures of 50°F, Gr (Equation (7)) is 0.850×10^6, Pr (Equation (6)) is 0.72 and $\bar{h}_c$ estimated is about 1 BTU/h ft² °F. This is in the low range of the values given for free convection of air in Table II. During broiling, if gas flame is used, the flame temperature can easily reach above 3500°F [8], and if an electric oven is used, the elective coil may reach above 2000°F. The temperature factor for radiation, F_t (Equation (17)) is around 35 as taken from a graph [6]. This is obtained by using 2000°F as an approximation for the temperature of the radiating surface. The surface temperature of steak would be usually above refrigeration temperature initially and certainly should not reach above 200°F if the steak remains edible, and this is taken into consideration. Assuming the view factor to be unity and emissivity of beef 0.7 (Table III), the radiation heat transfer coefficient is about 25 BTU/h ft² °F. Therefore in broiling a steak, convective heat transfer is negligible and radiation predominates. A steak, during the broiling under this condition, according to Equation (6), gets approximately 50000 BTU/h ft².

Another example for the combined mode of heat transfer is the thermal processing of food, which is actually a series of heat transfer steps (see Figure 13). When solid food is packaged in a flexible package [8] heat transfers from the heating media, through water in the retort, to the surface of the flexible package, through the lamination of the packaging materials, then from the inside of the packaging by some mechanisms to the surface of the solid foods and finally into the food.

The overall heat transfer coefficient, U, for the process, is reciprocal of the sum of all the resistances for heating the solid food in the flexible package.

$$U = \frac{1}{\dfrac{1}{h_{\text{out}} A_p} + \dfrac{x_p}{K_p A_p} + \dfrac{1}{h_{\text{in}_1} A_p} + \dfrac{1}{h_{\text{in}_2} A_s} + \dfrac{x_s}{K_s A_s}},$$

where

A_p = heat transfer area of the package;

h_{out} = heat transfer coefficient to the package wall; from heating medium, i.e. water in a retort;

x_p = thickness of the package film;

K_p = heat conductivity of the package film;

h_{in_1} = heat transfer coefficient inside of the package between package inner wall and the inside gap;

h_{in_2} = heat transfer coefficient between the inner gap
  of the pouch and the surface of the solid food;

x_s = half thickness of solid food;

A_s = area of heat transfer in solid food.

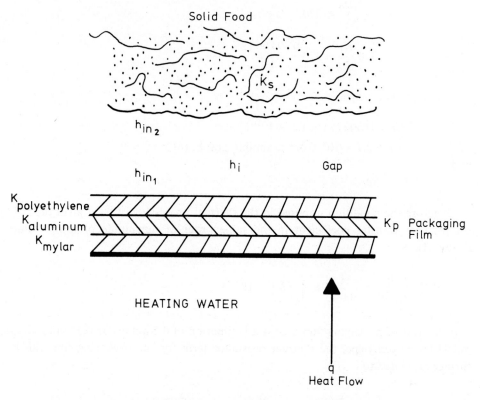

Fig. 13. Heat transfer into flexibly packaged solid food.

The order of magnitude of heat transfer coefficient of water heating is $300 \sim 9000$ BTU/h ft² °F [7] and 500–10000 BTU/h ft² °F [6].

In the case when the heating medium is water at 250°F, temperature of the wall is 150°F and for the plate having 4.5 in. length,

$$Pr = 1.88$$

$$g\beta\varrho^2/\mu^2 = 1.11 \times 10^9 \; °F^{-1} \, ft^{-3} \; [6]$$

$$\log_{10} \frac{(g\beta\varrho^2)}{\mu^2} (\Delta T \, L^3) \left(\frac{C_{p\mu}}{K}\right) = \log_{10}(1.11 \times 10^9) \times \left(100 \left(\frac{4.5}{12}\right)^3\right)(1.88)$$

$$\log_{10}(1.1 \times 10^{10}) = 10.04.$$

From this value and using extrapolation of the Nu versus $(Gr) \, (Pr)$ graphs for the horizontal plate given in a heat transfer text [6] gives

$$\log_{10}\left(\frac{h_c D_0}{K_f}\right) \approx 2.24$$

$$\frac{h_c D_0}{K_f} \approx 370$$

$$h_c = 370 \, \frac{K_f}{D_0} = 370 \, \frac{0.394 \times 12}{4.5} = 389 \,.$$

For vertical plate, when

$$Gr = (Gr)\,(Pr) = 1.1 \times 10^{10} \,,$$
$$Nu = 2.7 \times 10^2 \text{ from a similar graph, [6]}$$

and

$$h = 282 \,.$$

Therefore the condition imposed in the thermal processing of food gives the convective heat transfer coefficient in the same order of magnitude of lower values (Table II). Now

$$\frac{1}{h_{out} A_p} \approx \frac{1}{300 A_p} \approx \frac{1}{A_p}\,(3.3 \times 10^{-3})$$

If the flexible packaging film used is a lamination of 0.5 mil mylar/0.35 mil foil/3.0 mil H.D. polyethylene, the thermal resistance term for the packaging film can be further expended to

$$\frac{x_p}{K_p A_p} = \frac{x_{mylar}}{K_{mylar} A_{mylar}} + \frac{x_{foil}}{K_{foil} A_{foil}} + \frac{x_{polyethylene}}{K_{polyethylene} A_{polyethylene}}$$

but,

$$A_p = A_{mylar} = A_{foil} = A_{polyethylene}$$

$$\frac{x_p}{K_p A_p} = \frac{1}{A_p}\left(\frac{x_{mylar}}{K_{mylar}} + \frac{x_{foil}}{K_{foil}} + \frac{x_{polyethylene}}{K_{polyethylene}}\right)$$

Thermal conductivity of aluminum is given as 119 BTU/h ft °F at 212°F, and 117 BTU/h ft °F at 32°F [6].

Thermal conductivity of H.D. polyethylene, $K_{polyethylene}$, is given as 11×10^{-4} gm cal/s cm °C [24], or 0.267 BTU/h ft °F. The thermal conductivities of the majority of the polymer films range between 1.0×10^{-4} to 15×10^{-4} cal/s cm °C [24] or 0.0242 to 0.363 BTU/h ft °F. The thermal conductivity of mylar (polyester polymer of ethylene glycol and teraphthalic acid) is 3.63×10^{-4} cal/s cm °C [49] or 0.0878 BTU/h ft °F. The resistant term due to aluminum foil is negligible compared with the plastic layers and can be neglected. Therefore, the combined resistant of flexible term becomes

$$\frac{x_p}{K_p A_p} = \frac{1}{A_p} (1.400 \times 10^{-3}).$$

An estimate of heat transfer from the inner surface of the package requires more consideration. It would be logical to visualize heat transferring from the inside surface of the package at the gap which exists between the packaging film and food and from the gap to the surface of the food by a series of two convections. In actual cases, usually when food is packaged in a flexible pouch, the vacuum is drawn to remove air. Therefore, heat transfer by convection is minimized. However, as the heating progresses, gas and vapor are evolved from food and the mass evolved may contribute to convection heating. Because of these phenomena, which require involved analytical data, the actual condition that exists in the gap is difficult to quantize. It is possible that at least initially this gap, especially when well evacuated, can offer considerable resistance to heat transfer by minimizing convection. However, even with the most effective air removal, because of the non-smooth surface of most of the food and the rigidity of the laminated film, perfect contact between the packaging film and food to allow direct conduction of heat is difficult. Taking the average gap between the package and food as 1/32 in. if it is assumed that air is present between this gap and heat transfers by conduction, due to the small dimension and the temperature gradient which is not so steep, the resistance term for the gap becomes

$$\frac{x_i}{K_i A_p} = \frac{1}{A_p} (2.604 \times 10^{-3}).$$

In the case when the food to be processed is solid, if the conductivity of the food material is assumed the same as that of water, 0.364 BTU/h ft °F at 100°F and 0.394 BTU/h ft °F at 200°F, then the overall heat transfer coefficient can be expressed as

$$U = \frac{1}{\dfrac{1}{h_{out} A_p} + \dfrac{x_p}{K_p A_p} + \dfrac{x_i}{K_i A_i} + \dfrac{x_s}{K_s A_s}},$$

$$= A_p \frac{1}{(3.3 \times 10^{-3}) + (1.4 \times 10^{-3}) + (2.604 \times 10^{-3}) + (5.3 \times 10^{-2})}$$

$$\text{BTU/h °F}$$

when the thickness of solid food is 0.5 in. The above equation indicates that the heat transfer during the thermal processing of food is limited by the heat flow within the solid food itself, and also for all practical purposes the resistant terms for the convection outside of the package film can be neglected, leaving only the resistant term for

TABLE XVII[a]

Thermal conductivity, specific heat, apparent density and thermal diffusivity
of container materials for foods

	Thermal conductivity (BTU/h ft °F)	Specific heat (BTU/lb °F)	Apparent density (lb/ft^3)	Thermal diffusivity (ft^2/h)	References
Aluminum	117–155	0.21–0.26	2.6–2.8	208[c]	[7]
Glass (borosilicate)	0.65	0.20	140	0.023[b]	[23]
Nylon (type 6/6)	0.14	0.40	70	0.0050[b]	[24]
Polyethylene (high density)	0.28	0.55	60	0.0085[b]	[24]
Polyethylene (low density)	0.19	0.55	58	0.0060[b]	[24]
Polypropylene	0.068	0.46	57	0.0026[b]	[24]
Polytetrafluoroethylene (Teflon)	0.15	0.25	130	0.0046[b]	[24]
Stainless steel (302) Type 18-8 austenitic	9.17	0.118	494	0.157[b]	[22]
Steel	21–26	0.12	7.0–7.2	28.2[c]	[7]
Tin	33–36		6.4–7.5		[7]

[a] Compiled from References:

  [7] McAdams, William H.: 1954, *Heat Transmission*, McGraw-Hill Book Company, Inc., New York, 3rd ed.

 [23] Lange, N. A.: 1961, *Handbook of Chemistry*, McGraw-Hill Book Co., New York, N.Y., 10th ed.

 [24] 'Plastics Properties Chart, Part I, *Modern Plastics Encyclopedia*, Breskin Publications, New York (1963).

[b] Calculated from the given thermal conductivity, specific heat, and apparent density.

[c] Calculated from the average values of thermal conductivity, specific heat and apparent density.

conduction through solid food material. Therefore, in processing solid food or conduction heating food

$$U \approx \frac{A_s K_s}{x_s}.$$

This again examplifies the importance, or controlling effect of thermal properties of
food in many cases.

    Thermal properties of some packaging material which are useful in estimation of
heat transfer is given in Table XVII.

2.7. Methods of determination of thermal properties of foods

In determination of thermal properties, essentially, the known quantity or level of
heat is supplied to the system and change in temperatures of the system is measured.
On this basis, construction of the apparatus for determination of thermal properties is
simple, and many of the versions are constructed and reported in the literature
[6, 8, 25, 26, 36, 46, 49 through 63]. Steady state methods are simpler but more tedious,
while transient methods are faster but require more skill. Commercial equipment for
determining thermal properties is also available [64]. At present the Differential
Scanning Calorimeter is being used for the determination of specific heat, enthalpy of

reaction or phase change and even emissivities of many polymeric and other materials [65 through 72], although it has not yet been widely applied to food materials. Such commercial equipment, although costly, is convenient and can rapidly and accurately obtain the desired data and avoid inherent particular error sometimes associated with a laboratory fabricated system.

In experimental determination of the thermal properties of food materials, it is necessary to select representative and homogeneous samples, as in any other experiment with the biological materials. A good contact between the apparatus and samples is essential to eliminate the interfacial resistance which may contribute to significant error. Accuracy of the method, apparatus, and analysis of data need to be determined and precision of the method should be checked with known methods and standard sample.

Consideration of the components of food, such as water, fat, air content or porosity, and behavior of the component over the temperature range is required to have some knowledge of the range of values expected. This is necessary in order to impose the right boundry condition for the experiment to get meaningful data.

References

[1] Moore, Walter J.: 1955, *Physical Chemistry*, Prentice-Hall, Inc., Englewood Cliffs, N.J., 2nd ed.

[2] Stumbo, C. R.: 1973, *Thermobacteriology in Food Processing*, Academic Press, New York, 2nd ed.

[3] Nickerson, John T. and Sinskey, Anthony J.: 1972, *Microbiology of Foods and Food Processing*, American Elsevier Company, New York.

[4] Harris, Robert S. and von Loesecke, Harry (eds.): 1960, *Nutritional Evaluation of Food Processing*, John Wiley & Sons, Inc., New York.

[5] Dwivedi, Basant K. and Arnold, Roy G.: 1972, 'Chemistry of Thiamine Degradation – Mechanisms of Thiamine Degradation in a Model System', *J. Food Sci.* **37**, 886.

[6] Kreith, Frank: 1963, *Principles of Heat Transfer*, International Textbook Company, Scranton, Pa., 7th Printing.

[7] McAdams, William H.: 1954, *Heat Transmission*, McGraw-Hill Book Company, Inc., New York, 3rd ed.

[8] Perry, John H. (ed.): 1950, *Chemical Engineers Handbook*, McGraw-Hill Book Company, Inc., New York, N.Y. 3rd ed.

[9] Charm, Stanley E.: 1963, *Fundamentals of Food Engineering*, The Avi Publishing Company, Inc., Westport, Conn.

[10] Riedel, L.: 1956, *Kaltetechnik* **8**, 374–377.

[11] Riedel, L.: 1957, *Kaltetechnik* **9**, 38–40.

[12] Dickerson, R. W.: 1968, 'Thermal Properties of Foods', in D. K. Tressler, W. B. Van Arsdel, and M. J. Copley (eds.), *The Freezing Preservation of Foods*, Avi Publishing Co., Westport, Conn., 4th ed., Vol. 2.

[13] Lentz, C. P.: 1961, 'Thermal Conductivity of Meats, Fats, Gelatin, Gels, and Ice', *Food Technol.* **15**, 243–247.

[14] Miller, H. L. and Sunderland, J. E.: 1963, 'Thermal Conductivity of Beef', *Food Technol.* **17**, 124–126.

[15] Hill, J. E.: 1966, 'The Thermal Conductivity of Beef', M.S. Thesis, Georgia Institute of Technology, Atlanta, Ga.

16. Jason, A. C. and Long, R. A. K.: 1955, 'The Specific Heat and Thermal Conductivity of Fish Muscle', *Intern. Congr. Proc., 9th, Paris.* **1**, 2160–2169.

[17] Smith, J. G., Ede, A. J., and Gane, R.: 1952, 'The Thermal Conductivity of Frozen Foodstuffs', *Mod. Refrig.* **55**, 254–257.

[18] Moline, S. W., Sawdye, J. A., Short, A. J., and Rinfret, A. P.: 1961, 'Thermal Properties of Foods at Low Temperatures', *Food Technol.* **15**, 228–341.

[19] *Handbook of Chemistry and Physics*. Chemical Rubber Co., Cleveland, Ohio. 46th ed.

[20] Editors of *Food Engineering*: 1962, 'Special Report: Plant Handbook Data', *Food Engineering* **34**, 89–104.

[21] Lagoni, H. and Merton, D.: 1956, 'Wärmeleitsfähigkeitsmessungen an Butter', *Kiel. Milchwirtsch. Forschungsber.* **8**, 85.

[22] *Metals Handbook*, American Society for Metals, Cleveland, Ohio (1948).

[23] Lange, N. A.: 1961, *Handbook of Chemistry*, McGraw-Hill Book Co., New York, N.Y., 10th ed.

[24] 'Plastics Properties Chart, Part I', *Modern Plastics Encyclopedia*, Breskin Publications, New York (1963).

[25] Sharma, D. K. and Thompson, T. L.: 1973, 'Specific Heat and Thermal Conductivity of Sorghum', *Trans. ASAE* **16**, 114–117.

[26] Keppeler, R. A. and Boose, J. R.: 1970, 'Thermal Properties of Frozen Sucrose Solutions', *Trans. ASAE* **13**, 335–339.

[27] Riedel, L.: 1959, *Kaltetechnik* **11**.

[28] Riedel, L.: 1954, *Kaltetechnik* **6**.

[29] Riedel, L.: 1957, *Kaltetechnik* **9**.

[30] Riedel, L.: 1958, *Kaltetechnik* **10**.

[31] Riedel, L.: 1960, *Kaltetechnik* **12**.

[32] Fruton, Joseph S. and Simmonds, Sofia: 1959, *General Biochemistry*, John Wiley & Sons, Chapman & Hall, Ltd., London, 2nd ed.

[33] Rha, ChoKyun: 1974, 'Utilization of Single Cell Protein for Human Food', in Steven Tannenbaum and Daniel I. C. Wang (eds.), *Single Cell Protein*, The M.I.T. Press, Cambridge, Mass. (in press).

[34] Balmaceda, E. and Rha, C. K.: 1973, 'Rate of Coagulation of Single Cell Protein Concentrate', *Biotechnol. Bioengng.* **15**, 819.

[35] Keenan, Joseph H. and Keyes, Frederick G.: 1963, *Thermodynamic Properties of Steam, Including Data for the Liquid and Solid Phases*, John Wiley & Sons, Inc., New York, 1st ed., 35th Printing.

[36] Jasansky, A. and Bilanski, W. K.: 1973, 'Thermal Conductivity of Whole and Ground Soybeans', *Trans. ASAE* **16**, 100–103. Published by the American Society of Agricultural Engineers, St. Joseph, Michigan.

[37] Wang, R. H. and Knucken, J. G.: 1958, 'Thermal Conductivity of Liquid-Liquid Emulsions', *Ind. Eng. Chem.* **50**, 11, 1667.

[38] Maxwell, J. C.: 1954, A Treatise on Electricity and Magnetism, vol. I. and II., ch. 9; vol. I, article 314, Dover, New York, 3rd ed.

[39] Hamilton, R. L. and Crosser, O. K.: 1963, 'Thermal Conductivity of Heterogeneous Two-Component Systems', *Ind. Engng. Chem.* **1**, 187–191.

[40] Cheng, S. C. and Vachon, R. I.: 1969, 'The Prediction of the Thermal Conductivity of Two and Three Phase Solid Heterogeneous Mixtures', *Int. J. Heat Mass. Transfer* **12**, 249–264.

[41] Cheng, S. C., Law, Y. S., and Kwan, C. C. Y.: 1972, 'Thermal Conductivity of Two and Three Phase Solid Heterogeneous Mixtures', *Int. J. Heat Mass. Transfer* **15**, 355–358.

[42] Harper, J. C.: 1962, 'Transport Properties of Gases in Porous Media at Reduced Pressures with Reference to Freeze Drying', *Am. Inst. Chem. Eng. J.* **8**, 298–302.

[43] Harper, J. C. and El Sahrigi, A. F.: 1964, 'Thermal Conductivities of Gas-Filled Porous Solids', *Ind. Eng. Chem. Fundamentals* **3**, 318–324.

[44] Lusk, G. Karel, M., and Goldblith, S. A.: 1964, 'Thermal Conductivity of Some Freeze-Dried Fish', *Food Technol.* **18**, 121–124.

[45] Leitman, J.: 1967, M.S. Thesis, Georgia Institute of Technology, Atlanta, Ga.

[46] Walters, Robert E. and May, K. N.: 1963, 'Thermal Conductivity and Density of Chicken Breast Muscle and Skin', *Food Technol.* **17**, 808.

[47] Dickerson, R. W. and Read, R. B.: 1973, 'Cooling Rates of Foods', *J. Milk Food Technol.* **36**, 167–171.

[48] Rha, ChoKyun: 1967, 'Thermal Sterilization of Flexibly Packaged Foods', D. Sc. Thesis, Massachusetts Institute of Technology, Cambridge, Mass.

[49] Amborski, L. E. and Flierl, D. W.: 1953, 'Physical Properties of Polyethylene Terephthalate Films', *Industrial Engng. Chem.* **45**, 2290–2295.

[50] Pelanne, C. M. and Bradley, C. B.: 1962, 'A Rapid Heat-Flow Meter Thermal-Conductivity Apparatus', *Materials Res. Standards* **2**, 549–552.

[51] Dickerson, Roger W., Jr.: 1965, 'An Apparatus for the Measurement of Thermal Diffusivity of Foods', *Food Technol.* **19**, 198–204.

[52] Saravacos, George D. and Pilsworth, Malcolm N., Jr.: 1965, 'Thermal Conductivity of Freeze-Dried Model Food Gels', *J. Food Sci.* **30**, 773–778.

[53] Morley, M. J.: 1966, 'Thermal Conductivities of Muscles, Fats and Bones', *J. Food Technol.* **1**, 303.

[54] Gentzler, G. L. and Schmidt, F. W.: 1972, 'Determination of Thermal Conductivity Values of Freeze-Dried Evaporated Skim Milk', *J. Food Sci.* **37**, 554–557.

[55] Poppendiek, H. F., Randall, R., Breeden, J. A., Chambers, J. E., and Murphy, J. R.: 1966, 'Thermal Conductivity Measurements and Predictions for Biological Fluids and Tissues', *Cryobiology* **3**, 318–327.

[56] Roth, Douglas D., Tsao, George T. and Lancaster, Earl B.: 1970, 'Thermal Conductivity of Starch Granules', *International Journal for Research, Processing, and Use of Carbohydrates and their Derivatives*, **2**, 40–41.

[57] Moench, A. F. and Evans, D. D.: 1970, 'Thermal Conductivity and Diffusivity of Soil Using a Cylindrical Heat Source', *Soil Sci. Soc. Amer. Proc.* **34**, 377–381.

[58] Rolinski, Edmund J. and Sweeney, Thomas L.: 1968, 'Thermal Conductivity of Fibrous Silica', *J. Chem. Engng. Data* **13**, 203–206.

[59] Pasquino, Anne D. and Pilsworth, Malcolm N., Jr.: 1964, 'The Thermal Conductivity of Polystyrene, Oriented and Unoriented, with Measurements of the Glass-Transition Temperature', *Polymer Letters* **2**, 253–255.

[60] Benneth, A. H., Chace, W. G., Jr., and Cubbedge, R. H.: 1970, 'Thermal Properties and Heat Transfer Characteristics of Marsh Grapefuit', Technical Bulletin No. 1413, U.S. Department of Agriculture.

[61] Agrawal, K. N. and Verma, V. V.: 1969, 'An Apparatus for the Study of Heat and Mass Transfer Properties of Insulating Materials at Low Mean Temperatures', *Indian J. Technol.* **7**, 4–8.

[62] Qashou, M., Nix, G. H., Vachon, R. I., and Lowery, G. W.: 1970, 'Thermal Conductivity Values for Ground Beef and Chuck', *Food Technol.* **24**, 493.

[63] Spells, K. E.: 1960, 'The Thermal Conductivities of Some Biological Fluids', *Phys. Medicine Biol.* **5**, 139–153.

[64] Bull. # 141, 151, 241, 341, Thermatest Laboratories, Inc., 1225 Elko Drive, Sunnyvale, California.

[65] Berger, K. G. and Akehurst, E. E.: 1966, 'Some Applications of Differential Thermal Analysis to Oils and Fats', *J. Food Technol.* **1**, 237–247.

[66] Stafford, Bill B.: 1965, 'Application of Differential Thermal Analysis to Polyethylene Blends', *J. Appl. Polymer Sci.* **9**, 729–737.

[67] Watson, E. S., O'Neill, M. J., Justin, J., and Brenner, N.: 1964, 'A Differential Scanning Calorimeter for Quantitative Differential Thermal Analysis', *Anal. Chem.* **36**, 1233–1238.

[68] Rogers, R. N. and Morris, E. D.: 1966, 'Determination of Emissivities with a Differential Scanning Calorimeter', *Anal. Chem.* **38**, 412.

[69] Rogers, R. N. and Morris, E. D.: 1966, 'On Estimating Activation Energies with a Differential Scanning Calorimeter', *Anal. Chem.* **38**, 412.

[70] Antila, V.: 1966, 'Fatty Acid Composition, Solidification and Melting of Finnish Butter Fat', *Meijeritieteellimen Aikakaushiya* **27**, No. 1.

[71] Barrall, E. M. II, Porter, R. S., and Johnson, J. F.: 1967, 'Heats of Transition for Some Cholesteryl Esters by Differential Scanning Calorimetry', *J. Phys. Chem.* **71**, 1223.

[72] de Waal, H.: 1965, 'Quantitative Differential Thermal Thermal Analysis with an Isothermal Microcalorimeter', *Instrument Practice* **19**, Nov., 1022–1027.

CHAPTER XVIII

SOME PHYSICAL PROPERTIES OF FOODS

STEVENSON W. FLETCHER

Food and Agricultural Engineering Department, University of Massachusetts, Amherst, Mass., U.S.A.

1. Aerodynamic and Hydrodynamic Properties

The food industry has long utilized air and water to transport products from one location to another, especially raw products. Likewise, a gas or fluid can be utilized to separate a desirable product or products from undesirable materials. When designing systems to work with a specific food material, one must know about the aero and hydrodynamic characteristics of the material. The characteristics, or primary properties, that govern the product behavior in air or water are the drag coefficient and the terminal velocity.

Considering a drag coefficient which relates to the forces acting on the object in the medium stream, in general, with air or water moving around it, a body will orient itself so that the pressures on it can be divided into a force that tends to lift the object and a force that will retard its movement or drag its movement down (Figure 1).

Ignoring friction, the resultant force on the object causing movement can be given by the formula

$$F_r = \tfrac{1}{2} C A_p \varrho_f V^2,$$

where
   C is the dimensionless overall drag coefficient;
   A_p is the projected area in the direction of movements in ft^3;
   ϱ_f is the mass density of the fluid;
   V is the velocity of the fluid in ft s^{-1}

The force F_r is the net force in pounds or, more specifically, the weight of the particle at terminal velocity. The common way to use this formula is to assume the body to be a sphere and to calculate the sphere's diameter by any one of the shape characteristic methods. The drag coefficient can be obtained from standard charts based on the Reynolds number (Figure 2). Charts are also available for other regular shapes such as discs, flat plates and cylinders. These charts are developed from equations that consider frictional effects and pressure drag on the body and are available in most fluid dynamics textbooks.

Terminal velocity relates to the maximum velocity that an object will obtain in a free fall situation. Theoretically, due to gravitational acceleration, a body will speed up at a rate of 32.2 ft s^{-1} s^{-1} for an infinite time, but actually it will speed up until

ChoKyun Rha (ed.), Theory, Determination and Control of Physical Properties of Food Materials, 357–365.

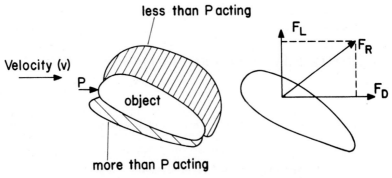

Velocity (v)

less than P acting

F_L

F_R

P

object

F_D

more than P acting

Fig. 1. Pressures of an object in a moving field.

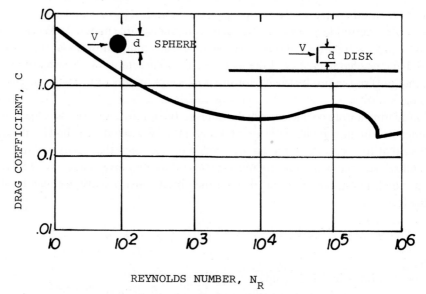

REYNOLDS NUMBER, N_R

Fig. 2. Drag coefficient for particles with regular geometric shapes.

the gravitational force equals the resultant drag force at which time its velocity will reach a peak and become constant. In the absence of a medium velocity, the product would fall if its density were higher than that of the medium; if the product density were less than the medium, the product would rise. Thus, it is with the tennis ball which falls in the air but rises in water.

Thus, to determine the terminal velocity of an item one might say that $F_g = F_r$ or

$$m_p g \left[\frac{(\varrho_p - \varrho_f)}{\varrho_p} \right] = \tfrac{1}{2} C A_p \varrho_f V_T^2$$

or

$$V_T = \left[\frac{2mg (\varrho_p - \varrho_f)}{\varrho_p \varrho_f A_p C} \right]^{1/2}$$

F_g = gravitational force;

F_r = net resistance force;

V_T = terminal velocity;

ϱ_p = density of product.

This basic formula can be simplified for known mediums and shapes depending on the application. Experimentally, terminal velocity is normally obtained by recording the distance-time relationship of a free falling object in a still medium and when the resulting curve becomes linear, the slope yields the terminal velocity.

Most of the theoretical studies in aerodynamic or hydrodynamic properties have been difficult to apply to actual conditions. There are many reasons for this, but perhaps the most significant include the constant rotation of the product, the difficulty in assuming a regular shape of a product, the varying moisture contents, and the particles hitting one another during movement, thus causing secondary velocity due to impact, changes in orientation, and changes in the lift and drag pressures on the product (Bilanski, 1971).

As a result of increased research efforts since World War II, the usuage of aerodynamic and hydrodynamic systems in the food industry is now great, ranging from fluidized freezing to pneumatic handling to separation processes. The processing of products in a fluidized process such as quick-freezing blueberries and peas is a direct application of aerodynamic properties. High velocity air suspends the particles by overcoming the gravitational forces, thus having the advantages of damage free movement, increased heat transfer and even heat transfer. Irregularly shaped items are being processed in fluidized systems. The development of a workable, economic system that has the proper design to give optimal air flow and velocity is dependent on aerodynamic properties.

Pneumatic transport and handling, either in air or water, has for a long time utilized aerodynamic and hydrodynamic properties and is the area in which the greatest amount of properties research has been done. Using a fluid as a carrying medium and a cushioning material serves to deliver products great distances quickly, economically, clean and with minimum damage. Information has been developed that enables optimal systems to be designed based on specific aerodynamic and hydrodynamic properties of the material being handled. Now, rather than to use trial and error methods, the aerodynamic properties of the material can be used to determine such things as airflow rate, tube diameter, pipe bend size and sweep radius, and feeder mechanization requirements (Ghosh and Kalyanaraman, 1970).

In hydraulic handling the terminal velocity of the particle being moved will determine whether the particle is suspended and is conveyed or sinks and collects. That is, when the gravitational forces plus the buoyancy equals the drag forces, the product settles out. Perhaps the significant equation in this area is the one proposed by Condolios and Chapus (1963) which is similar to our previous formulas only it is for settling velocity and takes into consideration deviations from a theoretical sphere.

$$V_T = \left[\left(\tfrac{4}{3} \right) g d_e \frac{(\varrho_p - \varrho_f)}{\varrho_f} \left(\frac{\psi}{C} \right) \right]^{1/2}$$

d_e diameter of a sphere of equivalent volume;

ϱ_p and ϱ_f = density of particle and fluid;

C drag coefficient;

$\psi = A_e/A_1$, where A_e is area of equivalent sphere and A_1 is area of largest cross section. From this simple formula the velocity of the water can be calculated.

Two other problems in the food industry have lent themselves to solution through knowledge of aerodynamic properties. One, which is an old practice, is the cleaning, separation or grading by flotation or air blasts. All of this commonly known method, is based on the drag coefficients and terminal velocity of the material, which, of course, are directly related to shape, size and weight. Thus, separation can be done according to shape, size or weight or a combination of these three.

Finally, aerodynamic and hydrodynamic properties are important when an operation requires that the product not be moved, yet a gas or liquid is required to move around it. In air drying, the higher the air velocity the faster the drying, yet the velocity can not be of such a force that it will literally blow the product around.

Theoretical research in the aerodynamic and hydrodynamic properties of food products, other than fresh fruits and vegetables, is limited. Bilanski (1971) has recently presented an excellent review of the current research in the areas of pneumatic handling and distribution, aerodynamic parameters, pneumatic separation and agricultural sprays.

2. Electrical Properties

The electrical properties in biological materials are perhaps the least studied and little understood of the physical properties. However, they still have some significance to the food industry, thus discussed here are some of the properties and their applications. Electrical properties are important, either as techniques of measuring some other parameter of a food or as a means of altering the properties of a product. Probably the three most common uses of electric properties has been in moisture content determination, quality of product evaluation and dielectric heating.

The electrical resistance or conductivity of a product is directly related to the moisture content along with the temperature. Many of the so-called instant moisture meters place a fixed voltage through a given quantity of the material being held at a constant temperature. The resistance to the flow of this voltage is measured by an ohm meter which is calibrated in terms of moisture content. Conductivity has also been measured for many of our meats, fresh fruits and vegetables, dairy products and vegetable oils. Conductivity has been correlated to lactose content and freezing temperature of milk by Pinkerton and Peters (1958) and used as a mastitis infection indicator by Sharma (1971).

Perhaps more significant has been the results of relating the dielectric constant or capacitivity, dielectric loss factor and dielectric loss angle to other parameters. A simple but imprecise definition of the dielectric constant is the ratio of the capacitance of a capacitor with the material as the dielectric (non-conducting current) to its capacitance with vacuum as the dielectric.

$$\varepsilon = \frac{C_m}{C_0}$$

C_0 = capacitance with air as dielectric;
C_m = capacitance with some material as dielectric;
   ε = dielectric constant.

The dielectric loss angle is the tangent of the phase angle δ between the changing current and the total current, where I_c is changing current or active current.

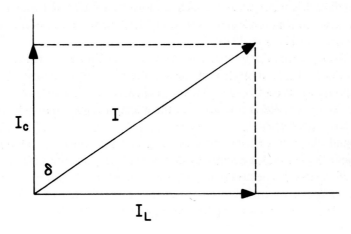

Fig. 3.

The dielectric loss factor or permittivity is the dielectric constant times the dielectric loss angle or

$$\varepsilon'' = \varepsilon' \tan \delta .$$

This term is useful in the determination of the power absorbed during dielectric heating by the relationship:

$$P = 55.61 \times 10^{-14} f E^2 \varepsilon''$$

where P is the available power given in watts/m^3, f is the frequency in Hz, E is the voltage gradient in volts/m^2 (Mohsenin, 1963).

Dielectric properties are fairly easy to measure with our existing instrumentation, although precision is not easy. They are dependent on the frequency of the electric field, the temperature and the moisture content.

The food industry has used dielectric properties extensively to measure moisture content or quality of the product and to heat the products. Dielectric heating is popular in many situations due to the uniformity of heating and high rates of temperature rise in poor heat conducting materials that are difficult to heat by normal heat transfer methods.

An extensive list of references and dielectric properties is again reported by Nelson (1971). Significant among the reported results is the work on moisture content determination through dielectric property measurement for dehydrated products such as dehydrated carrots (Dunlap and Makower, 1945) and dehydrated potatoes (Danial and White, 1967). Attempts have been made to relate optimal maturity of fresh fruits and vegetables such as potatoes (Pace et al., (1968) and apples (Thompson and Zachariah, 1971) with apparently successful correlations being obtained.

Butter has had a lot of effort placed on it relative to various properties measurements using dielectric methods. Items such as moisture content, salt content, moisture dispersion, quality after storage and temperature dependency (Parkash and Armstrong, 1970; Klepacki and Kurpisz, 1970).

Products being processed by dielectric or microwave heating in freeze drying or other operations has, of course, been the subject of much investigation, especially with meat products. Probably the most comprehensive work with meat and fish was done (Bengtsson et al., 1963) primarily on the effect of temperature and type of cut on the dielectric properties.

Other significant work in dielectric heating includes work on potato chips by Goldblith and Pace (1967); mixtures by de Loor (1968); and frozen foods in general by Harper et al. (1962). The most complete set of dielectric measurements for various products at various temperatures and moisture contents is the work by Risman and Bengtsson (1971) and includes graphic and tabular presentation.

3. Physical Characteristics

In many food processing applications and research problems, it is necessary to have an accurate estimate of shape and size of the material. The design of grading and sizing machinery, hydrodynamic and aerodynamic handling equipment design and the determination of heat or cooling times of a product are all common examples where size and shape of the product need be determined. For describing the shape of a food product there are several methods that can be considered.

First, the shape of an object can be determined by comparing the object with a set of standard shapes. Normally, the comparison is made of the projected images with a descriptive term assigned to each standard image. Mohsenin (1968) has proposed a set of descriptive terms that can be used to describe many fruits and vegetables but the use of this method to determine shape has had very limited usage.

Another method is to determine the degree of roundness or the roundness ratio. This is a measure of the degree of sharpness of a corner in relation to the whole solid. The method of determining the roundness ratio that is most commonly used is to take the ratio of the projected area of the whole to the area of the circle that will circum- scribe the objects area.

$$\text{Roundness ratio} = \frac{\text{Area (projected)}}{\text{Area (circumscribed circle)}}.$$

The term roundness ratio is used a lot in geology, but has limited use in physical property investigations primarily because it is a two dimensional term and with most food products a three dimensional term must be used.

A better term to use is sphericity or a measure of how close an object comes to the shape of a sphere. The most practical way to define sphericity or degree of sphericity is by the cube root of the volume of the solid divided by the volume of a sphere that will circumscribe the object. For a near perfect sphere, this will of course have a value of 1.

In some special applications the body being studied can be approximated as a common geometric shape. If this can be done then standard formulas can be used to determine surface areas and volume. The ones that are commonly used include the prolate spheroid (rotation around major axis), oblate spheroid (rotation around minor axis) and the right circular cone.

Another way to show how a body relates to a sphere is by the use of a shape factor. This is a dimensionless ratio that describes the relationship between the characteristic dimensions of a body. There are numerous ways of showing this relationship but the most common method is to show the ratio of the minor axis to the major axis. Using this concept b/a (with b the minor axis and a the major axis) would equal 1 for a sphere, 0 for a line or disc and any number between 0 and 1 would indicate how close the shape approaches a sphere.

Attempts have been made to dimensionally describe the product by using three mutually perpendicular diameters. The basic diameter will be the intermediate diameter which is the diameter of the smallest possible circle through which the object can be passed. This is set as some basis, such as 100. Thus, if you had an object that had diameters measuring 10 cm, 0.25 cm and 6.0 cm, you would have a dimensional description of 40/100/240.

Finally, the average projected area can be used to define shape or size. This simply averages the three projected areas computed from three mutually perpendicular axes. This is taken as the true average area or the so called criterion area.

Of more importance than size and shape are the physical characteristics of volume, density and specific gravity. With engineering materials such as wood or metal, a standard shape can be obtained and dimensions and weight taken fairly accurately. Some food materials can be shaped, such as cutting potatoes into plugs, but then the sample is not representing the whole produce but rather a distorted portion. Food materials often have irregular shapes, which do not have simple dimensions that can be measured easily with rulers, dividers or calipers.

The most commonly used method of determining the volume and specific gravity is by the use of the platform scale. This is one of three ways available where both the volume and specific gravity can be determined from the same data. The information needed includes the weight of the product in a container, the weight of the container and water and finally, the weight of the container, water and submerged product. If the product floats it should be forced under the water by a rod. If the product sinks it must be supported in the water and not allowed to touch the container bottom or sides. The volume of the product is

$$\text{Volume (in.}^3) = \frac{\text{Weight of displaced water (lb)}}{\text{Weight density of the water (lb/in}^{-3})}.$$

The weight of displaced water is the total weight of product water and container less the weight of the water and container. The only problems with this method is the limited accuracy and the problem of it being trapped in products such as lettuce.

Another method is the so called specific gravity balance which has the additional advantages of increased accuracy and can use liquids other than water. In this method the product is suspended from the balance and submerged in the liquid. If the product is heavier than water

$$\text{Volume} = \frac{\text{Weight in air} - \text{weight in water}}{\text{Weight density of water}}$$

$$\text{SG} = \left[\frac{\text{Weight in air}}{\text{weight in air} - \text{weight in water}} \right] \text{SG}_{\text{water}}$$

If the product floats, a sinker is added and the formulas adjusted accordingly.

When the liquid is replaced with toluene and evacuated you have the pycnometer. This has the advantage of being able to use it on a wide range of products and the elimination of the problem or an entrainment in the product.

There is a whole group of instruments available that measure density or SG quickly and fairly accurately. They simply relate the rate of radiation, light, electricity or sound transmission to density or specific gravity. All of them are basically similar, in that the degree or rate of transmission through a specific distance is in direct proportion to the density. Many instruments have been developed for particular products which is a requirement of these instruments due to the requirement of fixed distances.

Finally, let us take a brief look at surface area. Studies on methods dealing with surface area fall into three types of groups. The first general method assumes that the product can be described as a standard geometric shape and the surface area calculated from standard geometry equations.

The second group involves empirical methods and the most commonly used approach is to relate weight to surface area. Equations are available on apples, pears, plums, eggs, grain, citrus and peas, although once they are reported they are rarely used, primarily due to the fact that every situation is different and has its own empirical relationship.

The last group is used only with flat objects. The measurements can be directly obtained and related to a standard formula for a similar shape of if the product is irregular you can obtain a trace of the object through X-rays, photographs or light sensitive paper. The surface area can be calculated by weight relationships, counting squares or with the aid of a planemeter.

References

Bengtsson, N. E., Melin, J., Remi, K., and Soderlind, S.: 1963, 'Measurements of the Dielectric Properties of Frozen and Defrosted Meat and Fish in the Frequency Range 10–200 MH$_2$', *J. Sci. Food Agr.* **14**, 592–604.

Bilanski, W. K.: 1971, 'Aerodynamic Properties of Agricultural Products Research, Past and Present', ASAE Paper 71–846.

Condolios, E. and Chapus, E. E.: 1963, 'Transporting Solid Materials in Pipelines', *Chem. Enging.* **40**, 93–98.

Danial, D. N. and White, G. M.: 1967, 'Dielectric Properties of the Potato', ASAE Paper 67–808.

de Loor, G. P.: 1968, 'Dielectric Properties of Heterogeneous Mixtures Containing Water', *J. Microwave Power* **3**, 67–73.

Dunlop, W. C. and Makower, B.: 1945, 'Radio-Frequency Dielectric Properties of Dehydrate Carrots', *J. Phys. Chem.* **49**, 601–622.

Ghosh, D. P. and Kalyanaraman, K.: 1970, 'Pressure Drops Due to Solids around Horizontal Elbow Bends Pneumatic Conveyance', *J. Ag. Eng. Res.* **15**, 117–121.

Goldblith, S. A. and Pace, W. E.: 1967, 'Some Considerations in the Processing of Potato Chips', *J. Microwave Power* **2**, 95–98.

Harper, J. C., Chichester, C. O., and Roberts, T. E.: 1962, 'Freeze-Drying of Foods', *Agr. Eng.* **43**, 78–81, 90.

Klepacki, J. and Kurpisz, W.: 1969, 'Correlation between Dielectric Constant of Butterfat and Organoleptic Properties of Butter', *Roczniki Inst. Przemyslu Mleczarskiego* **11**, 205–218.

Mohsenin, N. N.: 1963, 'Engineering Approach to Evaluating Textural Factors in Fruits and Vegetables', *ASAE Trans.* **6**, 85–88, 92.

Mohsenin, N. N.: 1970, *Physical Properties of Plant and Animal Materials*, vol. 1: *Structure, Physical Characteristics, and Mechanical Properties*, Gorden and Breach Science Publ., New York.

Nelson, S. O.: 'Electrical Properties of Agricultural Products – A Review', ASAE Paper 71–847.

Pace, W. E., Westphal, W. B., Goldblith, S. A., and Van Dyke, D.: 1968, 'Dielectric Properties of Potatoes and Potato Chips', *J. Food Sci.* **33**, 37-42.

Parkash, S. and Armstrong, J. G.: 1970, 'Measurement of Dielectric Constant of Butter', *Dairy Ind.* **35**, 688–689.

Risman, P. O. and Bengtsson, N. E.: 1971, 'Dielectric Properties of Food at 3 GH$_2$ as Determined by a Cavity Perturbation Technique. I. Measuring Technique', *J. Microwave Power* **6**, 101–106.

Thompson, D. R. and Zachariah, G. L.: 1971, 'Dielectric Theory and Bioelectrical Measurements (Part II. Experimental)', *Trans. ASAE* **14**, 214–215.

CHAPTER XIX

COMPILATION, RECALL AND UTILIZATION OF DATA
ON THE PHYSICAL PROPERTIES OF FOOD MATERIALS

RONALD JOWITT

National College of Food Technology, University of Reading,
Weybridge, Surrey, England

1. Introduction

At the 3rd European Federation of Chemical Engineering Symposium 'Food-Recent Development in Food Preservation' at Bristol, England in April 1968, Jason and Jowitt (1969) voiced the widely held view that the dearth of information on physical properties of foodstuff (ppfs) could only be remedied by international collaboration and the offer by Jowitt on that occasion to provide at the National College of Food Technology, Weybridge, England, some kind of centre for this work was supported by delegates from several countries and organizations.

Although hindered by inadequate finance and, consequently, insufficient staff, some progress in this task has been made and some aspects are reported here. Despite disappointingly slow growth at the Centre, it is gratifying to record a total of 44 collaborators and correspondents in 18 different countries and the help and support of all these co-workers is gratefully acknowledged. The project at Weybridge has been described on other occasions (Jowitt *et al.*, 1971; McCarthy *et al.*, 1972; Jowitt, 1970). Here, the *system* and its operation will be discussed.

2. Sources of Information

It was clear from the outset that a primary task was to secure any and all information on ppfs from whatever source. These sources can be classified as:
  (1) the world literature on food science, technology and engineering;
  (2) other world literature – mainly that cited in 1;
  (3) public records, university theses and the like;
  (4) private records such as data in company files, and
  (5) experimental.
  Anticipating the setting up of food science and technology abstracts, Mann in 1966 surveyed the world literature on food science and technology and provided a convenient tabulation of sources indicated in (1) above. At Weybridge we have used the FSTA index as a primary source of references which might contain ppfs information, to considerable benefit.

Two years ago, the British Office for Scientific and Technical Information (OSTI) sponsored an exercise involving FSTA in which the National College of Food Technology at Weybridge took part. Over a period of six months, the International Food Information Service (IFIS) conducted searches of the world literature for references to thermo-physical properties of foodstuffs using their (FSTA) system of keywords, indexing, searching and publication listing, and then their findings for relevance and comprehensiveness were checked.

In any system for literature searching, two aims (or criteria) are paramount: First the relevance to the purpose of the search of any publication retrieved, and secondly, the completeness with which *all* relevant publications are retrieved.

Any system of searching, particularly those using mechanical or mechanistic methods and 'uninvolved' searchers will be imperfect in one, and usually in both, these respects. The aim is for 100% relevance and 100% comprehensiveness, but it will be evident that these goals are to some extent mutually exclusive, since the more the system is designed to exclude irrelevant material the greater the risk of relevant material being overlooked – and vice versa, just as in any physical separation process.

In this exercise IFIS retrieved a total of 602 references of which 397, or 66%, were relevant. Our own search of the same abstracts confirmed the comprehensiveness of the IFIS search in that although our initial searches produced a larger number of references, the per cent relevance was less at 46% and we retrieved very little of value which they had not.

Our conclusion is that this source (FSTA) provides a high degree of comprehensiveness (so far as the literature covered by FSTA is concerned) but a rather low percent relevance. On the whole, this is the preferred way for the purpose.

However, it was found increasingly, lately, that retrieval of *all* published ppfs data is by no means achieved in this way, because of the publication of relevant information in apparently quite irrelevant articles or journals, or the omission by abstractors of such information from their abstracts of papers which one would not infer to be relevant, and the considerable increase in yield if references cited in the primary references are themselves searched.

It is true that a very great deal of useful information exists in the individual theses and dissertations of higher degree candidates in the universities of the world and in the private files and records of commercial companies and research organizations. However, although it is important to collect such information as already exists, 'ubique' = 'wheresoever', it is increasingly apparent that the value of much published information is doubtful because of the absence of adequate contextual information – such as the history, the state and composition of the material, methods of measurement, number of replicate observations, spread of values and so on. So the policy is tending towards *assiduousness* rather than *obsessiveness* in the search for existing data, either published or unpublished.

So far as experimental determinations are concerned, the special consideration at Weybridge at present is to help in the information process whereby would-be experimenters can be informed of other establishments and persons in other countries (or in their own) in which such work is, or might be, going on, to make sure one is not inadvertently duplicating someone else's work.

It would be possible in due course and with the cooperation of others in this field, to operate an awareness service so that, by and large, it is possible to find out reasonably quickly and reliably what projects and what special expertise is current on a world-wide basis. There is no doubt that the correspondents and collaborators are keen to join in. What lacks at the moment are the hands and the funds to implement the scheme properly.

3. Classification of Foodstuffs and Properties

A system for the classification of foodstuffs, treatments and physical properties was proposed at the same European conference referred to above (Jason *et al.*, 1969). This has recently been revised and is appended. The basic system is unchanged but helpful suggestions from correspondents in many countries have been incorporated in this version which might not need further revision and can be used unchanged for some time to come. This does not mean that the classification is complete – it can never be that, and it will be added to as occasion demands, but such additions (not amendments) will not affect its continuing use.

An example of the use of the classification is given on page 1 of Appendix I from which the important features of the dendritic form may be noted, viz., the possibility to ascribe precise coordinates to any specific combination of foodstuff, treatment and physical property, and the ability to widen one's search for information on successively more general groups of foodstuff, or treatment or property, should data be unavailable for the particular combination of interest.

Intended to be logical, self-consistent and unambiguous the system is still arbitrary in some respects. Some foodstuffs such as fish, or fruit, being biological entities, are classified elsewhere according to more-or-less established systems such as that of Linnaeus and the question may be asked why such existing classification has not simply been followed here. The answer is that where such earlier systems have proved suitable they have been used, where not, not. The important criterion – the only one in fact – is: does this classification reflect the fact that all these materials are included as *foodstuffs* – not as members of this or that biological group or subgroup. Piscatologist friends have questioned the major subdivision of fish into fatty and non-fatty, because some species may have widely differing fat contents in different seasons, or stages of growth, or regions. Despite the truth of this observation, it was felt that for fish as food, the fat content has important and far-reaching significance, not least so far as some of its physical properties are concerned, and the separation of fish into 'fatty' (>2.0 wt $\%$ fat) and 'non-fatty' ($\not> 2.0$ wt $\%$ fat) in the condition most likely to be encountered in practice is, I believe a useful one. The fish section of the classification has received an invaluable contribution from Dr Louis Fedele of the University of Milan, Italy, who has been attached to the ppfs project at Weybridge for a period of 6 months.

4. Indexing

When a reference to a publication of potential value is retreived from the literature at Weybridge, an initial index card is made out for it bearing details of the formal reference and author(s) only.

Next, the publication or a copy of it is obtained for the file. Then, if relevant, the reference indexing procedure is followed and when the proper index is achieved, the initial index card is destroyed.

Appendix II is a copy of the reference-indexing form which is filled out from the information in the initial index card and the article itself (except for 'Reference Number'), down as far as 'Article Title in English'. Except for Russian, non-Roman texts are not reproduced and the Russian Cyrillic titles are transliterated using the 'British' System (British Standard 2979: 1958). Foreign titles are translated into English when English abstracts are not provided.

Next, the article is read and the foodstuffs, treatments and physical properties referred to are noted down along with reference to the nature of the information contained – the so-called 'master card reference(s)' which are

/M – contains information on method of measurement

/D – contains *quantitative data*

/T – contains *theory* related to ppfs

/E – comprises a *survey* or *review* of other work

Also noted are cited references of possible value. These foodstuffs, treatments and physical properties are then allotted the correct classification coordinates which are entered in the reference-indexing form along with the master card reference letter(s) and cited references to be followed up. Some practice and care in allocating new coordinates is necessary in order to preserve the 'dendritic' character of the system and to avoid ambiguities, inconsistencies or conflicts with other combination of letters.

5. Optical Coincidence Index Cards

The indexing system in use is known in the USA as 'peak-a-boo', in England as 'optical coincidence' and it is the optical equivalent of the old 'rod and hole' system. Optical index cards of proprietary manufacture may be hand or machine perforated in any of 10 000 'numbered' locations on the card and each *location* corresponds to a particular 'publication' containing relevant information. The capacity is thus 10 000 papers. Each *card* corresponds to a coordinate in the classification and is perforated at all locations which correspond to publications with information relevant to that coordinate, be it foodstuff, property, treatment or *group* of same.

For example, a reference containing information on the thermal conductivity of freeze-dried plaice (UWI ABAD-MNC) would be indexed under:

UWI A   non-fatty fish
UWI AB   non-fatty, flat fish
UWI ABA   non-fatty, flat, bony fish

– M   preserved
– MN   dehydrated
– MNC   freeze dried
  2   thermal properties

and

  2.2   thermal conductivity

In addition, there are four so-called 'master cards' each corresponding to the four categories listed above. These are perforated when a publication contains information of that general kind and are used to screen out information of that *kind* only during the subsequent retrieval procedure (see later).

Reverting to the indexing procedure, the Reference Indexing form is attached to the copy of the publication which is allotted a Reference Number – usually simply the next in sequence – and all the appropriate optical coincidence cards are drilled at this location.

Next, an accession card is typed (see Appendix III) as in the example, and also an author card, using information from the Reference Indexing Form, and both are stamped with the Reference Number, back and front.

At this point, the Reference Indexing form is detached from the publication and destroyed. The publication is filed in numerical order in a lateral filing cabinet. The Accession Card is filed in numerical order and the Author Card in alphabetical order.

6. Recall

The procedure for gaining access to information in the 'store' is as follows:

An optical Coincidence Card, or Cards (OCC) for the property (or group of properties) and a card (or cards) for the foodstuff (or group of foodstuffs) along with cards for the treatment of interest (if any) are taken from their respective filing locations and superimposed together on the optical reading unit plate. The numbers of any unobscured holes are noted. These are the reference numbers of publications containing the information on the subjects of the OCC's used.

To ascertain which of these contain say, numerical data, one superimposes also the /D Master OCC and the holes remaining unobscured indicate the reference numbers of publications containing data.

An unavoidable source of 'misinformation' in the present system takes the form of 'spurious relationships'. If a publication contains information on property '*a*' of foodstuff '*b*' and also on property '*x*' of foodstuff '*y*', it will be recalled as though it also contained information on property '*a*' of foodstuff '*y*' and property '*x*' of foodstuff '*b*', i.e., as though both properties were associated with both foodstuffs. This is more an interesting feature than a serious problem area.

Figure 1 is a flow diagram of these operations.

It will be noted that the preparation of bibliographies is a part of this project. A bibliography and a supplement to it on physical properties of fish has been prepared and copies are available at a nominal cost of $ 10 from the Centre. A bibliography on optical properties and one on thermal conductivity are planned.

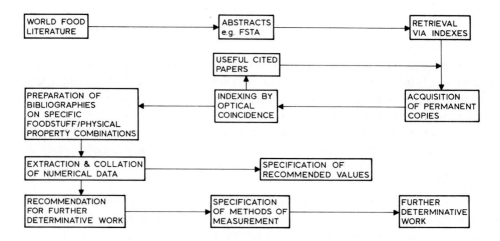

Fig. 1. Retrieval and processing of published papers on the physical properties of foodstuffs.

7. Utilization of Data

We distinguish five different areas of application of ppfs data and believe their identification is of importance as they each have their own combinations of special requirements of ppfs information, and applications for it. They are applications:

(1) of direct importance as attributes of particular foodstuffs;

(2) of indirect index of 'inaccessible' attributes;

(3) in the control of food processing operations;

(4) in the design of processes for foods;

(5) in the design of plant.

For example –

(1) The texture of foods is ultimately describable in physical terms and although we are a long way from doing that, there are many physical properties which correspond to textural properties or are of importance in themselves such as chewiness in candy, toughness in meat (shear resistance), jelly strength, firmness vs. crumbliness in cheese, crispness in apples and in potato chips.

(2) Experience has shown that there are simple relationships between such things as maturity in peas and their resistance to shear in a shear cell, between color and the ripeness of fruits and so on.

(3) Similarly, although we often may wish to cause (or prevent) changes in foods of a compositional, chemical or biological nature, we are often unable to monitor those changes except by some change in the physical nature of the material under

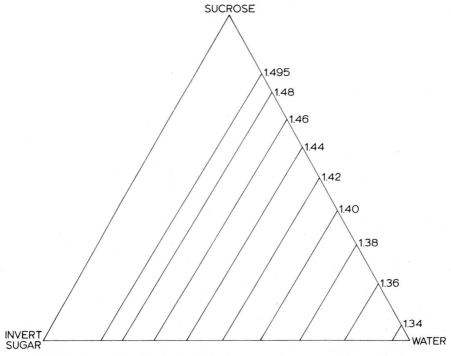

Fig. 2. Lines of equal refractive index for sucrose-invert sugar-water mixtures.

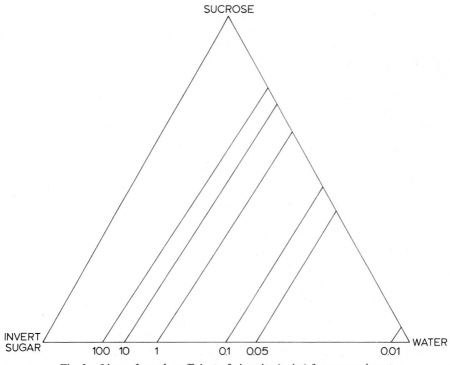

Fig. 3. Lines of equal coefficient of viscosity (poise) for sucrose-invert sugar-water mixtures.

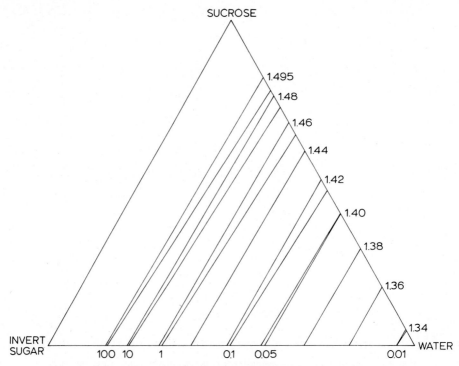

Fig. 4. Lines of equal refraction index and coefficient of viscosity for sucrose-invert sugar-water mixtures.

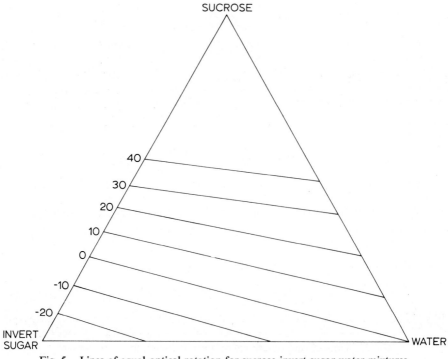

Fig. 5. Lines of equal optical rotation for sucrose-invert sugar-water mixtures.

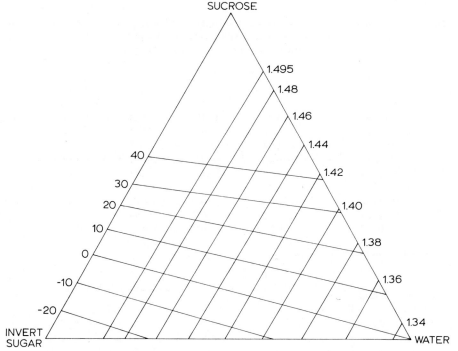

Fig. 6.   Lines of equal refractive index and optical rotation for sucrose-invert sugar-water mixtures.

process. These include viscosity, refractive index, microwave attenuation and infra-red reflectance changes – all used to measure, e.g., moisture content.

A recent project at the N.C.F.T. illustrates this. In a study of the use of refractive index as a foodstuff physical property, an investigation was made into the possibility of characterizing a three-component system by two different physical properties. Figure 2, shows the variation in refractive index with concentration of a sucrose-invert sugar-water mixture and Figure 3, the variation in viscosity for the same system. Figure 4, shows these two physical properties superimposed and it is clear that to-gether these two properties are no more use than either alone.

However, Figure 5, shows the angle of optical rotation of the system and Figure 6, the combination of rotation and refractive index. It is equally clear that values of refractive index and optical rotation for a sample of the three-component mixture should enable its composition to be determined with fair accuracy.

Since continuous measurement of refractive index and angle of rotation may be made in flow systems this should lend itself to in-line control of continuous processes for such systems.

Using this method to analyze samples of 'Golden Syrup' resulted in the errors shown in Table I. The error is small and most probably due to loss of some of the fructose from the invert sugar formed earlier in the process. If constant, such an error could readily be allowed for in calibration.

TABLE I

Composition of golden syrup (Tate & Lyle Ltd.): comparison of physical analysis (by optical rotation. and refractive index) with conventional analysis[a]
Values for five different retail samples.

Sample			% By weight				
			1	2	3	4	5
Sucrose	(a)	Physical	33.14	33.19	33.24	33.07	31.63
	(b)	Chemical	31.50	31.45	31.63	31.72	30.51
		(a)–(b)	+1.64	+1.74	+1.61	+1.35	+1.13
Invert	(c)	Physical	50.13	49.77	49.87	50.11	50.42
Sugar	(d)	Chemical	51.68	51.46	51.43	51.47	51.51
		(c)–(d)	−1.55	−1.69	−1.56	−1.36	−1.09
Water	(e)	Physical	16.73	17.04	16.89	16.82	17.95
	(f)	Chemical	16.82	17.09	16.94	16.81	17.98
		(e)–(f)	−0.09	−0.05	−0.05	+0.01	−0.03

[a] Clerget-Herzfeld double polarimetry method.

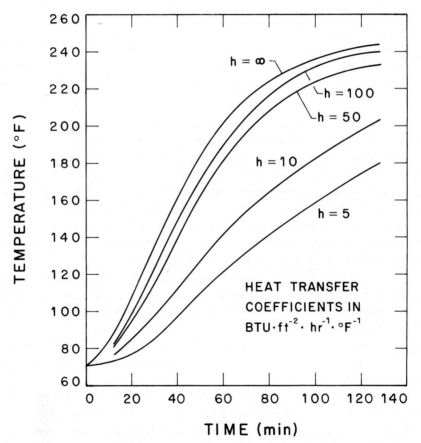

Fig. 7. Effect of variation of heat transfer in the rate of heating of a can of food.

(4) To design a process for food conversion or preservation requires an understanding of the physical basis of such a process. Although inadequate at present, a knowledge of the physical properties of the materials in process is essential for the proper design and prediction of the outcome of a food process. An example from the Jowitt-Thorne Fluidized Bed Canning Process is the influence of the physical properties of the canned food and of the fluidized bed on the rate of heating and of thermal destruction relative to conventional heating in steam.

Figure 7, (Jowitt and Thorne, 1971, 1972), shows the influence of outside heat transfer coefficient on the heating of the can contents in the case of a conductive – or solid-pack. Even though heat transfer coefficients in fluidized beds can be as low as one tenth to one fifth of that from condensing steam, it will be seen that there is no material difference in heating rate. The difference in process time (Figure 8) for equivalent sterility is even less significant. However, for convection packs the lower heating rate in a fluidized bed than in steam at the same temperature is more marked and it

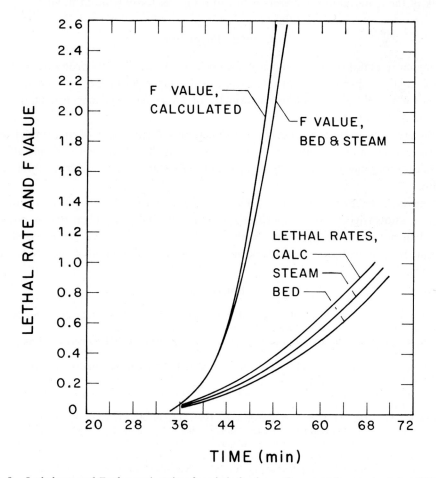

Fig. 8.   Lethal rate and *F* value against time for a babyfood can of cream style corn, heated at 250 °F.

is intended, in the near future, to investigate this in some detail with particular reference to the influence of the physical properties of the canned foodstuff and to possible design and operation changes to offset the disadvantage in practice.

(5) It is in plant design that the dearth of physical data on foods has perhaps been most serious and most evident. Recent work at the N.C.F.T. by Mr S. J. Higgs and coworkers has highlighted the importance of the wall-effect exhibited by some food liquids and semi-liquids containing large quantities of solid, fibrous material in suspension. Such materials tend to 'slip' past surfaces in relative motion and exhibit much lower resistance to flow than would be deduced from the shear resistance of the bulk material.

In the case of one sample of tomato paste – a material which exhibits this effect to a marked extent – which was pumped through a glass tube 1 cm diameter, it was estimated that 77% of the shear flow took place at the wall due to the slip effect.

Similarly, stirrer and mixer design for such materials requires special care if the bulk of the material in the mixers is not to rotate uselessly without mixing.

8. Future Possibilities

Many more examples of the value of a knowledge of the physical properties of foodstuffs could be given and with more work in this field and proper coordination and cooperation nationally and internationally of effort, even greater benefit to food technology will result.

But even now patterns are beginning to emerge and I believe that of equal importance to the collection and tabulation of ppfs data across the whole field of foodstuff and physical properties will be to distinguish and understand these patterns and relationships.

One day it might be possible to *predict* the value of a foodstuff's physical property from a knowledge of its composition, structure and history. When that is possible we shall not need the ppfs data banks which we are now engaged in assembling.

But that state will only be possible as a *result* of the assembling of such collections of data.

References

Jason, A. C. and Jowitt, R.: 1969, 'Physical Properties of Foodstuffs in Relation to Engineering Design', *Dechema-Monographien* **63**, 21–72.

Jowitt, R.: 1970, 'Fyzikalni vlastnosti potravin arskyeh materialu', CSVT Symposium Proceedings, Prague 1970.

Jowitt, R. and Thorne, S. N.: 1971, 'Evaluates Variables in Fluidised-Bed Retorting', *Food Enging.* **43**, 60–64.

Jowitt, R. and Thorne, S. N.: 1972, 'The Continuous Heat Processing of Canned Foods in Fluidized Beds', *Food Tr. Rev.* **42**, 9–13.

Jowitt, R., McCarthy, O. J., and Bimbenet, J. J.: 1971, 'Propriétés physiques des produits alimentaires. Proposition de collaboration internationale', *Ind. Aliment. et Agric.* **88**, 1217–1220.

Mann, E. J.: 1966, 'Report on International Survey of the World Literature on Food Science and Technology', *Dairy Sci. Abstr.* **28**, 603–606.

McCarthy, O. J., Jowitt, R., and Hallstrom, B.: 1972, 'Klassificerung av livsmedels fysikaliska egenskaper', *Livsmedelsteknik* **14**, 69–70.

Appendix I

The Classification of Foodstuffs and Their Physical Properties*

National College of Food Technology, (University of Reading), St. George's Avenue, Weybridge, Surrey

(Second revision of the Classification published in Jowitt, R. (1968) *Food Tr. Rev.* **38,** 55–64)

EXAMPLE OF THE USE OF THE CLASSIFICATION

What would be the coordinates of: 'thermal conductive of freeze-dried plaice'?

Foodstuff

Page 388: UWI specifies Fish, shellfish and products
UWI A specifies Non-fatty fish
UWI AB specifies Non-fatty, flat fish
UWI ABA specifies Non-fatty, flat, bony fish
and UWI ABAD are the coordinates of 'Plaice'.

Treatment

Page 394: – M specifies Preservation
– MN specifies Dehydration
– MNC specifies Freeze drying
and therefore – UWI ABAD-MNC are the coordinates of 'Freeze-dried plaice'.

Physical property

Page 397: 2. specifies Thermal properties
2.4 specifies Thermal conductivity

Therefore UWI ABAD-MNC 2.4 are the coordinates of 'thermal conductivity of freeze-dried plaice'.

FOODSTUFF CLASSIFICATION

		Page
UWA	Dairy products	380
UWB	Sugar and sugar confectionery, polysaccharides (including starches)	381
UWC	Cereals and cereal products	382
UWD	Bakery products	383
UWE	Edible oils and fats	383
UWF	Vegetables and vegetable products	383
UWG	Fruit, nuts and products	385
UWH	Meat, eggs and products	387
UWI	Fish, shellfish and products	388
UWN	Soft drinks	390
UWO	Alcoholic beverages	391

* Copyright reserved: R. Jowitt, June 1972.

NOTES ON FOODSTUFF CLASSIFICATION

1. Synonyms are given in parenthesis, e.g.: – UWG XL Peanut (groundnut).
2. Explanatory information is also given in parenthesis, e.g.: – UWH WA Intact egg (shell and contents).
3. Foodstuffs which are closely similar but not synonymous are classified under the same coordinates, separated by commas, e.g.: – UWF ID Cucumber, gherkin.
4. Foodstuffs which could be classified in two or more places in the classification are given coordinates in only one place. They are listed in all other appropriate places without coordinates, but with a reference to the one set of coordinates used, e.g. salt is given the coordinates UWR TAR; it is also listed under UWR V with a reference to UWR TAR.

UWA	DAIRY PRODUCTS
UWA A	Cow's milk and its components
UWA AA	Raw milk
UWA AB	Market milk
UWA AC	Skim milk
UWA AD	Buttermilk
UWA AE	Cream
–	Butter (Classified under UWE FE)
UWA B	Processed cow's milk and its components
UWA BA	Cultured milks
UWA BAA	Cultured buttermilk (streptococcal)
UWA BAB	Acidophilus milk (lactobacillary)
UWA BAC	Yogurt (streptococcal/lactobacillary)
UWA BAD	Kefir, leben (yeast fermented)
UWA BB	Processed milks (traditional)
UWA BBA	Rabbri, khurchan (condensed milks)
UWA BBB	Khoa, kheer (dried milks)
UWA BC	Curd, cheese
UWA BCA	Milk cheese
UWA BCAA	Bacterial cheese
UWA BCAAA	Surface slime
UWA BCAAB	No surface slime
UWA BCAABA	Hard
UWA BCAABB	Semi-hard
UWA BCAABC	Soft
UWA BCAB	Mould cheese
UWA BCABA	Surface mould
UWA BCABB	Internal mould, blue veined
UWA BCAC	Acid-coagulated cheese
UWA BCD	Cream cheese
UWA BD	Casein
UWA BE	Whey
UWA BF	Milk shakes
–	Lactose (Classified under UWB ACB)
UWA C	Reconstituted milks
UWA CA	Filled milks
UWA D	Milk of other mammals (divide like UWA A to UWA C)
UWA DA	Goat's milk
UWA DB	Indian buffalo's milk and components

	Ghee (Classified under UWE FF)
UWA DC	Ewe's milk
UWA DD	Mare's milk
UWA DDB	Processed mare's milk and components
.UWA DDBA	Cultured mare's milk
UWA DDBAA	Kumiss
UWA E	Ice cream
UWA F	Milk surrogates
UWA FA	Vegetable protein milks
UWA FAA	Soya milk
UWA FAB	Leaf protein milk
UWA FB	Microbial protein milks
UWA FC	Fish protein milks
UWA FD	Non-dairy 'whiteners'
UWB	SUGARS AND SUGAR CONFECTIONERY, POLYSACCHARIDES (INCLUDING STARCHES)
UWB A	Sugars
UWB AB	Monosaccharides
UWB ABA	Dextrose
UWB ABB	Fructose
UWB ABC	Invert sugar
UWB ABE	Galactose
UWB AC	Disaccharides
UWB ACA	Maltose
UWB ACB	Lactose
UWB ACC	Sucrose
UWB AD	Trisaccharides
UWB AE	Manufactured sugars
UWB AEA	Icing sugar
UWB AEB	Caster sugar
UWB AEC	Granulated sugar
UWB AED	Lump sugar
UWB AEE	Brown sugar
UWB AEF	Demerara sugar
UWB AEG	Loaf sugar
UWB AF	Syrups
UWB AFA	Cane syrups
UWB AFB	Beet syrups
UWB AFC	Refinery syrups
UWB AFD	Glucose syrups (liquid glucose)
UWB AFE	Maple syrup
UWB AFF	Palm syrup
UWB AFG	Date syrup
UWB AG	Molasses
UWB AGA	Treacle
UWB AH	Honey and honey products
UWB AHC	Honeycombs
UWB AHG	Clear honey
UWB AHK	Opaque honey
UWB AHR	Honey and invert sugar blends
UWB B	Sweets, confectionery
UWB BA	Chocolate and chocolate confectionery
UWB BAC	Whole chocolate (bars, blocks, liquid chocolate)
UWB BACA	Plain chocolate
UWB BACB	Milk chocolate
UWB BACC	Flavoured chocolate
UWB BACD	With other ingredients (e.g. fruit, nuts, etc.)

UWB BAD	Chocolate coated confectionery
UWB BADA	Soft centres
UWB BADB	Hard centres
UWB BADC	Liquid centres
UWB BC	Boiled sweets
UWB BD	Gums, jellies
UWB BDA	Gelatine base gums, jellies
UWB BDB	Starch base gums, jellies
UWB BE	Fondants
UWB BF	Caramels, toffees
UWB BH	Nougat
UWB BH	Barley sugar
UWB BI	Liquorice confectionery
UWB BJ	Marshmallows
UWB BK	Marzipan
UWB BZ	Medicated confectionery
UWB D	Polysaccharides and derivatives
UWB DA	Natural edible gums
UWB DAA	Pectin
UWB DB	Synthetic gums
UWB DC	Starch and starch products
UWB DCA	Corn starch (maize starch)
UWB DCB	Potato starch
UWB DCC	Sago
UWB DCD	Cassava (manioc) starch (tapioca, manioca)
UWB DCE	Rice starch
UWB DCF	Fruit starches (e.g. banana, soya)
UWB DCG	Nut starches (e.g. chestnut)
UWB DCH	Arrowroot, canna starch
UWB DC	Yam sItarch
UWB DCJ	Custard, blancmange powders
UWB DDP	Pentosans
UWC	CEREALS AND CEREAL PRODUCTS
UWC A	Rice
UWC B	Wheat
UWC C	Oats
UWC D	Rye
UWC E	Maize, corn, sweet corn
UWC F	Barley, pearl barley
UWC G	Millet
UWC H	Sorghum
UWC I	Cereal flours (divide like UWC A to UWC H)
UWC U	Breakfast cereals
UWC UA	Porridge oats
UWC UB	Other breakfast cereals (divide like UWC A to UWC H)
UWC W	Pasta products
UWC WA	Extruded pasta, noodles
UWC WAA	Vermicelli
UWC WAB	Macaroni
UWC WAC	Spaghetti
UWC WB	Sheet, laminated, rolled pasta
UWC WBA	Lasagne
UWC WBB	Tagliatelle
UWC WC	Stuffed pasta
UWC WCA	Ravioli
UWC WCB	Tortellini

UWD	**BAKERY PRODUCTS**
UWD A	Leavened breads
UWD AB	Wholemeal breads
UWD AC	White bread
UWD D	Fancy breads
UWD DA	Fruit breads
UWD DB	Malt breads
UWD DC	Spiced breads
UWD F	Dietetic breads
UWD G	Rusks
UWD I	Cakes
UWD IA	Sponge cakes, sponge puddings
UWD K	Dry pastries
UWD L	Puff pastries
UWD M	Tarts
UWD N	Puddings, dumplings
UWD O	Meat pies (divide like UWH)
UWD P	Fish pies (divide like UWI)
UWD Q	Fruit pies (divide like UWG)
UWD R	Unleavened breads, biscuits
UWD RA	Water biscuits, crackers
UWD RC	Sweet biscuits
UWD RE	Ship's biscuits
UWD RG	Wafers
UWD RGA	Communion wafers
UWD S	Doughs
UWD T	Batters
UWD U	Gluten, zein
UWE	**EDIBLE OILS AND FATS**
UWE A	Oils and fats (by function)
UWE AC	Cooking oils
UWE AF	Salad oils
UWE AH	Cooking fats, shortening
UWE D	Vegetable oils and fats
UWE DA	Cocoa butter
UWE DF	Other vegetable oils and fats (divide like UWF B to UWF T and UWG A to UWG D)
UWE DX	Nut butters and oils (divide like UWG X)
UWE F	Animal oils and fats
UWE FA	Lard, pork fat
UWE FB	Dripping, tallow, oleo fat, beef fat
UWE FD	Horse fat
UWE FE	Butter
UWE FF	Ghee
UWE FH	Other animal oils and fats (divide like UWH A to UWH S)
UWE I	Fish oils (divide like UWI A to UWI D)
UWE M	Margarine
UWE MA	Vegetable margarine
UWE MMA	Soft margarine
UWE P	Margarine and butter mixtures, oleomargarine
UWE PA	Vanaspati
UWF	**VEGETABLES AND VEGETABLE PRODUCTS**
UWF B	Leaf, stem and flower crops
UWF BC	Cabbage
UWF BCP	Pickling cabbage (red cabbage)

UWF BD	Kale
UWF BE	Brussels sprout
UWF BF	Spinach
UWF BG	Broccoli
UWF BH	Cauliflower
UWF BJ	Asparagus
UWF BK	Globe artichoke
UWF BL	Leek
UWF BM	Kholrabi
UWF BN	Celery
UWF BO	Chicory
UWF BP	Lettuce
UWF BQ	Cress
UWF BR	Endive
UWF BS	Pumpkin flower
UWF I	Fruit crops
UWF IA	Egg plant (aubergine)
UWF IB	Capsicum (red peppers, green peppers)
UWF IC	Tomato
UWF ID	Cucumber, gherkin
UWF IF	Dish-cloth gourd
UWF IG	Squash
UWF IH	Bitter squash, bitter melon
UWF II	Pumpkin
UWF IJ	Marrow
UWF J	Seed crops
UWF JA	Oil seeds
UWF JAA	Cotton seed
UWF JAB	Rape seed
UWF JAC	Sunflower seed
UWJ JAD	Safflower seed
UWJ JAE	Palm kernel
UWJ JAF	Sesame seed
UWJ JB	Pulses
UWJ JBA	Lentil
UWJ JBB	Pea
UWJ JBC	Bean
UWJ JBCA	Runner bean, French bean, snap bean
UWF JBCB	Dwarf bean
UWF JBCC	Soya bean
UWF JBCD	Broad bean
UWF JBCE	Haricot bean
UWF JBCF	Butter bean
UWF JBCG	Lima bean
UWF T	Root crops, tubers, bulbs
UWF TA	Potato
UWF TB	Beet
UWF TC	Carrot
UWF TD	Parsnip
UWF TE	Turnip (rutabaga, swede)
UWF TF	Onion
UWF TG	Jerusalem artichoke
UWF TH	Salsify
UWF TI	Sweet potato
UWF TJ	Radish
UWF TK	Horseradish
UWF V	Edible fungi

UWF VM	Mushrooms
UWF VR	Yeasts
UWF VT	Truffle
UWF W	Marine vegetation, plankton
UWF WA	Algae
UWF WB	Seaweed, laver
UWF X	Products
UWF XA	Juices
UWF XB	Mixes
UWF XBA	In salad dressing
UWF XD	Pastes, purees
UWF XH	Jellies
UWF XK	Soups
UWF XL	Stews
UWF XM	Vegetable protein products
UWF XMA	Protein concentrates
UWF XMB	Protein isolates
UWF XMC	Protein hydrolysates
UWF XMD	Textured, 'structured', vegetable protein products

UWF XA to UWF XM may be divided further like UWF B to UWF W to specify products from individual species, e.g. UWF XDIC – Tomato puree.

UWG	FRUIT, NUTS AND PRODUCTS
UWG A	Fruits grown only in tropical climates
UWG AA	Pineapple
UWG AB	Jack fruit
UWG AC	Tamarind
UWG AD	Langsat
UWG AE	Mango
UWG AF	Rambutan
UWG AG	Durian
UWG AH	Mangosteen
UWG AI	Rose apple
UWG AJ	Guava
UWG AK	Papaya
UWG AL	Custard apple
UWG B	Fruits grown in tropical and subtropical climates
UWG BA	Date
UWG BB	Fig
UWG BC	Avocado pear
UWG BD	Olive
UWG BE	Pomegranate
UWG BF	Banana
UWG BG	Plantain
UWG BH	Melon
UWG BI	Cape gooseberry
UWG BJ	Citrus fruits
UWG BJA	Orange
UWG BJB	Tangerine, clementine
UWG BJC	Grapefruit
UWG BJD	Lemon
UWG BJE	Lime
UWG C	Fruits grown in temperate climates
UWG CA	Pomaceous fruits
UWG CAA	Apple
UWG CAB	Pear

UWG CABA	Quince
UWG CABB	Medlar
UWG CB	Drupaceous fruits
UWG CBA	Peaches
UWG CBAA	Apricot
UWG CBAB	Nectarine
UWG CBB	Plums
UWG CBBA	Prune
UWG CBBB	Damson
UWG CBBC	Gage plum
UWG CBBCA	Greengage
UWG CBBD	Golden plum
UWG CBBE	Red plum
UWG CBBF	Victoria plum
UWG CBC	Cheery
UWG CC	Berries and soft fruits
UWG CCA	Blackberry
UWG CCB	Loganberry
UWG CCC	Raspberry
UWG CCD	Strawberry
UWG CCE	Redcurrant
UWG CCF	Blackcurrant
UWG CCG	Gooseberry
UWG CCH	Mulberry
UWG CCI	Cranberry
UWG CCJ	Grapes
UWG CCJA	Raisins, sultanas, currants
UWG CCK	Blueberry
UWG CCL	Dewberry
UWG D	Fruit surrogates
UWG DA	Rhubarb
UWG E	Liquid fruit products, purees, pastes (divide like UWG A to UWG D)
UWG EA	Natural strength juices
UWG EB	Squashes, crushes, cordials
UWG EC	Comminuted products
UWG ECA	Comminuted fruit drinks
UWG ECB	Purees
UWG ECC	Pastes
UWG ED	Aroma concentrates
UWG W	Other fruit products (divide like UWG A to UWG D)
UWG WA	Jams
UWG WB	Marmalades
UWG WC	Curds
UWG WD	Mixes, mincemeat
UWG WE	Jellies
UWG WF	Stews
UWG X	Nuts
UWG XA	Walnut
UWG XB	Pecan (hickory) nut
UWG XC	Pig nut
UWG XD	Chestnut
UWG XE	Hazel nut (filbert, cobnut)
UWG XF	Almond
UWG XG	Cashew nut
UWG XH	Pistachio
UWG XK	Brazil nut
UWG XL	Peanut (groundnut)

UWG XM	Coconut
UWG XZ	Nut products (divide like UWG XA to UWG XM)
UWG XZA	Juices
UWG XZAM	Coconut juice, coconut milk, coconut cream
UWG XZB	Mixes
UWG XZC	Pastes, purees
UWG XZCL	Peanut butter
UWG XZD	Jellies
UWG XZE	Soups
UWG XZF	Nut protein products

UWH	Meat, eggs and products
UWH A	Beef
UWH AA	Veal
UWH B	Venison
UWH C	Buffalo
UWH D	Mutton, lamb
UWH E	Horseflesh
UWH F	Goat
UWH G	Pork
UWH H	Bacon, ham
UWH I	Rabbit, hare
UWH K	Chicken
UWH L	Duck (domestic)
UWH M	Goose
UWH N	Turkey
UWH O	Guinea fowl
UWH P	Pigeon
UWH R	Game birds
UWH RA	Pheasant
UWH RB	Partridge
UWH RC	Grouse
UWH RD	Wild duck
UWH S	Miscellaneous vertebrate meat
UWH SA	Whalemeat
UWH SB	Sealmeat
UWH SC	Turtlemeat
UWH SD	Snakemeat
UWH SE	Frogmeat
UWH W	Eggs
UWH WA	Intact egg (shell and contents
UWH WB	Whole egg (yolk and white)
UWH WC	Egg yolk
UWH WD	Egg white
UWH WE	Egg shell
UWH Z	Products
UWH ZA	Sausages
UWH ZB	Minced meat
UWH ZD	Mixes
UWH ZH	Pastes, pâtés, purees
UWH ZK	Jellies
UWH ZP	Soups
UWH ZS	Stews
UWH ZT	Extracts

UWH ZA to UWH ZT may be divided further like UWH A to UWH W to specify individual products, e.g. UWH ZAG – Pork sausages, UWH ZPK – Chicken soup.

UWI	FISH, SHELLFISH AND PRODUCTS
UWI A	Non-fatty fish
UWI AA	Round
UWI AAA	Bony
UWI AAAA	The cod family
UWI AAAAA	Cod
UWI AAAAB	Haddock
UWI AAAAC	Pollack
UWI AAAAD	Coalfish (saithe, coley, green cod)
UWI AAAAE	Whiting
UWI AAAAF	Burbot (pout, bib)
UWI AAAAG	Hake
UWI AAAAH	Ling
UWI AAAB	Smelt (sparling)
UWI AAAC	Pike
UWI AAAD	Perch-pike, pike-perch
UWI AAAE	Weakfish (sea trout – N. America)
UWI AAAF	Croaker
UWI AAAG	Cunner
UWI AAAH	Lutjanus spp.
UWI AAAHA	Red snapper
UWI AAAHB	Muttonfish
UWI AAAI	Flying fish
UWI AAAJ	Rock fish
UWI AAAK	Japanese hokke
UWI AAAL	Bullhead (freshwater catfish)
UWI AAAM	Pufferfish (globefish)
UWI AAB	Cartilaginous
UWI AB	Flat
UWI ABA	Bony
UWI ABAA	Dab
UWI ABAB	Lemon sole
UWI ABAC	Witch
UWI ABAD	Plaice
UWI ABAE	Flounder (fluke)
UWI ABAF	Megrim (whiff)
UWI ABAG	Sole
UWI ABB	Cartilaginous
UWI ABBA	Skate, ray
UWI B	Fatty fish
UWI BA	Round
UWI BAA	Bony
UWI BAAA	The herring family
UWI BAAAA	Herring
UWI BAAAB	Sprat (whitebait – N. America)
UWI BAAAC	Shad, alewife
UWI BAAAD	Sardine, pilchard
UWI BAAAE	Anchovy
UWI BAAAF	Menhaden
UWI BAAB	The salmon
UWI BAABA	Salmon
UWI BAABB	King (Chinook, black, chub) salmon
UWI BAABC	Sockeye (red) salmon
UWI BAABD	Chum (dog) salmon
UWI BAABE	Pink (humpback) salmon
UWI BAABF	Silver (coho, jack) salmon
UWI BAAC	The trout

UWI BAACA		Brown trout
UWI BAACB		Rainbow trout
UWI BAACC		Sea trout
UWI BAAD		Koayu (ayu)
UWI BAAE		Coregonus spp.
UWI BAAEA		Whitefish (lake herring, whitebait – Australia)
UWI BAAEB		Cisco
UWI BAAF		Capelin
UWI BAAG		The perches
UWI BAAGA		Perch (lake or river perch)
UWI BAAGB		Bass (sea perch – N. America)
UWI BAAH		The sea breams
UWI BAAHA		Common sea bream (red bream, snapper – Australia)
UWI BAAHB		Porgy
UWI BAAI		The mullets
UWI BAAIA		Gray mullet
UWI BAAIB		Red mullet
UWI BAAJ		The mackerels
UWI BAAJA		Mackerel
UWI BAAJB		Spanish mackerel
UWI BAAK		Scad (horse mackerel, jack mackerel)
UWI BAAL		Bluefish (elf – S. Africa)
UWI BAAM		Butterfish
UWI BAAN		Wolf fish (catfish)
UWI BAAO		The tunny
UWI BAAOA		Tuna (bluefin tuna)
UWI BAAOB		Albacore (long finned tuna)
UWI BAAOC		Skipjack
UWI BAAOD		Bonito
UWI BAAP		John Dory
UWI BAAQ		Garfish (garpike)
UWI BAAR		Sheatfish (som – Russia)
UWI BAAS		The eels
UWI BAASA		Eel
UWI BAASB		Conger eel
UWI BAAT		The carp family
UWI BAATA		Carp
UWI BAATB		Roach
UWI BAAU		Scorpeniform spp.
UWI BAAUA		Redfish (Norway haddock, ocean perch)
UWI BAAUB		Gurnard
UWI BAB	Cartilaginous	
UWI BABA		Shark, dogfish
UWI BABB		Sturgeon
UWI BABC		Monkfish (frogfish, angler fish)
UWI BABD		Ratfish
UWI BB	Flat	
UWI BBA	Bony	
UWI BBAA		Turbot
UWI BBAB		Halibut
UWI BBB	Cartilaginous	
UWI C	Molluscs	
UWI CA	Oyster	
UWI VB	Mussel	
UWI CC	Clam	
UWI CD	Cockle	
UWI CE	Scallop	

UWI CF	Whelk
UWI CG	Edible snail
UWI CH	Winkle
UWI CI	Octopus, squid, cuttlefish
UWI D	Crustaceans
UWI DA	Shrimp
UWI DD	Prawn
UWI DC	Crayfish
UWI DD	Lobster
UWI DE	Crab
UWI DF	Scampi
UWI X	Products (divide like UWI A to UWI D)
UWI XA	Mixes
UWI XB	Pastes, purees
UWI XC	Jellies
UWI XD	Soups
UWI XE	Stews
UWI XF	Fish cakes
UWI XG	Fish protein products
UWI XGA	Edible fish meal
UWI XGB	Fish protein concentrate
UWI XGC	Fish protein isolate
UWI XGD	Fish protein hydrolysate
UWI XH	Fish sausage
UWN	Soft drinks
UWN B	Natural mineral waters, spa waters
UWN C	Synthetic and flavoured mineral waters (suffix -LQ if carbonated, see p. 394)
UWN CA	Soda water
UWN CB	Ginger beer
UWN CC	Tonic water
UWN CF	Fruit flavoured
UWN CFA	Orange
UWN CFB	Lemon
UWN CFC	Lime
UWN CFD	Grapefruit
UWN CFG	Cherry
UWN CG	Cola types
UWN CH	Sarsaparilla, root beer
UWN CI	Cream soda types
UWN J	Essences, concentrates
UWN JA	Plant syrups
UWN JAA	Orange blossom
UWN JAB	Marshmallow
UWN JAC	Rose hip
UWN JAD	Peppermint
UWN K	Leaf products
UWN KT	Teas
UWN KTA	Black, fully fermented
UWN KTAB	China, Japan
UWN KTAC	Other sources
UWN KTB	Oolong, semi fermented
UWN KTC	Green; unfermented
UWN KTV	Concentrates
UWN KTVA	Tableted
UWN KTVB	Powdered
UWN KTX	Tea surrogates

UWN N	Seed products
UWN NA	Coffee
UWN NAC	Special coffees
UWN NACA	Blended
UWN NACB	Decaffeinated
UWN NAD	Concentrates
UWN NADA	Liquid
UWN NADC	Tableted
UWN NADD	Powdered
UWN NL	Coffee surrogates
UWN NLA	Chicory
UWN NLB	Fig
UWN NLC	Barley
UWN NLD	Rye
UWN NLK	Acorn
UWN NB	Cocoa and chocolate drinks
UWN NBA	Cocoa powder
UWN NBB	Chocolate powder
UWN NBC	Chocolate-based beverages
UWO	Alcoholic beverages
UWO A	Fermented beverages
UWO AB	Beers
UWO AC	Ciders
UWO AD	Perry
UWO AE	Grape wines
UWO AEA	Fortified wines (sherry, port, madeira)
UWO AF	Wines of other fruits (divide like UWG
UWO AG	Mead
UWP B	Distilled beverages
UWO BA	Brandy, cognac
UWO BB	Apricot brandy
UWO BC	Peach brandy
UWO BD	Cherry brandy
UWO BE	Blackcurrant liqueurs, Cassis
UWO BF	Gin
UWO BG	Whisky
UWO BGA	Scotch
UWO BGB	Irish
UWO BGC	Rye
UWO BGD	Bourbon
UWO BJ	Vodka
UWO BK	Rum, arrack
UWO BL	Ouzo
UWO BM	Pisco
UWO BN	From spices and herbs (divide like UWR TA and UWR TB)
UWO Z	Other alcoholic beverages
UWP	Pickles and sauces
UWP A	Vinegars
UWP C	Sauces
UWP CA	Thick sauces
UWP CAB	Fruit and vegetable sauces
UWP CAF	Vegetable sauces (divide like UWF)
UWP CAFTK	Horseradish sauce
UWP CAG	Fruit sauces (divide like UWG)
UWP CB	Thin sauces

UWP CBA	Worcestershire sauce
UWP CBB	Yorkshire relish
UWP CBC	Soya sauce
UWP CBD	Fish sauce
UWP D	Pickles
UWP DA	In vinegar
UWP DB	In sauce
UWP DBA	In mustard sauce, piccalilli
UWP K	Chutneys
UWP KA	Tomato chutneys
UWP KB	Mango chutneys
UWP L	Sauerkraut
UWP U	Salad cream
UWP V	Mayonnaise
UWR	ADDITIVES (INCLUDING COLOURS AND FLAVOURS), TRACE CONSTITUENTS
UWR A	Nutrients
UWR AA	Vitamins
UWR AB	Minerals
UWR G	Gelling agents
UWR GG	Gelatin
–	Pectin (Classified under UWB DAA)
UWR GL	Agar-agar
UWR GM	Irish moss, carrageenan
UWR GN	Alginates
UWR K	Emulsifying and stabilising agents
UWR KA	Glycerol
UWR KB	Glycerol monostearate
UWR KD	Esters of polyhydric alcohols
UWR KF	Cellulose derivatives
UWR KH	Lecithin
–	Edible Gums (Classified under UWB DA)
–	Starches (Classified under UWB DC)
UWR L	Bread improvers
UWR N	Bleaching agents
UWR R	Colouring substances
UWR RC	Coal-tar dyes
UWR RCA	Water soluble
UWR RCB	Oil soluble
UWR RS	Other synthetic dyes
UWR RV	Vegetable dyes
UWR T	Flavouring substances
UWR TA	Spices and condiments
UWR TAA	Vanilla
UWR TAB	Mint
UWR TAC	Ginger
UWR TAD	Turmeric
UWR TAE	Curry powders
UWR TAF	Allspice (pimento)
UWR TAG	Chilli
UWR TAH	Cloves
UWR TAJ	Cinnamon
UWR TAK	Nutmeg
UWR TAM	Capers
UWR TAO	Peppers
UWR TAP	Mustards
UWR TAR	Salt

UWR TB	Herbs
UWR TBA	Parsley
UWR TBB	Wintergreen
UWR TBC	Fennel
UWR TE	Flavour concentrates
UWR TEA	Essential oils
UWR TEB	Aqueous aroma concentrates
UWR TG	Aromatic chemicals
UWR TS	Sweetening agents
–	Sugars (Classified under UWB A)
UWR TSB	Saccharin
UWR TSC	Cyclamates
UWR TT	Hydrolised protein
UWR TU	Flavour enhancers
UWR TUA	Monosodium glutamate
UWR TUE	Ribonucleotides
UWR V	Preservatives
UWR VA	Saltpetre
–	Salt (Classified under UWR TAR)
UWR W	Processing aids
UWR WA	Yeast
UWR WB	Monoglycerides, diglycerides and derivatives
UWR WC	Mineral oils
UWR WD	Baking powder
UWR Z	Residues, trace constituents
UWR ZA	Antibiotics
UWR ZE	Pesticides
–	Mineral oils (Classified under UWR WC)
UWR ZP	Trace compounds
UWR ZR	Trace elements
UWT	WATER, ICE
UWT A	Water
UWT AA	Sea water
UWT B	Ice

TREATMENT CLASSIFICATION

Suffixes to indicate the treatment to which the foodstuff is being or has been subjected.

-D	Preliminaries, preparation
-L	Processing
-M	Preservation
-N	Filling and closing conditions
-P	Pack type
-S	Storage atmosphere
-T	Storage temperature
-W	Distribution stage
-Y	Preparation stage

-D	PRELIMINARIES, PREPARATION
-DB	Transport, reception, unloading, unpackaging
-DF	Sorting, grading, screening, classifying
-DH	Trimming, peeling, skinning, dehairing, defeathering
-DK	Seed removal, stone removal, coring, eviscerating, boning
-DM	Cleaning, washing
-DP	Cutting, slicing, filleting, dicing, shredding

-L	PROCESSING
-LA	Grinding, milling, comminution, mincing
-LB	Filtration
-LC	Expressing
-LD	Homogenisation, emulsification
-LE	Mixing, kneading
-LF	Fermentation, leavening
-LG	Centrifugation, clarification, cyclone separation
-LH	Precipitation, flocculation
-LI	Concentration
-LIA	Evaporation, distillation
-LIB	Freeze concentration
-LIC	Reverse osmosis, ultrafiltration
-LJ	Agglomeration, granulation, pelleting, tableting
-LK	Crystallisation
-LL	Solvent extraction
-LM	Rendering
-LMA	Heat rendering
-LMB	Impulse rendering
-LN	Bleaching
-LO	Hydrogenation
-LP	Deionisation
-LQ	Aeration, carbonation
-LR	Moulding, shaping, extruding, pressing, dividing
-LS	Decoration, coating, enrobing, encapsulation
-LT	Holding, conditioning, tempering, proving
-LU	Heating, blanching
-LV	Thawing, melting
-LVA	Dielectric thawing
-LW	Cooking
-LWA	In fat
-LWAA	Frying
-LWAB	Roasting
-LWB	In water, steam
-LWBA	Pressure cooking
-LWBB	Steaming
-LWBC	Boiling, coddling, poaching
-LWC	In air
-LWCA	Baking
-LWCB	Toasting
-LWCC	Grilling, barbecuing, broiling
-LX	Melangeuring
-LY	Conching
-LZ	Hydrolysis
-LZA	Acid hydrolysis
-LZB	Enzymic hydrolysis
-LZC	Acid-enzymic hydrolysis
-M	PRESERVATION
-MA	Fermentative preservation
-MAA	Alcoholic
-MAB	Acetous
-ME	Pasteurisation, sterilisation
-MEA	In-container
-MEAA	Cans
-MEAB	Bottles, jars
-MEAC	Pouches

-MEB	Before filling
-MEBA	Batch
-MEBB	Continuous
-MEBBA	Indirect heating
-MEBBAA	HTST
-MEBBAB	UHT
-MEBBB	Direct heating
-MEBBBA	HTST
-MEBBBB	UHT
-MI	Cooling
-MK	Chilling
-ML	Freezing
-MLA	Blast freezing
-MLB	Plate freezing
-MLC	Fluidised bed freezing
-MLD	Immersion freezing
-MLDA	In brines, syrups
-MLDB	In cryogenic liquids
-MLE	Cryogenic sray freezing
-MN	Dehydration
-MNA	Spray drying
-MNB	Roller (drum) drying
-MNC	Freeze drying
-MND	Tray, tunnel drying
-MNE	Pneumatic drying
-MNF	Bin drying
-MNFA	Fixed bed drying
-MNFB	Spouted bed drying
-MNFC	Fluidised bed drying
-MNG	Atmospheric drying (traditional)
-MNH	Foam mat drying
-MNI	Puff drying
-MT	Curing
-MTA	Smoking
-MTB	Salting
-MTC	Pickling
-MTD	Nitrate/nitrite treatment
-MTE	Kippering, bloating
-MW	Chemical preservation
-MZ	Irradiation
-MZA	Infra red irradiation
-MZB	Ultra violet irradiation
-MZC	Gamma irradiation, X-irradiation
-MZD	Beta irradiation
-MZE	Alpha irradiation
-N	FILLING & CLOSING CONDITIONS
-NA	Unsterile
-NB	Aseptic
-P	PACK TYPE
-PA	Bulk
-PB	Domestic/catering pack, consumer unit
-PBA	Cans
-PBB	Bottles, jars
-PBC	Pouches
-PBD	Cartons

-PBE	Shrink wraps
-S	STORAGE ATMOSPHERE
-SA	Ambient
-SB	Gas
-SBA	Carbon dioxide
-SBB	Nitrogen
-SC	Vacuum
-T	STORAGE TEMPERATURE
-TA	Ambient
-TB	Refrigerated
-TBA	Chilled
-TBB	Frozen
-W	DISTRIBUTION STAGE
-WA	Manufacturer's stock
-WB	Distributor's stock
-WC	Retailer's stock
-WD	Consumer's stock
-WDA	Institutional
-WDB	Domestic
-Y	PREPARATION STAGE
-YA	Unpacking
-YB	Thawing
-YC	Rehydrating
-YD	Cooking (divide like -LW)
-YE	Serving
-YF	Eating

PHYSICAL PROPERTY CLASSIFICATION

1. Mechanical properties
2. Thermal properties
3. Diffusional and related properties
4. Electromagnetic, electrostatic properties

1	MECHANICAL PROPERTIES
1.1	Specific gravity, density, specific volume, specific weight
1.2	Bulk density
1.3	Porosity, void fraction, aeration
1.4	Particle size, particle size distribution
1.5	Specific surface
1.6	Morphometric properties
1.7	Friction properties, coefficients of static and kinetic friction
1.8	Aerodynamic and hydrodynamic characteristics
1.9	Surface properties, interfacial tension
1.9.1	Surface tension, interfacial tension
1.9.2	Adhesive properties
1.10	Acoustic properties
1.10.1	Velocity of sound
1.10.2	Sound absorption coefficient
1.10.3	Acoustic resistance
1.11	Elastic properties
1.11.1	Young's modulus; apparent moduli in uniaxial tension or compression
1.11.2	Shear modulus; apparent moduli in shear

Appendix II

Reference-Indexing Form

REFERENCE NUMBER:	FORM OF COPY (P, R, O, Mfilm, Mfiche):
AUTHORS:	
REFERENCE:	
LANGUAGE OF ARTICLE OR BOOK AND LANGUAGES OF SUMMARY(S), ABSTRACTS NO.:	
ARTICLE TITLE (ORIGINAL LANGUAGE):	
ARTICLE TITLE IN ENGLISH:	
INDEX TERMS: Foodstuffs:	
Treatments:	
Physical Properties: & Master Card references	
REFERENCES	

Appendix III

Design of Accession Card

The design of the Accession Card is shown by the following example:

---

                                    1   **21**   P

2 Lempka, A. and Wojciak, W.

3 *Die Nahrung* **13** (1) 27–31 (1969)

4 (De, en, ru) (*FSTA* **1** (5) K33)

5 Einfluss lyophilisierter Pektinpräparate auf gewisse physikalische und physikalische-chemische Eigenschaften der Schokolade.

6 (The effect of lyophilised pectin preparations on several physical and physico-chemical properties of chocolate).

                                        ○

Front of card:

1 Unique reference number; reference is indexed under this number in the optical coincidence index. 'P' indicates that the reference is stored in the form of a photocopy.
   Similarly, R = Reprint
             O = Original
             Mfilm = Microfilm
             Mfiche = Microfiche
2 Name(s) of author(s) and initials.
3 Journal title in full followed by **volume number**, (issue number), inclusive page numbers, and (year). References to books, proceedings, etc., will be given in the recognised forms.
4 First parentheses give the language of the reference and the language(s) of summary(ies) in a different language. Abbreviations for languages are those used in *Food Science & Technology Abstracts*. In this example the reference is in German and it contains a summary in English and a summary in Russian.
   Second parentheses give a reference to an English abstract (if known) of a foreign-language reference.
5 Title of the paper in the original language. No accents are used except the Umlaut. Cyrillic titles are given in the transliterated form. (Transliteration is carried out using the 'British' System B.S.2979: 1958).
6 Title of the paper in English (where the reference is in a foreign language).

---

7 **21** ○

8 UWB (B, BA, BAC)

9 -L(X)

10 1(.10, .11, .13, .13.4, .13.4.6)

11 M/D, M/M

Reverse of the card:

Note that the information on the reverse of the card is upside-down in relation to that on the front; this makes it possible to read this information without removing the card from the captive index. (Information on the front of the card is continued on the reverse if space on the front runs out).

7 Reference number repeated

8 Classification coordinates of foodstuffs. To ease typing load and to enable easier reading of the coordinates for each main foodstuff group, these coordinates are placed in parentheses after the main coordinate.
   Hence:

   UWB (B, BA, BAC) = UWB, UWB B, UWB BA, UWB BAC.

9 Classification coordinates of treatments; (see remarks for 8).

10 Classification coordinates of physical properties; (see remarks for 8).

11 Master cards.

DESIGN OF AUTHOR CARDS

Sufficient information to completely identify the reference is given on the Author Card; Title and classification coordinates are not given.

Lempka, A. and Wojciak, W.

Die Nahrung **13** (1) 27–31 (1969)
(De, en, ru) (*FSTA* **1** (5) K33)

21

○

INDEX OF NAMES

INDEX OF SUBJECTS